Moving Loads –
Dynamic Analysis and Identification
Techniques

Structures and Infrastructures Series

ISSN 1747-7735

Book Series Editor:

Dan M. Frangopol

Professor of Civil Engineering and
Fazlur R. Khan Endowed Chair of Structural Engineering and Architecture
Department of Civil and Environmental Engineering
Center for Advanced Technology for Large Structural Systems (ATLSS Center)
Lehigh University
Bethlehem, PA, USA

Volume 8

Moving Loads –
Dynamic Analysis and Identification Techniques

Siu-Seong Law[1] and Xin-Qun Zhu[2]

[1] Civil and Structural Engineering Department, Hong Kong Polytechnic University, Kowloon, Hong Kong
[2] School of Engineering, University of Western Sydney, Australia

CRC Press
Taylor & Francis Group
Boca Raton London New York Leiden

CRC Press is an imprint of the
Taylor & Francis Group, an **informa** business

A BALKEMA BOOK

Colophon

Book Series Editor:
Dan M. Frangopol

Volume Authors:
Siu-Seong Law and Xin-Qun Zhu

First issued in paperback 2017

Taylor & Francis is an imprint of the Taylor & Francis Group,
an informa business

© 2011 Taylor & Francis Group, London, UK

Typeset by MPS Ltd, a Macmillan Company, Chennai, India

British Library Cataloguing in Publication Data
A catalogue record for this book is available from the British Library

Library of Congress Cataloging-in-Publication Data

Law, Siu-Seong.
  Moving loads : dynamic analysis and identification techniques / Siu-Seong Law,
Xin-Qun Zhu.
    p. cm. — (Structures and infrastructures series, ISSN 1747-7735 ; v. 8)
    ISBN 978-0-415-87877-7 (hardback) — ISBN 978-0-203-84142-6 (eBook) 1. Live
loads. 2. Structural dynamics. I. Zhu, Xin-Qun. II. Title.

    TA654.3.L39 2011
    624.1′72—dc22
                                                                    2010052748

Published by: CRC Press/Balkema
               P.O. Box 447, 2300 AK Leiden, The Netherlands
               e-mail: Pub.NL@taylorandfrancis.com
               www.crcpress.com – www.taylorandfrancis.co.uk – www.balkema.nl

ISBN 13: 978-1-138-11491-3 (pbk)
ISBN 13: 978-0-415-87877-7 (hbk)

Structures and Infrastructures Series: ISSN 1747-7735
Volume 8

Table of Contents

Part I – Moving Load Problems

Part II – Moving Load Identification Problems

Editorial

Welcome to the Book Series *Structures and Infrastructures*.

Our knowledge to model, analyze, design, maintain, manage and predict the life-cycle performance of structures and infrastructures is continually growing. However, the complexity of these systems continues to increase and an integrated approach is necessary to understand the effect of technological, environmental, economical, social and political interactions on the life-cycle performance of engineering structures and infrastructures. In order to accomplish this, methods have to be developed to systematically analyze structure and infrastructure systems, and models have to be formulated for evaluating and comparing the risks and benefits associated with various alternatives. We must maximize the life-cycle benefits of these systems to serve the needs of our society by selecting the best balance of the safety, economy and sustainability requirements despite imperfect information and knowledge.

In recognition of the need for such methods and models, the aim of this Book Series is to present research, developments, and applications written by experts on the most advanced technologies for analyzing, predicting and optimizing the performance of structures and infrastructures such as buildings, bridges, dams, underground construction, offshore platforms, pipelines, naval vessels, ocean structures, nuclear power plants, and also airplanes, aerospace and automotive structures.

The scope of this Book Series covers the entire spectrum of structures and infrastructures. Thus it includes, but is not restricted to, mathematical modeling, computer and experimental methods, practical applications in the areas of assessment and evaluation, construction and design for durability, decision making, deterioration modeling and aging, failure analysis, field testing, structural health monitoring, financial planning, inspection and diagnostics, life-cycle analysis and prediction, loads, maintenance strategies, management systems, nondestructive testing, optimization of maintenance and management, specifications and codes, structural safety and reliability, system analysis, time-dependent performance, rehabilitation, repair, replacement, reliability and risk management, service life prediction, strengthening and whole life costing.

This Book Series is intended for an audience of researchers, practitioners, and students world-wide with a background in civil, aerospace, mechanical, marine and automotive engineering, as well as people working in infrastructure maintenance, monitoring, management and cost analysis of structures and infrastructures. Some volumes are monographs defining the current state of the art and/or practice in the field, and some are textbooks to be used in undergraduate (mostly seniors), graduate and postgraduate courses. This Book Series is affiliated to *Structure and Infrastructure Engineering* (http://www.informaworld.com/sie), an international peer-reviewed journal which is included in the Science Citation Index.

It is now up to you, authors, editors, and readers, to make *Structures and Infrastructures* a success.

Dan M. Frangopol
Book Series Editor

About the Book Series Editor

 **Dr. Dan M. Frangopol** is the first holder of the Fazlur R. Khan Endowed Chair of Structural Engineering and Architecture at Lehigh University, Bethlehem, Pennsylvania, USA, and a Professor in the Department of Civil and Environmental Engineering at Lehigh University. He is also an Emeritus Professor of Civil Engineering at the University of Colorado at Boulder, USA, where he taught for more than two decades (1983–2006). Before joining the University of Colorado, he worked for four years (1979–1983) in structural design with A. Lipski Consulting Engineers in Brussels, Belgium. In 1976, he received his doctorate in Applied Sciences from the University of Liège, Belgium, and holds two honorary doctorates (Doctor Honoris Causa) from the Technical University of Civil Engineering in Bucharest, Romania, and the University of Liège, Belgium. He is an Honorary Professor at Tongji University and a Visiting Chair Professor at the National Taiwan University of Science and Technology.

Dan Frangopol is a Distinguished Member of the American Society of Civil Engineers (ASCE), a Fellow of the American Concrete Institute (ACI), the International Association for Bridge and Structural Engineering (IABSE), and the International Society for Health Monitoring of Intelligent Infrastructures (ISHMII). He is also an Honorary Member of both the Romanian Academy of Technical Sciences and the Portuguese Association for Bridge Maintenance and Safety. He is the initiator and organizer of the Fazlur R. Khan Distinguished Lecture Series (www.lehigh.edu/frkseries) at Lehigh University.

Dan Frangopol is an experienced researcher and consultant to industry and government agencies, both nationally and abroad. His main areas of expertise are structural reliability, structural optimization, bridge engineering, and life-cycle analysis, design, maintenance, monitoring, and management of structures and infrastructures. His work has been funded by NSF, FHWA, NASA, ONR, WES, AFOSR and by numerous other agencies. He is the Founding President of the International Association for Bridge Maintenance and Safety (IABMAS, www.iabmas.org) and of the International Association for Life-Cycle Civil Engineering (IALCCE, www.ialcce.org), and Past Director of the Consortium on Advanced Life-Cycle Engineering for Sustainable Civil Environments (COALESCE). He is also the Chair of the Executive Board of the International Association for Structural Safety and Reliability (IASSAR, www.columbia.edu/cu/civileng/iassar), the Vice-President of the International Society for Health Monitoring of Intelligent Infrastructures (ISHMII, www.ishmii.org), and the founder and current chair of the ASCE Technical Council on Life-Cycle Performance, Safety, Reliability and Risk of Structural Systems (content.seinstitute.org/committees/strucsafety.html).

Dan Frangopol is the recipient of several prestigious awards including the 2008 IALCCE Senior Award, the 2007 ASCE Ernest Howard Award, the 2006 IABSE OPAC Award, the 2006 Elsevier Munro Prize, the 2006 T. Y. Lin Medal, the 2005 ASCE Nathan M. Newmark Medal, the 2004 Kajima Research Award, the 2003 ASCE Moisseiff Award, the 2002 JSPS Fellowship Award for Research in Japan, the 2001 ASCE J. James R. Croes Medal, the 2001 IASSAR Research Prize, the 1998 and 2004 ASCE State-of-the-Art of Civil Engineering Award, and the 1996 Distinguished Probabilistic Methods Educator Award of the Society of Automotive Engineers (SAE). He has given plenary keynote lectures in numerous major conferences held in Asia, Australia, Europe and North America.

Dan Frangopol is the Founding Editor-in-Chief of *Structure and Infrastructure Engineering* (Taylor & Francis, www.informaworld.com/sie) an international peer-reviewed journal, which is included in the Science Citation Index. This journal is dedicated to recent advances in maintenance, management, and life-cycle performance of a wide range of structures and infrastructures. He is the author or co-author of more than 270 books, book chapters, and refereed journal articles, and over 500 papers in conference proceedings. He is also the editor or co-editor of more than 30 books published by ASCE, Balkema, CIMNE, CRC Press, Elsevier, McGraw-Hill, Taylor & Francis, and Thomas Telford and an editorial board member of several international journals. Additionally, he has chaired and organized several national and international structural engineering conferences and workshops.

Dan Frangopol has supervised 34 Ph.D. and 50 M.Sc. students. Many of his former students are professors at major universities in the United States, Asia, Europe, and South America, and several are prominent in professional practice and research laboratories.

For additional information on Dan M. Frangopol's activities, please visit www.lehigh.edu/~dmf206/

Preface

The interaction phenomenon is very common between different components of a mechanical system. It is a natural phenomenon and is found with the impact force in aircraft landing; the estimation of degree of ripeness of an apple from impact on a beam; the interaction of the magnetic head of a computer disk leading to miniature development of modern computer; etc. Uncertainty in some of them, e.g. the soil-structure interaction behavior of foundation and the water-structure interaction of hydraulic structures, would lead to inaccurate analysis results on the behavior of the structure. The interaction force is difficult to measure unless instruments have been installed during construction for this purpose. Some of the interaction problems are difficult to quantify due to the lack of thorough knowledge on the interaction behavior. Methods have been developed to estimate the interaction forces from the measured responses of the structure but they are confined to cases with an impact force on a beam or plate. Analytical skills are required to estimate interaction forces of the mechanical system in order to enable advanced developments in different areas of modern technology.

The wind–structure interaction and vehicle–bridge interaction are two common problems with modern large civil engineering structures. The stiffness of a structure would be time-varying with large deformation of the structure and the mass matrix would also be time-variant. The interaction forces cannot be measured accurately without plenty of anemometers and pressure gauges installed on the structure or with a fully instrumented vehicle. The interaction forces have been traditionally obtained from an iterative analysis of the wind–structure system or the vehicle–bridge system with the coupled set of equation of motion.

This book takes the vehicle–bridge system for an illustration of the moving load problem. Weigh-in-motion techniques based on strain measurements in bridge engineering have been successfully developed to estimate the equivalent axle loads of vehicles from strain response at a location. Several weigh-in-motion systems are available commercially and they have been used in many countries for monitoring the effect of over-weight vehicles on early degradation of highway pavement. However, they cost several million dollars a unit. The strain readings are basically static readings, which is not representative of the dynamic components of the loads. The complete interaction loading time history needs to be known for subsequent interactive analysis. This book presents different analytical methods for the estimation of these interaction forces, taking advantage of modern computer power in the numerical analysis. These methods have the following innovative features:

1. Measurements are required at only a few locations of the structure as input.
2. The measurements can be strain, displacement, velocity and accelerations. One type of measurement or a combination of them can be used for the moving load estimation.
3. The formulation of the moving load identification is innovative, with the forces to be identified expressed as a function of the system matrices and the structural responses.
4. The solution of the identification equation is formulated in the least-squares sense and it is best solved with the damped least-squares or regularization technique.
5. Some algorithms described in this book are for beam and plate problems whilst others can be applied to more complex structures, so long they can be discretely modelled with a finite element model.
6. The whole time–history of the interaction force is taken as the unknown to be identified. Therefore, knowledge of the vehicle system is not required, except its location relative to the bridge deck.
7. Unlike the weigh-in-motion techniques, the individual moving loads can be identified separately and their summation would give an estimate on the overall weight of the vehicle.
8. The issue of accuracy of identification and computation efficiency is emphasized throughout this book.

Chapter 1 introduces the background of the force identification and moving load identification problems and their main components. The computation techniques required for both the forward and inverse analysis are discussed with reference to some more advanced solution techniques for the identification.

Chapters 2 to 4 form Part I of this book on the forward moving load problems. The different analysis methods on the vehicle–bridge system are summarized with descriptions on the dynamic response analysis techniques of the mechanical system modelled with moving loads on top of a multiple span continuous beam or an orthotropic plate. The dynamic responses are obtained from the precise time step integration scheme or the Newmark-Beta method. The effect of support flexibility from elastic bridge bearings is also presented. The study on the dynamic responses of a structure under moving loads is concluded with a full description on the vehicle–bridge interaction problem in Chapter 4, with the first example a vehicle moving on top of a planar multi-span continuous bridge deck with different road surface roughness and the second example with multiple vehicles moving over a multi-lane continuous bridge deck to study on problems with industrial application.

Chapters 5 to 10 form Part II of this book. They are devoted to the studies on the inverse problem of moving load identification. Chapter 5 describes the technique in the frequency–time domain where the identification equation is formulated in the frequency domain and the solution of the equation is in the time domain. This technique involves the inversion of large system matrices demanding high computation effort and a long computation period. This method has been validated with laboratory tests on the identification of two moving loads. The regularized solution is a great improvement over the least-squares solution, using different combinations of measured responses. The time domain method described in Chapter 6 is found to yield better results than the frequency–time domain method using the same responses from the laboratory tests.

Chapter 6 presents the well-developed moving load identification techniques in the time domain with beams and plates. The first method is formulated based on the modal superposition method and the moving loads are expressed explicitly as a function of the measured responses. The solution is obtained using the least-squares method. The second method improves on the first method by having the system matrices divided into smaller matrices and reducing the computation effort with the matrix inversion significantly. The solution is obtained via regularization improving the least-squares solution, which is subject to measurement noise effect. Different types of measured responses, their combinations, location of measurements, incomplete measurement, different amount of measured information, travelling speed of the interaction force, effect of measurement noise, single or multiple forces, etc. are studied via examples of beam and plate. Laboratory experimental verification of the techniques is also presented with the moving loads passing over the beam and plate structures. A comparison on the computation accuracy from the frequency–time domain method in Chapter 5 and the two time domain methods in Chapter 6 is given in Section 6.3.3.

Chapter 7 summarizes the developments of the more general state space method with the time-varying system matrices in the moving load problem. The first method based on dynamic programming has a natural smoothing effect on the solution resulting in an effectively damped solution. The second method is solved with regularization technique. The second method also has the flexibility of taking in different types of responses as input. Numerical and experimental studies are presented for both axle and wheel load identification, with good results. This chapter is concluded with a comparison of results from the two State Space approaches and from the Exact Solution Technique (EST) Method in Chapter 6.

Chapter 8 presents a technique for enhancing the accuracy of identification with the unknown interaction forces, modelled with function bases such as series expansion, generalized orthogonal function and wavelet deconvolution. The number of unknowns to be identified is significantly reduced and the measured data is smoothed with this approximation, leading to improved results with a much over-determined set of identification equations. Simulation and experimental studies with moving forces on a beam and plate verified the validity of this improved formulation of moving load identification technique.

Chapter 9 presents the moving load identification methods with the structure represented by its finite element model for easy and refined modeling particularly for large scale complex structures. The analysis techniques range from the early approximate interpretive method to the general regularization method and they are different in the ways used to solve the identification equation. The latter method is more accurate and computationally effective as compared to existing methods in Chapters 5 and 6, and comparative studies with the EST Method are shown to illustrate the benefits of the finite element formulated method.

Chapter 10 presents examples of industrial applications of the moving load identification techniques in assessing the axle weights of travelling vehicles. Strategies on how to access a set of accurate measured data for the identification are discussed. The comparative studies of the different algorithms in this book are highlighted and the last set of comparisons between the two broad categories of method, based on the modal superposition technique and the finite element method, is given. This chapter is closed with further discussion of the more advanced techniques of the response analysis and moving load identification.

Chapter 11 recapitulates the main features and behaviors of the different moving load identification algorithms presented in this book with highlights on the different beneficial properties. This book is concluded with a section on future research directions in the moving load identification problem.

Dedication

This book is dedicated to our wives, Connie Lam and Yan Wang, and our families for their support and patience during the preparation of this book. Also to all of our students and colleagues who, over the years, have contributed to our knowledge of dynamic behavior of structures under moving load and the moving load identification problem.

Acknowledgements

A special acknowledgement to the American Society of Mechanical Engineers for their permission to use some of the material originally published in the following journal articles:

- Law S.S., Chan T.H.T. and Zeng Q.H. (1999). "Moving force identification: A frequency and time domains analysis." *Journal of Dynamic Systems, Measurement and Control*, ASME, 12(3), 394–401.
- Zhu X.Q. and Law S.S. (2001). "Identification of moving loads on an orthotropic plate." *Journal of Vibration and Acoustics*, ASME, 123(2), 238–44.
- Zhu X.Q. and Law S.S. (2003). "Time domain identification of moving loads on bridge deck." *Journal of Vibration and Acoustics*, ASME, 125(2), 187–98.
- Zhu X.Q., Law S.S. and Bu J.Q. (2006). "A state space formulation for moving loads identification." *Journal of Vibration and Acoustics ASME*, 128(4), 509–520.
- Law S.S., Wu S.Q. and Shi Z.Y. (2008). "Moving load and prestress identification using wavelet-based method." *Journal of Applied Mechanics*, ASME, 75(2), 141–7.

About the Authors

Siu-Seong Law – Dr. Law, is currently a Professor of the Civil and Structural Engineering Department of the Hong Kong Polytechnic University. He received his doctorate in civil engineering from the University of Bristol, United Kingdom (1991). His main area of research is in the inverse analysis of force identification and condition assessment of structures with special application in bridge engineering. He has published extensively in the area of damage models and time domain approach for the inverse analysis of structure.

Xin-Qun Zhu – Dr. Zhu, who received his Ph.D. in civil engineering from the Hong Kong Polytechnic University (2001), is currently a lecturer in Structural Engineering at the University of Western Sydney. His research interests are primarily in structural dynamics, with emphasis on structural health monitoring and condition assessment, vehicle-bridge/road/track interaction analysis, moving load identification, damage mechanism of concrete structures and smart sensor technology. He has published over 100 refereed papers in journals and international conferences.

Chapter 1

Introduction

1.1 Overview

Moving load problems exist in a wide variety of applications: rail mechanics; dynamics of computer tapes; floppy and hard discs; automotive and aircraft braking systems; belt drives; web winding and conveyer systems. Typical moving load problems can also be found in the vibration of tracks; bridge or ground excitations from travelling vehicles; noise and vibration from spinning computer hard discs; and even in the machining processes in manufacturing. As modern lightweight and high strength materials are very popular in engineering, it is expected that vibration and instabilities problems due to moving loads will numerous.

Knowledge of the dynamic loads acting on a structure is very important for the design, control, diagnosis and life management of a structure. Techniques for force identification have been developed independent of the direct measurement and instrumentation but requiring only the measured responses of the structure. Problems where the force initiation site is known, or both the force and its location are not known, have been studied in the last two decades. Techniques that identify the forces in motion without disturbing the normal operation process have attracted interest from many engineers and researchers. A literature review is presented below on the various methods developed in this area including the analytical methods, numerical prediction models, laboratory measurements and numerical methods of simulation. The basics of the solution algorithm of the ill-posed problem are discussed with a brief introduction to the modal condensation techniques which are most useful for the problem solution with only limited measured information compared to the total degrees-of-freedom of the structure. Particular emphasis is placed on those different analytical techniques, with different types of measurements, which are successful in the identification of the moving interaction forces of a bridge–vehicle system.

1.2 Background of the Moving Load Problem

Many methods have been presented to predict the dynamic behavior of beams subject to various kinds of moving loads. Fryba (1972) presented in his monograph various analytical solutions for vibration problems of simple and continuous beams under moving loads. Three types of algorithms are used to analyze the interaction problem, which are, namely, the direct time integration method (Henchi et al., 1998); the iterative method (Green and Cebon, 1994) and the vehicle–bridge interaction element method (Yang and Yau, 1997). A set of coupled equations with a large number of degrees-of-freedom (DOFs) for both the beam and the load system are constructed in the first method. Two uncoupled sets of equations for the beam and the loads are formulated separately in the second method with geometrical compatibility conditions and equilibrium conditions for the interaction forces between the beam and the load system. An interaction element consisting of a beam element and the suspension units of the load resting on the element is defined in the third method, and all the degrees-of-freedom associated with the load system within each substructure are eliminated by dynamic condensation. The results obtained by all these methods are subject to modeling errors.

The above interaction problem is typically represented by the bridge–vehicle interaction in practice, and the following sections refer to this problem in the discussion of different computation methods in the moving force problem.

The importance of investigating the moving loads on top of the bridge deck was first recognized in the 19th century. Following the collapse of some railway bridges in Great Britain, engineers and researchers began to pay more attention to the dynamic behavior of bridges under moving vehicular loads and carry out research on new techniques for the bridge design and bridge condition assessment (Cantieni 1983; 1992; O'Connor and Chan, 1988; Chan and O'Connor, 1990). The structural conditions of the bridge will be affected by the operation loads including the dead load, live load, wind load and seismic load. Among these loads, the moving vehicular axle load has a large effect on the structural condition, especially for median span bridges. A moving truck generates a dynamic response which is greater than the static response because of the interaction between the moving vehicle and the supporting bridge structure. To evaluate the influence of a passing vehicle on a bridge deck, design codes express the dynamic problem as a pseudo-static problem and require that the weight of a specified vehicle be multiplied by a dynamic amplification factor (DAF). For example, the provisions in the Standard Specifications for Highway Bridges (AASHTO, 2002) specify dynamic load effects in terms of an impact factor in the form of a simple expression. It is empirically derived, based on experimental measurement on railway bridges, and is solely a function of the bridge span. These AASHTO provisions seem to have served well for many years. However, modern bridge design utilizes lighter materials and longer spans, so questions have been raised regarding the appropriateness of the AASHTO impact provisions. Research over the last 30 years has shown that the dynamic response of bridges under vehicular loading is influenced by many parameters other than just the bridge span, such as the travelling velocity and characteristics of the vehicle, the characteristics of the bridge, the bridge surface roughness and conditions of the approach roadways. Many new bridge codes (AASHTO, 2007) include provisions for a dynamic load allowance (DLA) to account for the vehicular load effects including those from impact. The DLA is presented in the form of

$$DLA = \left(\frac{R_{dyn} - R_{stat}}{R_{stat}} \right) \times 100\% \qquad (1.1)$$

where R_{dyn} = maximum dynamic response; and R_{stat} = maximum static response. However, the DLA may not always reveal the TRUE dynamic behavior of the bridge.

There are other factors which have great influence on the dynamic responses of a modern bridge which have not previously been studied. They are: vehicle moving at non-uniform speed, i.e. the braking and acceleration of a vehicle; vehicle moving on top of a modern continuous bridge deck; and multiple vehicles and their transverse positions on multi-lane bridge deck.

Though vehicle-induced bridge dynamics are important, major bridge failures are not normally caused by dynamic wheel loads (ANON, 1992). The bridge deck has damage accumulating with daily traffic loads leading to a greater need of regular maintenance. The wheel load causes more subtle problems which contribute to fatigue, surface wear, and cracking of concrete which leads to corrosion. Cebon (1988) concluded that dynamic wheel loads may increase the road surface damage by a factor of two to four over that due to static wheel loads. For this reason the studies of dynamic wheel loads and ways to measure them have always been of interest.

The bridge deck can be modeled either as a single or multi-span continuous beam or as an orthotropic rectangular plate. An orthotropic plate possesses different mechanical properties in two mutually perpendicular directions. There are two forms of orthotropy in practice: material orthotropy and shape orthotropy. Most bridge decks are orthotropic, such as an isotropic slab, a grillage, a Tee-beam bridge deck, a multi-beam bridge deck, a multi-cell box-beam bridge deck and a slab stiffened with ribs of box section. There are also four types of vehicle models (moving forces, quarter-truck, half-truck and full-truck), in different states of complexity that can be used for the moving load interaction study. The effects of the road surface roughness, the physical parameters of the bridge deck or the vehicle, braking and vehicle speed variation on the dynamic impact factor will be discussed in Part I of this book. Several new computational techniques will be employed for analyzing the bridge–vehicle interaction taking the above factors into consideration.

1.3 Models for the Vehicle–Bridge System

The vehicle–bridge interaction is often studied with a simplified model to establish a clear connection between the governing parameters and the bridge response.

1.3.1 Continuous Beam under Moving Loads

The bridge deck is modeled as a continuous beam and the vehicle is modeled as a moving force, a moving mass or a moving oscillator. The response of the idealized beam structure is governed by three characterizing parameters: the speed parameter, the frequency ratio and the mass ratio. Existing studies with this structural model are shown below.

1.3.1.1 Moving Force, Moving Mass and Moving Oscillator

The dynamic response of an elastic structure traversed by a moving mass or load is a very complicated function of the moving mass, the supporting structure and the speed of the moving load. In the case of a low moving speed or a low mass ratio, the simple moving force model can be a good approximation of the complex moving mass problem. In the case of a high moving speed or a high mass ratio, the moving force model is not accurate due to the strong interaction effect, and the complicated time-variant system analysis for the moving mass problem must be conducted to ensure an accurate investigation. The 'moving-mass moving-force' problem has been discussed in detail based on the finite element method (Lin and Trethewey, 1990). This problem has been summarized into two extreme cases of (a) when the ratio of the vehicle weight and the supporting beam weight is large, the system can be simplified to a single degree-of-freedom system with a concentrated weight moving on a massless beam; and (b) when the ratio of the vehicle weight and the supporting beam weight is small, the system can be treated as an elastic beam subject to a moving concentrated force.

The moving oscillator model is more realistic for some engineering applications (e.g. a softly sprung vehicle traversing a flexible structure, such that the interaction effect becomes important). The Eigen-function expansion method for calculating the dynamic response of a continuous elastic structure carrying a moving linear oscillator was developed by Pesterev and Bergman (1997). The method reduces the problem of the integration of a system of linear ordinary differential equations that governs the time-dependent coefficients of the series expansion of the response in terms of the eigen-functions of the continuous structure. An exact direct numerical procedure for the solution of the moving oscillator problem was presented by Yang et al. (2000). This method was extended to the calculation of the response of a spatially one-dimensional, non-conservative linear distributed parameter system under the action of a moving concentrated load using the complex eigen-functions of the continuous system by Pesterev and Bergman (1998). It was later extended to handle the problem of multiple moving oscillators by Pesterev et al. (2001).

1.3.1.2 Multi-span Beam

For a continuous bridge deck, the multi-span beam model is frequently used in the analysis of the bridge–vehicle interaction, and modal superposition method is commonly employed to analyze the vibration of the beam under moving loads. Hayashikawa and Watanabe (1981) and Wang (1997) used the eigen-stiffness matrix method for the analysis. Wu and Dai (1987) studied the response of a multi-span, non-uniform beam, subject to a series of loads moving with varying speed in identical and opposite directions. The transfer matrix method is used to determine the natural frequencies and mode shapes of the beam. Lee (1994) used the Euler-Bernoulli beam theory and the assumed mode method to analyze the transverse vibration of a beam with intermediate point constraints subject to a moving load. The vibration modes of a simply-supported beam are used as the assumed modes. The point constraints, in the form of supports, are assumed to be linear springs of large stiffness. Lin (1995) has mentioned that the selection of support stiffness is problem dependent, and it should be used with care if numerical stability in the solution has to be maintained. Zheng et al. (1998) used the modified beam functions as the assumed modes to analyze the vibration of a multi-span non-uniform Euler-Bernoulli beam subject to a moving load. The modified beam functions satisfied the zero deflection conditions at all the intermediate supports as well as the boundary conditions at the two ends of the beam. Henchi et al. (1997) used exact dynamic stiffness elements under the framework of finite element approximation to study the dynamic response of multi-span structures under a convoy of moving loads.

1.3.1.3 Timoshenko Beam

Based on the Timoshenko beam theory, the dynamic response of a simply supported beam excited by moving loads has been studied by Mackertich (1990) including the shear deformation and rotary inertia effects. Lee (1996) studied the dynamic response of a Timoshenko beam subject to a concentrated mass moving with a constant speed using the Lagrangian approach and the assumed mode method. The assumed functions are the vibration modes of a beam supported at both ends. Wang (1997) investigated the forced vibration of multi-span Timoshenko beams to a moving force using the transfer matrix method together with the modal analysis. The transfer matrices are formed by adopting the conditions of the rotation angle continuity and the balance of moment at the junction between two adjacent spans. The modal frequencies and their corresponding sets of mode shape functions of the multi-span Timoshenko beam are then determined by the transfer matrix method.

1.3.1.4 Beam with Crack

Despite the ever increasing number of research publications on the dynamic response of structures with moving loads, there are few publications on the dynamic response of beams with inherent cracks under the action of moving loads. Lee and Ng (1994) used the assumed mode method to analyze the dynamic response of a beam with a single-sided crack subject to a moving load on the top. The beam is modeled as two segments separated by the crack. Two different sets of admissible functions satisfying the respective geometric boundary conditions are then assumed for these two fictitious sub-beams. The rotational discontinuity at the crack is modeled by a torsional spring with an equivalent spring constant for the crack. The equality of transverse deflection at the crack is enforced by a linear spring of very large stiffness. Parhi and Behera (1997) utilized an analytical method along with the experimental verification to investigate the behavior of a cracked beam with a moving mass. A local stiffness matrix was used to model the crack section. Mahmoud (2001) has studied the effect of transverse cracks on the dynamic response of simply supported undamped beams subject to a moving load or mass. Majumder and Manohar (2003) developed a time–domain approach to detect damage in beam structures using a moving oscillator as the excitation source and the local damage is simulated as a reduction in the flexural rigidity of the finite element of the beam. Mazurek and Dewolf (1990) and Lee et al. (2002) investigated the feasibility of detecting structural deterioration in highway bridges using vehicular excitation in a laboratory. All the above studies are related to a simple open crack model.

Research work on the dynamic behavior of a damaged bridge under moving forces has been carried out by many scholars. Mahmoud and Abou Zaid (2002) proposed a rotational spring model to simulate the effect of transverse cracks in simple supported undamped Euler-Bernoulli beams for the study of dynamic response under a moving mass. The presence of cracks resulted in higher deflections and altered the beam response patterns. In particular, the largest deflection in the damage beam at a given vehicle speed needs longer time to build up, and a discontinuity appeared in the slope of the deflected shape of the beam at the crack location. Experimental validation of damage beams modeled by rotational spring model under a moving mass was carried out by Bilello and Bergman (2004), in which good agreement with the theoretical predictions was shown. Moreover, the percentage of variation in the beam response due to damage was, in general, larger than those induced in the structural natural frequencies. Law and Zhu (2004) studied the dynamic behavior of damaged reinforced concrete bridge structures under a moving vehicle, modeled as either a moving mass or a moving mass-spring system. The damage function proposed by Abdel Wahab et al. (1999) was adopted in the study which is capable of representing either the open or breathing crack model. Both numerical simulation and the experimental verification were conducted. The relative frequency change (RFC) and absolute frequency change (AFC) of the beam varied when the vehicle was moving on the bridge. They are sensitive to the weight of vehicle, and the frequency ratio between vehicle and bridge also had some effect on the RFC and AFC. Yang et al. (2008) presented an analytical study on the free and forced vibration of inhomogeneous Euler-Bernoulli beams containing open edge cracks. The rotation spring model was adopted to model the cracks effect. Factors which may affect the dynamic deflection of the beam including the total number of cracks, slenderness ratio, boundary conditions and moving speed of force, were examined. Results showed that the natural frequencies are reduced with an increase in the dynamic deflection due to the presence of the edge crack and the axial compressive force. The natural frequencies were greatly influenced by the edge cracks while the dynamic deflection was not very sensitive to the presence and the location of crack. Both free vibration and dynamic response were much more affected by the axial compression than by the edge cracks.

1.3.1.5 *Prestressed Beam*

For an unbonded prestressed bridge, the prestressing force produces an axial force effect as well as a bending moment due to the eccentricity of prestressed tendons. The prestressed bridges are commonly modeled as axial loaded beams, though the measured modal frequencies from beams or bridges show an opposite trend to that from the axial loaded beam theory (Saiidi et al., 1994). However, the axial loaded beam model had been adopted in theoretical analyses. Law and Lu (2005) studied the time domain response of an unbonded prestressed beam modeled as an axial loaded beam. Kocaturk and Simsek (2006) used the Lagrange equations to study the dynamic response of eccentrically prestressed visco-elastic Timoshenko beams under a moving harmonic load, and the same problem was further studied using higher order shear deformation theory (Simsek and Kocaturk, 2006). Although the effect of prestress force in bonded prestress beams has been investigated in recent years (Hamed and Frostig, 2006), the dynamic analysis of such a bridge model can rarely be found.

1.3.2 *Continuous Plate under Moving Loads*

A beam model cannot truly represent the three-dimensional behavior of the bridge deck, particularly when a moving vehicle has a path that does not follow the centerline of the bridge deck. Many types of bridge decks, including those of slab bridges, hollow-core slab bridges, and deck and girder bridges can be effectively modeled as isotropic or orthotropic plates (Bakht and Jaeger, 1985). In modeling the highway bridge as a plate, the mathematical difficulties are compounded by the complicated vehicle–bridge dynamic interaction and the difficulty in representing the behavior of the bridge structure subject to the effect of multiple vehicles.

1.3.2.1   *Plate Models*

The isotropic plate model and the orthotropic plate model are normally used to model the bridge deck in the analysis of the bridge-vehicle interaction. The major problems involved are in the assumptions and the implementation procedure.

The vibration analysis of an isotropic plate subject to a moving load is based on the following assumptions (Fryba, 1972):

1. The small elastic strains occurring in the structure are within the scope of Hooke's law.
2. There exists in the plate the so-called neutral surface. The distances between points lying on that surface do not vary with the plate deflections.
3. Mass particles lying on the normal line to the neutral surface continue to lie on it even after the plate has been deformed.

When the bridge deck is simplified as an orthotropic plate, the flexural and torsional rigidities of the equivalent orthotropic plate need to be determined for subsequent dynamic analysis. It is noted that these equivalent rigidities can be approximately obtained (Bakht and Jaeger, 1985) for typical slab and pseudo-slab bridge sections, such as isotropic slabs, grillages, Tee-beam bridge decks, multiple beam bridge decks, multi-cell box-beam bridges and slabs stiffened with ribs of box-section.

Huffington and Hoppmann (1958) have determined the equation of motion and eigen-functions for the flexural vibrations of rectangular plates of orthotropic material. Leissa (1973) has presented some comprehensive and accurate analytical results on the free vibration of rectangular plates. For a broader class of boundary-value problem, an orthogonality criterion for the eigen-functions was established and relations on the kinetic and potential energies were derived. The free vibration of orthotropic plates with fixed-simply supported and free-free boundary conditions was investigated using orthotropic plate theory by Grace and Kennedy (1985). Natural frequency parameter for the rectangular orthotropic plates on a pair of simply supported parallel edges was studied numerically using analytical method for a variety of boundary conditions and material orthotropy by Jayaraman et al. (1990). Transverse free vibration of beam-slab type highway bridges was analyzed by Ng and Kulkarni (1972). A modified method based on the orthotropic plate theory for computing the natural frequencies of bridge slabs was presented through a set of empirical relationships between the plate parameters.

1.3.2.2   *Moving Forces*

The moving loads can be simplified into a moving force for vibration analysis. The plate is assumed to be thin with small deformation and the load is assumed to be in constant contact with the plate. The force is represented by step functions in small time intervals. Wu et al. (1987) analyzed the dynamic responses of non-uniform rectangular flat plates with various boundary conditions subject to different typical moving loads using the finite element method. The effects of eccentricity, acceleration and initial velocity of the moving load, and the effect of span length of the plate, are the main aspects of study. The dynamic behavior of a multi-span flat plate supported by beam members of a rigid plane grid and subject to the action of a series of moving loads was also investigated. The dynamic response of an infinite plate on elastic foundation subject to constant amplitude or harmonic moving loads was investigated by Kim and Roesset (1998). Larrondo et al. (1998) analyzed the transverse vibrations of simply supported anisotropic rectangular plates carrying an elastically mounted concentrated mass. Laura et al. (1999) analyzed the transverse dynamic response of simply supported rectangular plates with generalized anisotropy, subject to an uniformly distributed $P\cos(\omega t)$ – type excitation. Agrawal et al. (1988) also derived the equation of motion on the dynamic responses of orthotropic plates under moving masses using the Green's function.

When the effects of rotary inertia and shear deformation of the plate are taken into consideration, the Mindlin plate theory can be used to analyze the bridge–vehicle interaction. Wang and Lin (1996) made use of this theory to analyze the vibration of a homogenous and isotropic Mindlin plate on periodic supports, subject to a moving load on top. The component method was adopted to establish the transfer matrix and the modal frequencies, and their corresponding mode shape functions for the multi-span plate were calculated by the transfer matrix method. The effect of span number, rotary inertia and transverse shear deformation of the plate, the velocity of the load and the corresponding load position on the maximum response of the plate were investigated.

1.3.2.3 Quarter-truck Model

The vehicle is modeled as a single degree-of-freedom (DOF) or two DOFs sprung-mass-dashpot system which is very useful for modeling multiple vehicles. They are modeled as a set of independent discrete units moving with the same velocity (Taheri and Ting, 1989). This eliminates the inertia effects due to roll, pitch and yaw motions of the vehicle. The masses of the vehicles are lumped on the suspension systems which are modeled as linear springs and dashpots. All movements of the suspensions, except the vertical motions, are constrained. The structural impedance method (Taheri and Ting, 1989) and finite element method (Taheri and Ting, 1990) were developed to analyze the dynamic response of the plates subject to moving loads based on this model. The algorithms accounted for the complete dynamic interactions between the moving loads and the plate. The methods can be applied to general moving mass problems and to the simplified moving force and static problems. Humar and Kashif (1995) used this model to identify the parameters governing the response of isotropic and orthotropic plate models, and to examine the nature of dynamic response of the models subject to central and off-center moving vehicles and multiple vehicles. The conclusions show that the characterizing parameters governing the response of a bridge structure modeled as an isotropic plate are the aspect ratio of the plate, the speed parameter, the frequency ratio between the natural frequencies of the vehicle and the bridge, and the mass ratio. The parameters governing the response of an orthotropic plate model are the speed parameter, the frequency ratio, the mass ratio and two other factors related to the geometry and stiffness properties of the plate.

1.3.2.4 Half-truck Model

The half-truck model is usually represented by a planar, two-axle or three-axle, sprung mass system with frictional device. This model was used to study the effects of vehicle braking on the bridge (Gupta and Traill-Nash, 1980). Three bridge deck models (beam, beam with torsional DOFs and uniform orthotropic plate) were used in the analysis. The effect of the bridge transverse flexibility is considered firstly by including an additional degree-of-freedom in the simple beam representation. The response studies are extended into the braking of vehicle on the bridge approach as well as on the span. The effect of the bridge transverse flexibility on the bridge response is studied by obtaining the response for symmetric as well as eccentric vehicular loading on the bridge. Bridge idealization as an orthotropic plate was used to simulate the transverse flexibility of the bridge deck. The three-axle half-truck model was also used to identify the effect of various parameters on the dynamic load (Hwang and Nowak, 1991).

1.4 Dynamic Analysis of the Vehicle–Bridge System

Existing research on the bridge–vehicle interaction problem can mainly be grouped into two categories according to the method adopted to decompose the equation of motion of the bridge–vehicle system.

1.4.1 *Methods Based on Modal Superposition Technique*

Among various methods in the dynamic analysis of bridge structures under a moving vehicle, the modal superposition technique is typically used to decompose the equation of motion of a bridge–vehicle system in which the response of structure is represented in terms of a set of modal shapes with different amplitudes. The equation of motion of the dynamic system, which is a partial differential equation, is transformed into a set of ordinary differential equations which can be easily solved by numerical methods such as the Newmark-Beta method, etc. Existing research works where modal superposition technique is applied to a continuous bridge model are discussed below.

For bridge decks modeled as Euler-Bernoulli beams, Green and Cebon (1994) gave the solution on the dynamic responses in the frequency domain under a 'quarter-car' vehicle model using an iterative procedure. The algorithm was validated by extensive experiments on a typical highway bridge. Modal tests showed that beam and plate models of bridge dynamics gave reasonable predictions on the measured vibration mode shapes and natural frequencies of the bridge structure. Results from vehicle tests indicated that the method was accurate for predicting the dynamic response of short-span highway bridges due to heavy vehicular loads. An analysis method for a beam with non-uniform cross-section with a time varying concentrated force travelling on top was proposed by Gutierrez and Laura (1997). The vibrational behavior of an elastic homogeneous isotropic beam with different boundary conditions due to a moving harmonic force was studied by Abu-Hilal and Mohsen (2000). Law and Zhu (2005) investigated the influence of braking on a multi-span non-uniform bridge deck under moving vehicle axle forces. Results showed that vehicle braking generated an equivalent impulsive force covering a wide range of frequency spectrums. A large number of vibration modes are excited and they are required in the computation for a high accuracy in the dynamic response of the structure.

For bridge decks modeled as Timoshenko beams, the vibration of a continuous bridge deck modeled as a multi-span Timoshenko beam under a vehicle modeled by a mass-spring system with two DOFs was studied by Chatterjee et al. (1994).

Wang (1997) proposed a method for modal analysis to investigate the vibration of a multi-span Timoshenko beam under a moving force. The ratio of the radius of gyration of the cross-section to the span length was defined as a parameter and the effect of this parameter on the first modal frequency of the beam was studied. Wang and Chou (1998) employed the large deflection theory to derive the equation of motion of the Timoshenko beam due to the coupling effect of an external force with the weight of the beam. Results showed that the effect of weight of the beam increases the fundamental natural frequency of the structure. Both the dynamic deflection and moment of the beam predicated by the theory, including the effect of weight of the beam, are less than those predicated either by the small deflection theory or by the large deflection theory without including the effect of weight of beam. A dynamic analysis of a Timoshenko beam subject to a moving force was also investigated by Sniady (2008) in which a closed form solution was provided.

For bridge decks modeled as plates, Marchesiello et al. (1999) presented an analytical approach to the vehicle–bridge dynamic interaction problem with a seven DOFs vehicle system moving on a multi-span continuous bridge deck modeled as an isotropic plate. Both the flexural and torsional mode shapes were included in the study. An iterative method was adopted to calculate the responses of the bridge deck and vehicle separately, i.e. the equations of motion of the bridge and vehicle system respectively were not coupled. The theoretical modes, defined by means of the Rayleigh-Ritz approach, had been found to be in good agreement with that from finite element model. Zhu and Law (2003a) investigated the dynamic behavior of orthotropic rectangular plates under moving loads. Results showed that the impact factor of the orthotropic plate increased with the ratio between the flexural and torsional rigidities of the plate, and an equivalent beam model of the bridge deck could give an estimate on the impact factor along the centerline of the deck with an under-estimation of the dynamic response at the edge of the structure. A further study was carried out (Zhu and Law 2002a) on a more complex model with a two-axle three dimensional vehicle model with seven DOFs according to the H20-44 vehicle design loading (AASHTO, 2002) moving on a multi-lane continuous bridge deck. The dynamic behavior of the bridge deck under single and several vehicles moving in different lanes was analyzed using the orthotropic plate theory and modal superposition technique. The impact factor is found varying in an opposite trend with the dynamic responses in the different loading cases studied.

1.4.2 Methods Based on the Finite Element Method

In the modal superposition technique, mode shapes are required to decompose the system equation and they may be difficult to obtain for complex structures. The finite element method is capable of handling more complex vehicle–bridge models with complex boundary conditions in the dynamic analysis. Research work conducted on simple finite element model of the bridge–vehicle system had already been summarized in the monograph by Fryba (1972). Research works on more complex models will be presented in this section to show the advantage of the finite element method.

Wang et al. (1992; 1996) investigated the dynamic loading on girder bridges with different number of girders and span lengths due to several vehicles moving across rough bridge decks. The vehicle was simulated as a nonlinear model with 11 DOFs according to the HS20-44 truck in the AASHTO specifications (AASHTO, 2007). The maximum impact factors from different bridge girders were obtained for different numbers of loading trucks, road surface roughness, transverse loading positions and the vehicle speeds. Huang et al. (1998) developed a procedure for obtaining the response of thin-walled, curved box-girder bridges due to the HS20-44 truck model. The analytical results show that most impact factors of torsion and distortion are much larger than those of vertical bending response. The impact factors of normal stress at different points in the same cross section are quite different. Research work on a highway steel bridge with longitudinal gradients under an 11 DOFs HS20-44 truck was also carried out by Huang and Wang (1998). The dynamic responses of three steel multi-girder bridges with different span lengths due to multiple vehicles moving across rough bridge decks at different vehicle speeds are evaluated.

Henchi et al. (1998) proposed an efficient algorithm for the dynamic analysis of a bridge deck discretized into three-dimensional finite elements with a system of vehicles running on top at a prescribed speed. The vehicular axle loads acting on the bridge deck were represented as nodal forces using shape functions. The coupled equations of motion of the bridge and vehicle model were solved directly without the use of an iterative method. Numerical simulation showed that the proposed coupled method was much more efficient than the uncoupled iterative method. It also emphasized that there is no limitation concerning the complexity (number of degrees-of-freedom) of the bridge structure in this method if the stability criterion was satisfied. Lee and Yhim (2004) studied the dynamic responses of single and two-span continuous composite plate structures subject to multiple moving loads by using the third-order plate theory. Results showed that the maximum deflection of symmetric laminates to dynamic loading was higher than that of the anti-symmetric laminates. However the differences in dynamic resistance for anti-symmetric layup sequences were similar to that of the symmetric cases. Law and Zhu (2005) also investigated the dynamic behavior of long-span box-bridges subject to moving vehicles with numerical simulation and experimental verification. Similar work was done by Kim et al. (2005) and Li et al. (2008) with numerical simulation of vehicles travelling along a girder bridge and field test data was used to verify the proposed algorithm. Nallasivam et al. (2007) analyzed the impact effect on curved box-girder bridges due to moving vehicles. The impact factors corresponding to torsion and distortion have, in general, very high corresponding bimoments while those of the other responses were also relatively higher than that of corresponding straight box-girder bridge. The analysis of bridge–vehicle dynamic response under the effect of braking and acceleration was studied by Ju and Lin (2007) with a finite element model. Numerical examples indicated that the bridge longitudinal response was more sensitive than the bridge vertical response when the vehicle braking or acceleration was active, especially for higher piers.

There are other methods based on a finite element model of the structure, especially for the interaction problems, such as the 'moving element method' in which the beam model is discretized into elements that 'flow' with the moving vehicle (Lane et al., 2008). A vehicle–bridge interaction (VBI) element was firstly introduced by Yang and Wu (2001) to solve the bridge–vehicle interaction problem. This element is versatile to represent vehicles of various complexities, ranging from the moving load, moving mass, sprung mass to the suspended rigid bar. Pan and Li (2002) used the method to solve the transient response of a vehicle–structure interaction problem in time domain. This method was also employed to model a train traveling along the rail (Koh et al., 2003).

A car of the moving train was modeled as a three DOFs mass-spring-damper system. The rail was modeled as an infinite Euler-Bernoulli beam so that the 'element' would never reach the end of the beam. Contrasting with the finite element method, the moving vehicle always acts at the same point in the numerical model, thereby eliminating the need for keeping track of the contact point with respect to individual elements. A similar 'moving element' method was also proposed by Wu (2005a; 2005b; 2007; 2008) in which the method with 'moving distributed mass element' was adopted to solve the problem of dynamic analysis on beam, frame and plate structures under moving forces.

1.5 The Load Identification Techniques

1.5.1 The Weigh-In-Motion Technique

Accurate knowledge of dynamic loads acting on a structure is very important for the design, control, diagnosis and life management of a structure. The load determination is of special interest when the applied loads cannot be measured directly, while the responses can be measured easily (Yen and Wu, 1995a; 1995b; Choi and Chang, 1996; Moller, 1999; Rao et al., 1999). Traditionally the load was either measured directly with an instrument or computed from models of the structure–load system. It was very expensive and the results obtained are subject to bias in the first approach (Cantieni, 1992; Heywood, 1994), while the second approach is subject to modeling errors (Green and Cebon, 1994; Yang and Yau, 1997; Henchi et al., 1998). The first system for the Weigh-In-Motion of vehicles was possibly developed by Norman and Hopkins (1952). Moses (1984) has developed the weigh-bridge for weighing vehicles which is time consuming and subject to errors, and drivers of overweight vehicles tend to avoid the check points on the highway.

Techniques are required to determine the vehicular loads indirectly from the vibration responses of the bridge deck, in such a way that the different parameters of the bridge and vehicle system are accounted for in the measured responses with a cost less than direct measurement. Weigh-In-Motion (WIM) techniques are developed towards this need. Some of them use road-surface systems, which make use of piezo-electric (pressure electricity) or capacitive properties to develop plastic mat or capacitive sensors to measure axle weight (Davis and Sommerville, 1987). Under-structure WIM systems, represented by AXWAY (Peters, 1984) and CULWAY (Peters, 1986), are also used in which sensors are installed under a bridge or a culvert and the axle loads are computed from the measured responses. These systems can only give the equivalent static axle loads but not the peak dynamic wheel loads or the time series. Other systems use instrumented vehicles to measure dynamic axle loads (Cantieni, 1992; Heywood, 1994) but are subject to bias. O'Brien et al. (1999) has compared two Weigh-In-Motion systems based on dynamic measurement known to be used in Europe. One is a commercial system made by the Americans and the other was developed at Trinity College in Dublin, Ireland (Dempsey et al., 1995).

More recently, a cost-effective method for assessing the actual traffic loads on an instrumented railway bridge has been developed (Karoumi et al., 2005; Liljencrantz et al., 2007), and a WIM system was implemented with axle detection devices and four concrete strain transducers embedded in concrete. Successful application of the

WIM technique to a two-span continuous curved bridge with skew was also reported by Yamaguchi et al. (2009) even though the bridge is far from ideal for the Weigh-In-Motion technique.

The systems for most of the cable-supported bridges in the world are of the 'Plate bending' type based on the transient strain response from passing axle. Each of these systems cost several millions of dollars with sophisticated electronic components and high maintenance costs. Something cheaper and more reliable is required for monitoring the ever-increasing problem with over-weight trucks for highway maintenance and for improving bridge load design code. Since the dynamic vehicular axle load is known to increase the damage of the pavement, it should therefore be accurately and efficiently acquired for a proper design of the highway pavement or for the enforcement of vehicular weight limits on the highway.

1.5.2 *The Force Identification Techniques*

More advanced force identification techniques are needed for the estimation of the force or force reconstruction from measured structural responses. One major class of this problem has a known initiation site. Examples are the determination of the impact of aircraft on landing (Williams and Jones, 1948) and, in a more interesting case, the force on fruit (Schueller and Wall, 1991). Stevens (1987) gives an excellent survey of literature on the force identification problem as well as an overview of the subject. The other major class of problem is where both the force history and its location are unknown. Examples include using the modal response data to determine the location of impact forces on the read/write head of computer disks (Doyle 1984; Briggs and Tse, 1992). Doyle (1987a; 1987b; 1994) also developed a method for determining the location and magnitude of an impact force using the phase difference of the signals measured at two different locations straddling the impact point. Flexural wave propagation response was used to determine the location of the structural impacts in the frequency domain.

Whiston (1984) and Jordan and Whiston (1984) used the arrival-time difference between the maximum and minimum frequency components in the flexural wave in a beam to calculate a preliminary estimation of the force, and a further refinement process is conducted to reconstruct the impact force time history iteratively so that the force would not have a negative value. Michaels and Pao (1986) developed a method that determined an oblique dynamic force using wave motion displacement measurements.

The above works are based on the propagating-wave approach which relies on models in the frequency domain. Busby and Trujillo (1987) reconstructed the force history using a standing wave approach and Hollandsworth and Busby (1989) verified this experimentally with a force applied at a known location with accelerometers as sensors. Simonian (1981a; 1981b) used a dynamic programming filter to predict wind loads on a structure. Druz et al. (1991) formulated a nonlinear inverse problem and tried to find the location and magnitude of an external force. This force, however, was not general and is confined to a sinusoidal function defined by its amplitude and phase. More recently, Jiang and Hu (2008) present a new approach to reconstruct the distributed dynamic loads on an Euler-Bernoulli beam from the beam response. The concept of consistent spatial expression for the dynamic loads is put forward, and the Legendre polynomials are used as the consistent orthogonal basis functions to describe the distributed dynamic loads. The load reconstruction is realized with an optimal range of frequency and spatial modes.

1.5.3 The Moving Force Identification Techniques

This group of techniques addresses the more general approach of moving force identification from the measured responses of the structure, taking into account different influencing factors of the system, such as, the mechanical system (vehicle) generating the force, deceleration and acceleration of the force, force moving on top of a continuous beam and problems with multiple forces moving on top of a plate structure.

Hoshiya and Maruyama (1987) developed a system using weighted global iteration procedures associated with an extended Kalman filter to identify a moving force on a simply supported beam. O'Connor and Chan (1988) proposed an advanced weighing system (Interpretive method I) that can measure both static and dynamic axle loads from bridge strain responses. This category of methods was developed rapidly in the last two decades and most of the new developments take the vehicle–bridge interaction problem as an example of application. They can be broadly classified into the following two categories.

1. **Methods based on Modal Superposition Technique**

    In this group of methods, the modal superposition technique is firstly employed to decouple the equation of motion of the bridge structure to a set of ordinary differential equations, then the relationship between the moving forces and the bridge responses in each mode is established. Finally, the inverse problem can be solved by least-squares estimation with regularization or other optimization methods. Methods within this group include:

    - **Time Domain Method (TDM).** This method was first proposed by Law et al. (1997) in which the relationship of the moving axle force and response in each mode is established by integral convolution. The discrete form of equation of motion of the system in each mode can be obtained by assuming the time series of moving forces to be step functions in small time intervals. The time varying forces on a beam can be identified by solving the established discrete equations. Applications of this method to identify the moving forces on a multi-span continuous bridge were investigated by Zhu and Law (2000; 2001a; 2002b). The research was also extended to study the possibility of identifying axle loads when applied to the real vehicle–bridge system with road surface roughness and incomplete vehicle speed. Experimental tests show that the method can identify individual axle loads travelling at non-uniform speed with little error (Zhu and Law 2003c). The effect of support bearing stiffness was also included in this moving force identification procedure by Zhu and Law (2006).

    - **Frequency–Time Domain Method (FTDM).** This method was proposed by Law et al. (1999) in which the ordinary differential equations of the system, after applying the modal superposition technique, were Fast Fourier Transformed. A relationship between the moving forces and bridge responses in frequency domain can be established via the Frequency Response Function. The equations in frequency domain were discretized and the least-squares method was applied to solve the equations. The time history of moving forces can be obtained by performing the inverse Fourier transformation. The regularization method (Law et al., 2001) and the SVD technique (Yu and Chan, 2003) were adopted in the inverse procedure to improve the accuracy of this method.

- **State Space Method (SSM).** This method was firstly proposed by Zhu and Law (2001c) combined with a high-precision integration method (Zhong and Williams, 1994) in which the bridge was modeled as a non-uniform continuous Euler-Bernoulli beam. The Hamilton principle and modal superposition were employed to establish the system model including the mass, damping and stiffness matrices in state space, and the identification procedure is then performed with regularization. Application of this method to a plate model was investigated (Zhu et al., 2006; Law et al., 2007).
- **Method of Moments (MOM).** This method was proposed by Yu et al. (2008a; 2008b) in which the moving vehicle loads were described as a combination of whole basis functions, such as orthogonal Legendre or Fourier series, and the force identification problem was then transformed into a parameter identification problem.
- **Identification based on Genetic Algorithm.** This method was proposed by Jiang et al. (2003; 2004) in which the acceleration signals of a bridge structure at selected locations were adopted and the corresponding velocities and displacements were obtained by integration. Pseudo-inverse and singular value decomposition were employed to arrive at rough approximations of moving forces and a genetic algorithm is then used to find the best estimated value of forces by minimizing the errors between the measured and the reconstructed accelerations in each generation.

2. Methods based on Finite Element Method
    The finite element method is adopted to model the bridge structure. A location matrix (or vector for case with one moving force) for the moving forces is introduced with which the concentrate forces applied on the bridge deck can be transformed into nodal forces. Moving force identification can then be conducted with the least-squares estimation with regularization or other optimization methods. Methods of this group include:
    - **Interpretive Method I (IMI).** This method was proposed by O'Connor and Chan (1988) as the first method within this group. The bridge structure is modeled by an assembly of lumped masses interconnected by massless elastic beam elements. Thus the nodal responses of displacements and bending moments at any instant can be related to the moving forces (Chan and O'Connor, 1990).
    - **Interpretive Method II (IMII).** This method was first proposed by Chan et al. (1999) in which the bridge was modeled as a planar simply supported Euler-Bernoulli beam with a single force moving on top. The bridge responses at various locations, such as the vertical displacement or bending moments, were transformed to modal values. The central difference method is used to numerically differentiate the modal displacements to obtain the corresponding modal velocities and modal accelerations. Values of the axle load can then be obtained at any instant by solving the linear equations using the least-squares method. The IMII was extended to identify the moving forces on a multi-span continuous Timoshenko beam with non-uniform cross-section (Zhu and Law, 1999) and a generalized orthogonal function approach was proposed to obtain the

derivatives of bridge modal responses from the strain measurements instead of direct differentiation (Zhu and Law, 2001b).

- **Optimal State Estimation Approach (OSEA).** This method was first proposed by Law and Fang (2001), in which a location vector for the moving forces was introduced. The bridge–vehicle model was established in the state space and the moving forces were identified in time domain by adopting the dynamic programming optimization technique to have a set of effectively damped identification results. The method was used for the vehicle axle load identification of a three-dimensional bridge deck under a moving truck by Gonzalez et al. (2008) where the first-order regularization was adopted instead of the zero-order regularization to improve the accuracy of the identified results from the contaminated response data.

- **Finite Element Approach with Orthogonal Function Approximation.** The method was proposed by Law et al. (2004) where a location matrix was included for the moving forces same as the OSEA. Orthogonal functions were employed to fit the strain or displacement signals, and the velocities and accelerations can be accurately calculated through differentiation on the coefficients. Finally, the force identification was performed by solving a set of linear equations by least-squares method with regularization. The improved reduction system (IRS) scheme (O'Callahan, 1992) was employed to reduce the degrees-of-freedom of the bridge structure leading to enhanced identification accuracy.

- **Updated Static Component (USC) Technique.** This method was proposed by Pinkaew and Akarawittayapoom (2003). It is an iterative method based on finite element model in which the moving loads to be identified were separated into static and dynamic components. The moving forces were identified with updating the static component until convergence is achieved (Pinkaew, 2006). This method was further verified by experimental tests (Pinkaew and Asnachinda, 2007) and the application of which was extended to more complex bridge–vehicle models (Asnachinda et al., 2008). Results showed that the method is very accurate even with a relatively large level of noise in the measured bridge responses.

- **Wavelet-Based Method.** This method was first proposed by Wu and Shi (2006) in which the nodal responses and nodal forces were represented by wavelet basis. The relationship between the coefficients of responses and the coefficients of moving forces in wavelet domain can be established by employing a Galerkin procedure. The wavelet coefficients of moving forces can then be identified and the time history of forces can be reconstructed. The method was further applied on identifying vehicle axle loads on a prestressed bridge structure by Law et al. (2008).

A review of four of the above methods including IMI, IMII, TDM and FTDM with comparative studies and experimental tests was conducted by Chan, Yu and Law (2000). Parameters including the speed of vehicles, sampling frequency, axle-spacing-to-span ratio (ASSR) and the sensitivity to noisy data were studied. Numerical simulations and laboratory experiments showed that the accuracy of these methods was independent of sampling frequency and both IMII and TDM were further independent of the speed of vehicle. The accuracy of IMI was significantly affected by the ASSR and noise level. If the number of bridge modes was equal to the number of sensors, IMII would be good for applications with any ASSR and a low level of noise. TDM and FTDM were suitable for all cases studied, but time consuming.

A comparison between TDM and the finite element approach with orthogonal function approximation was given by Zhu and Law (2002c). Results showed that the latter gave consistently small errors in the results for all noise levels while the accuracy of TDM was affected by noise to a large extent. The orthogonal function approximation of response signals was shown to be effective in filtering the high-frequency noise components in the responses. It was also shown that the importance of having pre-processing of the measured data to remove the measurement noise before the identification is not over-emphasized.

It should be noted that two kinds of functions which are commonly used in moving force identification, i.e. the orthogonal functions and cubic spline interpolation, have different performances. Contrasted with the first technique, which is capable of smoothing signals, cubic spline interpolation does not have much benefit. However, it is a very accurate method for obtaining the differential of the interpolated signal and it can be used in IMII instead of the central difference method to achieve better performance.

Techniques based on finite element model are suitable for solving problems on complex bridge models with complex boundary conditions while the TDM and FTDM are not applicable due to the difficulty of obtaining the accurate mode shapes for such structures. Moreover, TDM and FTDM are very time consuming for complex structures with multiple forces.

It should be noted that the road surface roughness is a very important factor and has always been considered in most of the moving force identification procedures in which deterministic samples are adopted to represent the effect of the road surface roughness. It should also be noted that other methods of modeling the roughness (Thite and Thompson, 2003a; 2003b; Nordstrom and Nordberg, 2004; Liu and Shepard, 2005; Lu and Law, 2006, Nordberg and Gustafsson, 2006a; 2006b) which had been adopted in force identification may also be applicable for the moving force identification problem.

The different main stream moving force identification techniques will be discussed in Part II of this book. Techniques are categorized into those with the problem solution in the time domain, in the frequency–time domain and in the state–space domain. They can also be categorized into those with a finite element model of the structure and those with orthogonal function, approximating the measured responses.

1.6 Problem Statement on the Moving Load Identification

The moving load identification is not only an inverse problem but is also ill-posed, since the responses are typically continuous vector functions of the spatial coordinates, and they are defined at a few points only. Solutions are frequently unstable in the sense that small perturbations in the responses would result in large changes in the calculated load magnitudes. Lee and Park (1995) analyzed the characteristics of the error in the force determination in structural dynamic systems, and they used a regularization procedure to reduce the force determination error. Tikhonov's regularization method has been used by Busby and Trujillo (1987) in a modal-based load identification problem. Later they used a first-order regularization (Busby and Trujillo, 1997) where the penalty is in terms of the derivative of the force rather than the force itself, and the regularization parameter is determined by the L-curved method (Hansen, 1992) and the generalized cross-validation method (Golub et al., 1979). Also a time domain method was presented (Kammer, 1998) for estimating the discrete input forces acting on a structure based on system Markov parameters, and regularization techniques are employed to stabilize the computation.

The problem we have is one in which the system matrices K, C and M are known together with information on some of the displacements, velocities and accelerations. However the forcing term f is unknown. The goal of this problem is to find the forcing term f that causes the system to best match the measurement.

In practice it is not possible to measure all the dynamic responses and only certain combinations of the responses P are measured. The measurement equation is given as

$$d = AP \tag{1.2}$$

where d is a $(m \times 1)$ vector of computed responses; A is a $(m \times 2n)$ selection matrix relating the responses to the state variables, and P is of dimension $(2n \times 1)$. The actual measurements are represented by a vector r which is of the same dimension as d. The number of measured variable m is usually much less than the number of state variables (or n DOFs of the system) but greater than or equal to n_f, the length of vector f. In the case of a two-dimensional simply supported beam divided into L elements, $n = 2(L+1) - 2$ including all vertical displacements and rotational displacements at each of its nodes.

When the unknown force vector f is included in the equation of motion of the system, an exact match of the model with the measured data will usually not work. This is due to the fact that all measurements have some degree of noise. Even the least-squares criteria is not sufficient because a mathematical solution that will minimize the least-squares error E represented by:

$$E = (r - AP)^T R(r - AP) \tag{1.3}$$

will usually end up with the model exactly matching the data. This situation could be avoided by adding a smoothing term to the least-squares error (Busby and Trujillo, 1993; Santantamarina and Fratta, 1998) to become a non-linear least-squares problem

$$J(P, \lambda) = (r - AP, R(r - AP)) + \lambda(SP, SP) \tag{1.4}$$

where λ is the regularization parameter. (x, y) in Equation (1.4) denotes the inner product of two vectors x and y. R is an error-weighting matrix depending on the measured information. Some measured data are of small amplitude and therefore are greatly affected by systematic errors and background noise. Therefore the weighting matrix R is usually taken as the inverse of the covariance of the measured data. S is a smoothing matrix, which is typically either the identity matrix or a discrete approximation to a derivative operator (Santantamarina and Fratta, 1998). The first term in brackets on the right-hand side is the Euclidean scalar product. The second term with the regularization parameter λ has the effect of smoothing the identified forces. A small value of λ causes the solution to match the data closely but produces large oscillatory deviations. A large value of λ produces smooth forces that may not match the data well. When λ is zero, the solution becomes that for the least-squares problem.

The moving loads can be obtained from Equation (1.4) as:

$$P = (A^T R A + \lambda S^T S)^{-1} A^T R r \tag{1.5}$$

Both S and R are usually taken as the identity matrix, which means a minimum 2-norm in the solution and no priori information on the measured data is available respectively. Two terms can be derived from Equation (1.4) as follows using the singular value decomposition:

$$E_1^2 = (r - AP, r - AP) = \sum_{i=1}^{N_P * N_t} \left| \sigma_i(v_i^T P) - (u_i^T r) \right|^2 + \sum_{i=N_P * N_t + 1}^{N_S * N_t} \left| (u_i^T r) \right|^2 \tag{1.6}$$

$$E_2^2 = (P, P) = \sum_{i=1}^{N_p * N_t} \left| (v_i^T P) \right|^2 \tag{1.7}$$

where $A = U \Sigma V^T$, $U = \{u_1, u_2, \ldots, u_{N_S * N_t}\}$ and $V = \{v_1, v_2, \ldots, v_{N_P * N_t}\}$ are the right and left singular vectors respectively, and $\Sigma = diag\{\sigma_1, \sigma_2, \ldots, \sigma_{N_P * N_t}\}$ are the singular values with $(\sigma_1 \geq \sigma_2 \geq \cdots \geq \sigma_{N_P * N_t} \geq 0)$.

Substitute Equations (1.6) and (1.7) into Equation (1.4), the solution P_λ can be obtained as:

$$P_\lambda = \sum_{i=1}^{N_P * N_t} \frac{\sigma_i^2}{(\sigma_i^2 + \lambda)} \sigma_i^{-1} (u_i^T r) v_i \tag{1.8}$$

When $\lambda = 0$, P_λ is the least-squares solution and the noise effect will be amplified when $\sigma_i < u_i^T r$. When $\lambda > 0$, this formulation can reduce the influence of the components corresponding to those singular values σ_i^2 which are smaller than λ, so that the solution is less noise sensitive. Substitute Equation (1.8) into Equations (1.6) and (1.7), we have:

$$E_1^2 = \sum_{i=1}^{N_P * N_t} \frac{\lambda^2}{(\sigma_i^2 + \lambda)^2} \left| (u_i^T r) \right|^2 + \sum_{i=N_P * N_t}^{N_S * N_t} \left| (u_i^T r) \right|^2 \tag{1.9}$$

$$E_2^2 = \sum_{i=1}^{N_p * N_t} \frac{\sigma_i^2}{(\sigma_i^2 + \lambda)^2} \left| (u_i^T r) \right|^2 \tag{1.10}$$

where E_1^2 is an increasing function of λ, whereas E_2^2 is a decreasing function of λ, and thus an optimal regularization parameter exists with balance contribution from the two different errors.

If the true loads are known, the optimal regularization parameter can be obtained by minimizing the error between the true loads and the estimate loads. However, the true loads are not known in practice, and the generalized cross validation and L-curve method are employed to determine the optimal regularization parameter.

The Tikhonov regularization technique has always been successful for solving ill-posed problems. Examples of its application with the identification of interactions in medical science research can be found in (Kwon et al., 2005; Millan, 2005; Chang and Fischbach, 2006; Hanninen et al., 2007) and in the delay control of a nonlinear beam (Qian and Tang, 2008). The conventional Tikhonov regularization, however, suffers the disadvantage of losing its ability to regularize the solution with iterations, and an improved adaptive Tikhonov regularization has been proposed for a solution (Li and Law, 2010a; 2010b) where a changing reference limit is adopted to find the optimal regularization parameter λ in each iteration.

1.7 Model Condensation Techniques

Structural analysis of a large and complex structure is computationally expensive or sometimes impossible with the full degrees-of-freedom of the structure. In the inverse problem of force identification, the measured structural responses from field tests is limited compared to the total degrees-of-freedom of the structural system due to the limitations of measuring techniques and equipments. This problem can be solved with the transformation of the measured DOFs to the full DOFs of the structural system via the model condensation techniques. Since this type of technique plays a very important role in the structural dynamic analysis and the inverse analysis with a large-scale structural system, a brief review of this topic is given in this section.

The first model condensation method is called Guyan/Irons method (Guyan, 1965; Irons, 1965) which is a static condensation method without considering the inertia effects. Considering a system with only a static force F_m applied on some selected DOFs, the stiffness matrix and the response vector are partitioned such that:

$$\begin{bmatrix} K_{mm} & K_{ms} \\ K_{sm} & K_{ss} \end{bmatrix} \begin{Bmatrix} \mathbf{x}_m \\ \mathbf{x}_s \end{Bmatrix} = \begin{Bmatrix} F_m \\ 0 \end{Bmatrix} \tag{1.11}$$

where the subscript 'm' and 's' denote the master (selected) DOFs and slave (truncated) DOFs, respectively. From the second equation in Equation (1.11), the relationship between the selected response vector $\mathbf{x}_m$ and the truncated response vector $\mathbf{x}_s$ can be obtained as:

$$\mathbf{x}_s = -\mathbf{K}_{ss}^{-1} \mathbf{K}_{sm} \mathbf{x}_m \tag{1.12}$$

The response vector $\mathbf{x}$ is calculated from the selected response $\mathbf{x}_m$,

$$\mathbf{x} = \begin{Bmatrix} \mathbf{x}_m \\ \mathbf{x}_s \end{Bmatrix} = \begin{bmatrix} \mathbf{I} \\ -K_{ss}^{-1} K_{sm} \end{bmatrix} \mathbf{x}_m = \mathbf{T}_s \mathbf{x}_m \tag{1.13}$$

where $\mathbf{T}_s = \begin{bmatrix} \mathbf{I} \\ -\mathbf{K}_{ss}^{-1}\mathbf{K}_{sm} \end{bmatrix}$ is the static transformation matrix between the full state vector $\mathbf{x}$ and the master coordinates vector $\mathbf{x}_m$. The condensed mass and stiffness matrices $\mathbf{M}_{Rs}$, $\mathbf{K}_{Rs}$, respectively, can be expressed as:

$$\mathbf{M}_{Rs} = \mathbf{T}_s^T \mathbf{M} \mathbf{T}_s, \quad \mathbf{K}_{Rs} = \mathbf{T}_s^T \mathbf{K} \mathbf{T}_s \tag{1.14}$$

where $\mathbf{M}$ and $\mathbf{K}$ are the system matrices before reduction. The eigen-solution for the condensed system can be denoted as:

$$(\mathbf{K}_{Rs} - \omega^2 \mathbf{M}_{Rs})\mathbf{u}_m = 0 \tag{1.15}$$

where ω^2 and $\mathbf{u}_m$ are the eigenvalues and eigenvectors of the condensed system, respectively. It is noted that any frequency response functions generated from the reduced matrices in Equation (1.14) are exact only when the frequency is equal to zero. With the increase of the frequency of a dynamic system, the inertia effect ignored will become more significant which may cause condensation error in this method to a certain extent. Since improper selection of DOFs in model condensation may result in singularity of the eigenvalue problem, several selection schemes (Shan and Raymund, 1982; Matta, 1987) of master DOFs were developed to improve the accuracy. These schemes can also be adopted in other model reduction methods reviewed in this section.

In order to improve the Guyan/Iron method, the dynamic condensation method in which the inertia terms neglected in Equation (1.11) are included was developed (Kuhar and Stahle, 1974; Miller, 1980). The equation becomes:

$$\begin{bmatrix} \mathbf{M}_{mm} & \mathbf{M}_{ms} \\ \mathbf{M}_{sm} & \mathbf{M}_{ss} \end{bmatrix} \begin{Bmatrix} \ddot{\mathbf{x}}_m \\ \ddot{\mathbf{x}}_s \end{Bmatrix} + \begin{bmatrix} \mathbf{K}_{mm} & \mathbf{K}_{ms} \\ \mathbf{K}_{sm} & \mathbf{K}_{ss} \end{bmatrix} \begin{Bmatrix} \mathbf{x}_m \\ \mathbf{x}_s \end{Bmatrix} = 0 \tag{1.16}$$

The eigenvalue problem becomes:

$$\left(-\omega^2 \begin{bmatrix} \mathbf{M}_{mm} & \mathbf{M}_{ms} \\ \mathbf{M}_{sm} & \mathbf{M}_{ss} \end{bmatrix} + \begin{bmatrix} \mathbf{K}_{mm} & \mathbf{K}_{ms} \\ \mathbf{K}_{sm} & \mathbf{K}_{ss} \end{bmatrix} \right) \begin{Bmatrix} \mathbf{x}_m \\ \mathbf{x}_s \end{Bmatrix} = 0 \tag{1.17}$$

Thus the truncated response vector $\mathbf{x}_s$ can be obtained in terms of $\mathbf{x}_m$ as:

$$\mathbf{x}_s = -[\mathbf{K}_{ss} - \omega^2 \mathbf{M}_{ss}]^{-1}[\mathbf{K}_{sm} - \omega^2 \mathbf{M}_{sm}]\mathbf{x}_m = \mathbf{T}_d \mathbf{x}_m \tag{1.18}$$

where $\mathbf{T}_d = -[\mathbf{K}_{ss} - \omega^2 \mathbf{M}_{ss}]^{-1}[\mathbf{K}_{sm} - \omega^2 \mathbf{M}_{sm}]$ is the dynamic transformation matrix between the full state vector and the master coordinates. It is noted that the eigenvalue ω in $\mathbf{T}_d$ is unknown. This problem could be solved by selecting a system frequency $\overline{\omega}$ beginning with an initial zero value. A new transformation matrix based on the updated eigenvalue is then calculated, and the process is repeated until the eigenvalue no longer changes. After obtaining the dynamic transformation matrix, the condensed mass and stiffness matrices $\mathbf{M}_{Rd}$, $\mathbf{K}_{Rd}$, respectively, can be expressed as:

$$\mathbf{M}_{Rd} = \mathbf{T}_d^T \mathbf{M} \mathbf{T}_d, \quad \mathbf{K}_{Rd} = \mathbf{T}_d^T \mathbf{K} \mathbf{T}_d \tag{1.19}$$

To avoid the matrix inverse in calculating $\mathbf{T}_d$, matrix $\mathbf{T}_d$ can be obtained by applying the binomial theorem as:

$$
\begin{aligned}
\mathbf{T}_d &= -[\mathbf{K}_{ss} - \omega^2 \mathbf{M}_{ss}]^{-1}[\mathbf{K}_{sm} - \omega^2 \mathbf{M}_{sm}] \\
&= -\mathbf{K}_{ss}^{-1}[\mathbf{I} - \omega^2 \mathbf{M}_{ss}\mathbf{K}_{ss}^{-1}]^{-1}[\mathbf{K}_{sm} - \omega^2 \mathbf{M}_{sm}] \\
&= -\mathbf{K}_{ss}^{-1}[\mathbf{I} + \omega^2 \mathbf{M}_{ss}\mathbf{K}_{ss}^{-1}][\mathbf{K}_{sm} - \omega^2 \mathbf{M}_{sm}] \\
&= -\mathbf{K}_{ss}^{-1}[\mathbf{K}_{sm} + \omega^2 (\mathbf{M}_{ss}\mathbf{K}_{ss}^{-1}\mathbf{K}_{sm} - \mathbf{M}_{sm})]
\end{aligned}
\tag{1.20}
$$

Substituting Equations (1.18) and (1.20) into Equation (1.17), and neglecting the higher order terms of ω compared to ω^2, it gives:

$$
\begin{aligned}
\omega^2 (\mathbf{M}_{mm} - \mathbf{M}_{ms}\mathbf{K}_{ss}^{-1}\mathbf{K}_{sm} - \mathbf{K}_{ms}\mathbf{K}_{ss}^{-1}\mathbf{M}_{ss}\mathbf{K}_{ss}^{-1}\mathbf{K}_{sm})\mathbf{x}_m \\
= (\mathbf{K}_{mm} - \mathbf{K}_{ms}\mathbf{K}_{ss}^{-1}\mathbf{K}_{sm})\mathbf{x}_m
\end{aligned}
\tag{1.21}
$$

The eigenvalues can be calculated directly from Equation (1.21) without the need of iteration. In fact, the calculated eigenvalue equals to the one calculated from Guyan/Iron method which can be expressed as:

$$
\omega^2 \mathbf{M}_{Rs}\mathbf{u}_m = \mathbf{K}_{Rs}\mathbf{u}_m
\tag{1.22}
$$

where $\mathbf{M}_{Rs}$, $\mathbf{K}_{Rs}$ are the matrices as shown in Equation (1.14). Employing the relationship established in Equation (1.22), the dynamic transformation matrix $\mathbf{T}_d$ can be obtained from the following equation:

$$
\mathbf{T}_{id} = -\mathbf{K}_{ss}^{-1}[\mathbf{K}_{sm} + (\mathbf{M}_{ss}\mathbf{K}_{ss}^{-1}\mathbf{K}_{sm} - \mathbf{M}_{sm})\mathbf{M}_R^{-1}\mathbf{K}_R]
\tag{1.23}
$$

where $\mathbf{M}_R$ and $\mathbf{K}_R$ are the system mass and stiffness matrices after reduction. $\mathbf{T}_{id}$ is the transformation matrix of a new model condensation method called the Improved Reduction System (IRS) method. The formulation including Equations (1.20) to (1.23) was proposed by Gordis (1992).

The same dynamic transformation matrix $\mathbf{T}_{id}$ was also derived by O'Callahan (1992) with a different procedure. Considering a static force F applies on all the DOFs of a structure. By having the same partition as in Guyan/Iron method, Equation (1.11) becomes:

$$
\begin{bmatrix} \mathbf{K}_{mm} & \mathbf{K}_{ms} \\ \mathbf{K}_{sm} & \mathbf{K}_{ss} \end{bmatrix} \begin{Bmatrix} \mathbf{x}_m \\ \mathbf{x}_s \end{Bmatrix} = \begin{Bmatrix} F_m \\ F_s \end{Bmatrix}
\tag{1.24}
$$

where F_s is the force applied on the slave DOFs. The truncated set of equations in Equation (1.24) is:

$$
\mathbf{K}_{sm}\mathbf{x}_m + \mathbf{K}_{ss}\mathbf{x}_s = F_s
\tag{1.25}
$$

Solving Equation (1.25), $\mathbf{x}_s$ can be calculated as:

$$
\mathbf{x}_s = -\mathbf{K}_{ss}^{-1}\mathbf{K}_{sm}\mathbf{x}_m + \mathbf{K}_{ss}^{-1}F_s
\tag{1.26}
$$

Thus the full state vector $\mathbf{x}$ can be expressed as:

$$\mathbf{x} = \mathbf{T}_s \mathbf{x}_m + \mathbf{x}_{Fd} \tag{1.27}$$

where $\mathbf{T}_s$ has been shown in Equation (1.13) and $\mathbf{x}_{Fd}$ is the truncated distributed force adjustment which is defined as:

$$\mathbf{x}_{Fd} = \left\{ \begin{matrix} 0 \\ \mathbf{K}_{ss}^{-1} F_s \end{matrix} \right\} = \begin{bmatrix} 0 & 0 \\ 0 & \mathbf{K}_{ss}^{-1} \end{bmatrix} F = \mathbf{K}_s^{-1} F \tag{1.28}$$

According to Equations (1.22) and (1.24), and noting that the full modal vector $\mathbf{u}$ can be expressed by the selected vector $\mathbf{u}_m$ using the static reduction, i.e. $\mathbf{u} = \mathbf{T}_s \mathbf{u}_m$, $\mathbf{x}_{Fd}$ can be obtained as:

$$\mathbf{x}_{Fd} = \mathbf{K}_s^{-1} \mathbf{M} \mathbf{T}_s \mathbf{u}_m \omega^2 = K_{ss}^{-1} \mathbf{M}_{ss} \mathbf{T}_s \mathbf{M}_R^{-1} \mathbf{K}_R \mathbf{u}_m \tag{1.29}$$

Combine Equations (1.28) and (1.29),

$$\mathbf{x} = (\mathbf{T}_s + \mathbf{K}_{ss}^{-1} \mathbf{M}_{ss} \mathbf{T}_s \mathbf{M}_R^{-1} \mathbf{K}_R) \mathbf{u}_m = \mathbf{T}_{IRS} \mathbf{u}_m \tag{1.30}$$

where

$$\mathbf{T}_{IRS} = -\mathbf{K}_{ss}^{-1} [\mathbf{K}_{sm} + (\mathbf{M}_{ss} \mathbf{K}_{ss}^{-1} \mathbf{K}_{sm} - \mathbf{M}_{sm}) \mathbf{M}_R^{-1} \mathbf{K}_R] \tag{1.31}$$

It is noted that $\mathbf{T}_{IRS}$ in Equation (1.31) is identical to matrix $\mathbf{T}_{id}$ shown in Equation (1.23). The mass and stiffness matrices for the reduced system shown in Equation (1.23) or (1.31) have superior performance than that in Guyan/Iron method since the deleted inertia effect has been included. The improved mass and stiffness matrices for the reduced system then become:

$$\mathbf{M}_{RI} = \mathbf{T}_{IRS}^T \mathbf{M} \mathbf{T}_{IRS}, \quad \mathbf{K}_{RI} = \mathbf{T}_{IRS}^T \mathbf{K} \mathbf{T}_{IRS} \tag{1.32}$$

The IRS method is relatively insensitive to the number and location of the DOFs comparing to the two methods aforementioned. The IRS method was further improved and modified to become iterative by other researchers and it is called the Iterative IRS method (IIRS). There are two kinds of IIRS method. The first one was proposed by Blair et al. (1991). Since the transformation matrix for IRS utilized the reduced mass and stiffness matrices approximated by the Guyan/Iron method, an improvement can be made to use the newly approximated matrices described in Equation (1.32) and a better transformation matrix can be constructed by iterating on Equation (1.31) to become:

$$\mathbf{T}_{IRS,i+1} = -\mathbf{K}_{ss}^{-1} [\mathbf{K}_{sm} + (\mathbf{M}_{ss} \mathbf{T}_s - \mathbf{M}_{sm}) \mathbf{M}_{RI,i}^{-1} \mathbf{K}_{RI,i}] \tag{1.33}$$

and Equation (1.32) becomes:

$$\mathbf{M}_{RI,i+1} = \mathbf{T}_{IRS,i+1}^T \mathbf{M} \mathbf{T}_{IRS,i+1}, \quad \mathbf{K}_{RI,i+1} = \mathbf{T}_{IRS,i+1}^T \mathbf{K} \mathbf{T}_{IRS,i+1} \tag{1.34}$$

The second kind of IIRS method (Friswell et al., 1995) made further improvement by modifying Equation (1.33) as:

$$\mathbf{T}_{IRS,i+1} = -\mathbf{K}_{ss}^{-1}[\mathbf{K}_{sm} + (\mathbf{M}_{ss}\mathbf{T}_{IRS,i} - \mathbf{M}_{sm})\mathbf{M}_{RI,i}^{-1}\mathbf{K}_{RI,i}] \tag{1.35}$$

1.8 Summary

The structure–force interaction is typically discussed in terms of the bridge–vehicle interaction problem in practice with descriptions on the different models on the forces and the structure itself. Existing dynamic response analysis techniques are classified broadly into two areas of modal decomposition technique and finite element method with the problem of a vehicle moving across a bridge deck as the background.

Literature on the force identification and moving force identification are reviewed and grouped into two broad categories of methods with the solution algorithm based on the modal decomposition techniques and the finite element method. The solution technique to find the optimal regularized solution of the ill-posed moving load identification technique is discussed in terms of two types of errors. The basics on model condensation techniques which are very important for an accurate solution of the inverse problem are also given at the end of this Chapter.

Part I

Moving Load Problems

Chapter 2

Dynamic Response of Multi-span Continuous Beams under Moving Loads

2.1 Introduction

The dynamic behavior of a beam subject to moving loads has been extensively studied. The structural configuration can be categorized as single-span or multi-span uniform or non-uniform beam. Many of these reports are found in the monograph of Fryba (1972) and most of them are on a uniform, single span, simply supported beam.

This chapter presents the dynamic analysis of a multi-span continuous beam subject to moving loads. Three methods of analysis are introduced in Section 2.2: The Exact Solution Method, the Assumed Mode Shape Method and the Precise Integration Method. Discussions on the effect of elastic supports on the analyzed results can be found in Section 2.3.

2.2 Multi-span Continuous Beam

This Section summaries the more recent literature on the vehicle–bridge interaction with a multi-span continuous beam. Modal superposition method is commonly employed to analyze the vibration of a multi-span continuous beam under moving loads. Hayashikawa and Watanabe (1981) and Wang (1997) employed the eigen-stiffness matrix method. Chatterjee et al. (1994) investigated the dynamic response of the multi-span continuous bridge under a moving vehicle modeled as a single unsprung or sprung mass. Wu and Dai (1987) studied the response of a multi-span, non-uniform beam subject to a series of loads moving with varying speed in the same and opposite directions. The transfer matrix method is used to determine the natural frequencies and mode shapes of the beam. Lee (1994) used the Euler beam theory and the assumed mode method to analyze the transverse vibration of a beam with intermediate point constraints subject to a moving load. The vibration modes of a simply-supported beam are used as the assumed modes. The point constraints, in the form of supports, are assumed to be linear springs of large stiffness. Lin (1995) has mentioned that the selection of support stiffness is problem dependent, and it should be used with care if numerical stability in the solution has to be maintained. Zheng et al. (1998) used the modified beam functions as the assumed modes to analyze the vibration of a multi-span non-uniform Euler beam subject to a moving load. The modified beam functions satisfied the zero deflection conditions at all the intermediate supports as well as the boundary conditions at the two ends of the beam. Henchi et al. (1997) used an exact dynamic stiffness matrix under the framework of finite element approximation to

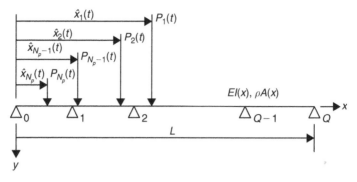

Figure 2.1 A continuous beam with $(Q-1)$ intermediate point supports under N_p moving forces

study the dynamic response of multi-span structures under a convoy of moving loads. Ichikawa et al. (2000) investigated the dynamic response of a symmetric three-span continuous Euler-Bernoulli beam subject to a moving load at time-dependent velocity using the method of eigenfunction expansion or the modal analysis, and estimated the effects of acceleration or deceleration of a moving load on the dynamic amplification factor. Khang et al. (2009) investigated the dynamic response of prestressed beams on rigid supports subject to moving concentrated loads. The effects of prestressing and the speed of moving loads on the dynamic response of the beam were discussed.

The following sections discuss the three latest techniques that are more suitable for the analysis of the vehicle–bridge interaction problem.

2.2.1 The Exact Solution

A continuous beam subject to a system of moving forces $P_l(t)(l=1,2,\ldots,N_p)$ is shown in Figure 2.1. The forces are assumed moving as a group at a prescribed velocity $v(t)$ along the axial direction of the beam from left to right. The beam is assumed to be an Euler-Bernoulli beam. The equation of motion can be written as

$$\rho A \frac{\partial^2 w(x,t)}{\partial t^2} + C \frac{\partial w(x,t)}{\partial t} + EI \frac{\partial^4 w(x,t)}{\partial x^4} = \sum_{l=1}^{N_p} P_l(t)\delta(x-\hat{x}_l(t)) \tag{2.1}$$

where L is the total length of the beam; A is the cross-sectional area; E is the Young's modulus; I is the moment of inertia of the beam cross-section; ρ, C and $w(x,t)$ are the mass per unit length, the damping and the displacement function of the beam respectively; $\hat{x}_l(t)$ is the location of moving force $P_l(t)$ at time t; $\delta(t)$ is the Dirac delta function and N_p is the number of forces.

2.2.1.1 Free Vibration

The eigenfunction of a Q span Euler-Bernoulli continuous beam as shown in Figure 2.2 can be written in the following form.

$$r_i(x_i) = A_i \sin \beta x_i + B_i \cos \beta x_i + C_i \sinh \beta x_i + D_i \cosh \beta x_i, \quad (i=1,2,\ldots,Q) \tag{2.2}$$

where $r_i(x_i)$ is the eigenfunction for the ith span, and β is the eigenvalue. Hayashikawa and Watanabe (1981) have presented the formulation of the eigenfunction with

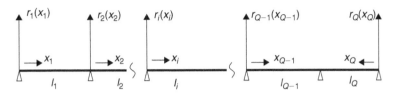

Figure 2.2 A Q-span continuous beam

arbitrary boundary conditions. The same problem can be solved for a simply supported beam with the following boundary conditions:

$$r_i(x_i)|_{x_i=0} = r_i(x_i)|_{x_i=l_i} = 0; \quad (i = 1, 2, \ldots, Q)$$

$$\left.\frac{\partial^2 r_1(x_1)}{\partial x_1^2}\right|_{x_1=0} = \left.\frac{\partial^2 r_Q(x_Q)}{\partial x_Q^2}\right|_{x_Q=0} = 0$$

$$\begin{cases} \left.\dfrac{\partial r_i(x_i)}{\partial x_i}\right|_{x_i=l_i} = \left.\dfrac{\partial r_{i+1}(x_{i+1})}{\partial x_{i+1}}\right|_{x_{i+1}=0} ; \\[4mm] \left.\dfrac{\partial^2 r_i(x_i)}{\partial x_i^2}\right|_{x_i=l_i} = \left.\dfrac{\partial^2 r_{i+1}(x_{i+1})}{\partial x_{i+1}^2}\right|_{x_{i+1}=0} ; \quad (i = 1, 2, \ldots, Q-2) \end{cases} \tag{2.3}$$

$$\left.\frac{\partial r_{Q-1}(x_{Q-1})}{\partial x_{Q-1}}\right|_{x_{Q-1}=l_{Q-1}} = -\left.\frac{\partial r_Q(x_Q)}{\partial x_Q}\right|_{x_Q=l_Q}$$

$$\left.\frac{\partial^2 r_{Q-1}(x_{Q-1})}{\partial x_{Q-1}^2}\right|_{x_{Q-1}=l_{Q-1}} = \left.\frac{\partial^2 r_Q(x_Q)}{\partial x_Q^2}\right|_{x_Q=l_Q}.$$

Substituting the boundary conditions in Equation (2.2), the mode shape of the continuous beam can be written as (Zhu and Law, 2001a):

$$\phi(x) = \begin{cases} A_1\left(\sin(\beta x) - \dfrac{\sin(\beta l_1)}{\sinh(\beta l_1)} \sinh(\beta x)\right), \quad 0 \leq x \leq l_1 \\[6mm] A_i\left(\sin\left(\beta\left(x - \displaystyle\sum_{j=1}^{i-1} l_j\right)\right) - \dfrac{\sin(\beta l_i)}{\sinh(\beta l_i)} \sinh\left(\beta\left(x - \displaystyle\sum_{j=1}^{i-1} l_j\right)\right)\right) \\[6mm] \quad + B_i\left(\cos\beta\left(\beta\left(x - \displaystyle\sum_{j=1}^{i-1} l_j\right)\right) - \cosh\left(\beta\left(x - \displaystyle\sum_{j=1}^{i-1} l_j\right)\right)\right. \\[6mm] \quad + \dfrac{\cosh(\beta l_i) - \cos(\beta l_i)}{\sinh(\beta l_i)} \left.\sinh\left(\beta\left(x - \displaystyle\sum_{j=1}^{i-1} l_j\right)\right)\right), \\[6mm] \qquad\qquad \displaystyle\sum_{j=1}^{i-1} l_j \leq x \leq \sum_{j=1}^{i} l_j \quad (i = 2, 3, \ldots, Q-1) \\[6mm] A_Q(\sin(\beta(L-x)) - \dfrac{\sin(\beta l_Q)}{\sinh(\beta l_Q)} \sinh(\beta(L-x))), \quad L - l_Q \leq x \leq L \end{cases} \tag{2.4}$$

where parameters $\beta, A_1, A_i, B_i (i = 2, 3, \ldots, Q-1)$ and A_Q are determined by solving the following set of equations (Gorman, 1975):

$$[F]\{A\} = 0 \qquad (2.5)$$

where

$$A = \{A_1, A_2, B_2, \ldots, A_{Q-1}, B_{Q-1}, A_Q\}.$$

The elements in matrix F are given by

$$f_{11} = \cos(\beta l_1) - \theta_1 {}^* \cosh(\beta l_1); \quad f_{12} = \theta_2 - 1; \quad f_{13} = -\phi_2;$$
$$f_{21} = \sin(\beta l_1); \quad f_{23} = -1;$$

$$\begin{cases} f_{2i-1,2(i-1)} = \cos(\beta l_i) - \theta_i {}^* \cosh(\beta l_i); \\ f_{2i-1,2i-1} = -\sin(\beta l_i) - \sinh(\beta l_i) + \phi_i \cosh(\beta l_i); \\ f_{2i-1,2i} = \theta_{i+1} - 1; \\ f_{2i-1,2i+1} = -\phi_{i+1}; \qquad\qquad\qquad\qquad (i = 2, 3, \ldots, Q-2) \\ f_{2i,2(i-1)} = -\sin(\beta l_i) - \theta_i {}^* \sinh(\beta l_i); \\ f_{2i,2i-1} = -\cos(\beta l_i) - \cosh(\beta l_i) + \phi_i \sinh(\beta l_i); \\ f_{2i,2i+1} = 2. \end{cases}$$

$$\begin{cases} f_{2Q-1,2(Q-2)} = -\cos(\beta l_{Q-1}) + \theta_{Q-1} \cosh(\beta l_{Q-1}); \\ f_{2Q-1,2Q-1} = \sin(\beta l_{Q-1}) + \sinh(\beta l_{Q-1}) - \phi_{Q-1} \cosh(\beta l_{Q-1}) \\ f_{2Q-1,2(Q-1)} = \theta_Q \cosh(\beta l_Q) - \cos(\beta l_Q); \\ f_{2(i-1),2(i-2)} = \sin(\beta l_{Q-1}) + \theta_{Q-1} \sinh(\beta l_{Q-1}) \qquad\qquad (2.6) \\ f_{2(i-1),2i-1} = \cos(\beta l_{Q-1}) + \cosh(\beta l_{Q-1}) - \phi_{Q-1} \sinh(\beta l_{Q-1}); \\ f_{2(i-1),2(i-1)} = -2 \sin(\beta l_Q) \end{cases}$$

where

$$\theta_i = \frac{\sin(\beta l_i)}{\sinh(\beta l_i)}; \qquad \phi_i = \frac{\cosh(\beta l_i) - \cos(\beta l_i)}{\sinh(\beta l_i)}; \qquad (i = 1, 2, \ldots, Q)$$

and the other coefficients f_{ij} equal to zero.

2.2.1.2   Dynamic Behavior under Moving Loads

Express the transverse displacement $w(x, t)$ in modal coordinates

$$w(x, t) = \sum_{i=1}^{\infty} \phi_i(x) q_i(t) \qquad (2.7)$$

where $\phi_i(x)$ is the mode shape function of the ith mode, which is determined from the eigenvalue and eigenfunction analysis as shown above; $q_i(t)$ is the ith modal amplitude. Substituting Equation (2.7) into Equation (2.1), and multiplying by $\phi_i(x)$, integrating with respect to x between 0 and L, and applying the orthogonality conditions, we obtain:

$$\frac{d^2 q_i(t)}{dt^2} + 2\xi_i \omega_i \frac{dq_i(t)}{dt} + \omega_i^2 q_i(t) = \frac{1}{M_i} \sum_{l=1}^{N_p} P_l(t) \phi_i(\hat{x}_l(t)) \tag{2.8}$$

where ω_i, ξ_i, M_i are the modal frequency, the damping ratio and the modal mass of the ith mode, with:

$$M_i = \int_0^L \rho A \phi_i^2(x) dx \tag{2.9}$$

The displacement of the beam at point x and time t can be found from Equations (2.7) and (2.8) as:

$$w(x, t) = \sum_{i=1}^{\infty} \frac{\phi_i(x)}{M_i} \int_0^t h_i(t - \tau) \sum_{l=1}^{N_p} P_l(\tau) \phi_i(\hat{x}_l(\tau)) d\tau \tag{2.10}$$

where

$$h_i(t) = \frac{1}{\omega_i'} e^{-\xi_i \omega_i t} \sin \omega_i' t; \quad \omega_i' = \omega_i \sqrt{1 - \xi_i^2} \tag{2.11}$$

2.2.2 Solution with Assumed Modes

There is a lot of multi-span continuous bridges with large cross-sections, and the effect of variation of the cross-sectional dimensions on the dynamic properties cannot be neglected. The effect of rotatory inertia and of shear deformation must be considered. In this section, the vibrational behavior of the multi-span non-uniform Timoshenko beam subject to a set of moving loads is analyzed basing on Hamilton's principle with the intermediate point constraints represented by very stiff linear springs.

The continuous Timoshenko beam in Figure 2.1 is constrained at $(Q-1)$ intermediate point supports under N_p number of moving loads. The loads $P_s(t)$ $(s = 1, 2, \ldots, N_p)$ are moving as a group at a prescribed velocity v along the axial direction of the beam from left to right. v is assumed constant in this study. The load locations are described by $\hat{x}_i(t)$ with $\hat{x}_i(t) = \hat{x}_{i0} + vt$ where $\hat{x}_{i0}$ is the initial location of the load, i.e. the loads can initially be at any positions on the beam. The bending moment and the transverse shear force of the beam are given as:

$$M(x) = EI(x) \frac{\partial \psi(x, t)}{\partial x}$$

$$V(x) = \kappa GA(x) \left[\frac{\partial w(x, t)}{\partial x} - \psi(x, t) \right] \tag{2.12}$$

where G is the shear modulus of the beam material and $A(x)$ is the cross-sectional area; E is the Young's modulus; $I(x)$ is the moment of the inertia of the beam cross-section;

κ is the shear coefficient; $w(x, t)$ is the transverse displacement function of the beam; $\psi(x, t)$ is the angle of rotation at a cross-section.

The kinetic energy T of the beam, the strain energy U_e, the potential energy due to point constraints U_Q, and the work done W due to the moving loads can be written for the Timoshenko beam as follows:

$$T = \frac{1}{2} \int_0^L \rho A(x) \left[\left(\frac{\partial w(x, t)}{\partial t} \right)^2 + \gamma^2(x) \left(\frac{\partial \psi(x, t)}{\partial t} \right)^2 \right] dx$$

$$U_e = \frac{1}{2} \int_0^L \left[EI(x) \left(\frac{\partial \psi(x, t)}{\partial x} \right)^2 + \kappa GA(x) \left(\frac{\partial w(x, t)}{\partial x} - \psi(x, t) \right)^2 \right] dx$$

$$(2.13)$$

$$U_Q = \frac{1}{2} k \sum_{i=1}^{Q-1} w(x_i, t)^2$$

$$W = \int_0^L \sum_{i=1}^{N_p} \delta(x - \hat{x}_i(t)) P_i(t) w(x, t) dx$$

where ρ is the mass density of material of the beam; $x_i (i = 0, 1, 2, \ldots, Q)$ are co-ordinates of the intermediate point supports and end supports and $\delta(t)$ is the Dirac delta function. $\gamma(x)$ is the radius of gyration of the beam cross-section. k is the stiffness of the point constraints. Expressing the vibration responses of the beam $w(x, t)$ and $\psi(x, t)$ in modal coordinates

$$w(x, t) = \sum_{i=1}^{n} q_i(t) W_i(x)$$

$$\psi(x, t) = \sum_{i=1}^{n} q_i(t) \phi_i(x)$$

$$(i = 1, 2, \ldots, n) \qquad (2.14)$$

where $W_i(x)$, $\phi_i(x)$ are the assumed vibration modes that satisfy the boundary conditions and $q_i(t)$ is the generalized co-ordinate, a set of such assumed vibration modes is given in Section 2.2.2.1.

Substituting Equation (2.14) into (2.13), we obtain:

$$T = \frac{1}{2} \int_0^L \rho A(x) \left[\sum_{i=1}^{n} \dot{q}_i(t) W_i(x) \sum_{j=1}^{n} \dot{q}_j(t) W_j(x) + \gamma^2(x) \sum_{i=1}^{n} \sum_{j=1}^{n} \dot{q}_i(t) \phi_i(x) \dot{q}_j(t) \phi_j(x) \right] dx$$

$$= \sum_{i=1}^{n} \sum_{j=1}^{n} \frac{1}{2} \dot{q}_i(t) m_{ij} \dot{q}_j(t)$$

$$U_e = \frac{1}{2} \int_0^L \left[EI(x) \sum_{i=1}^n q_i(t)\phi_i'(x) \sum_{j=1}^n q_i(x)\phi_j'(x) \right.$$

$$\left. + \kappa GA(x) \left(\sum_{i=1}^n q_i(t) W_i'(x) - \sum_{i=1}^n q_i(t)\phi_i(x) \right) \left(\sum_{j=1}^n q_j(t) W_j'(x) - \sum_{j=1}^n q_j(t)\phi_j(x) \right) \right] dx$$

$$= \frac{1}{2} \sum_{i=1}^n \sum_{j=1}^n q_i(t) k_{ij}' q_j(t)$$

$$U_Q = \frac{1}{2} k \sum_{s=1}^{Q-1} \sum_{i=1}^n q_i(t) W_i(x_s) \sum_{j=1}^n q_j(t) W_j(x_s) = \sum_{i=1}^n \sum_{j=1}^n \frac{1}{2} q_i(t) k_{ij}'' q_j(t)$$

$$W = \int_0^L \sum_{j=1}^{N_p} \delta(x - \hat{x}_j(t)) P_j(t) \sum_{i=1}^n q_i(t) W_i(x) dx = \sum_{i=1}^n q_i(t) f_i(t) \qquad (2.15)$$

where

$$m_{ij} = \int_0^L \rho A(x) [W_i(x) W_j(x) + \gamma^2(x)\phi_i(x)\phi_j(x)] dx$$

$$k_{ij}' = \int_0^L [EI(x)\phi_i'(x)\phi_j'(x) + \kappa GA(x)(W_i'(x) - \phi_i(x))(W_j'(x) - \phi_j(x))] dx$$

$$\qquad (2.16)$$

$$k_{ij}'' = k \sum_{l=1}^{Q-1} W_i(x_l) W_j(x_l)$$

$$f_i(t) = \sum_{l=1}^{N_p} P_l(t) W_i(\hat{x}_l(t)) \quad (i = 1, 2, \ldots, n; \ j = 1, 2, \ldots, n)$$

and $\dot{q}_i(t)$ and $\phi_i'(x)$ denote the first derivatives of $q_i(t)$ and $\phi_i(x)$, respectively; m_{ij} is the generalized mass, and $f_i(t)$ is the generalized force. Let $k_{ij} = k_{ij}' + k_{ij}''$ and k_{ij} is the generalized stiffness. The Lagrange equation may be written as follows:

$$\frac{d}{dt}\left(\frac{\partial L}{\partial \dot{q}} \right) - \frac{\partial L}{\partial q} + \frac{\partial U}{\partial q} - \frac{\partial W_c}{\partial q} = \frac{\partial W}{\partial q}$$

$$W_c = -q^T C \dot{q} \qquad (2.17)$$

where W_c is the work due to the viscous damping in the beam. Substituting Equation (2.15) into Equation (2.17), the equation can be written as:

$$\sum_{j=1}^n m_{ij}\ddot{q}_j(t) + \sum_{j=1}^n c_{ij}\dot{q}_j(t) + \sum_{j=1}^n k_{ij}q_j(t) = f_i(t), \quad (i = 1, 2, \ldots, n) \qquad (2.18)$$

and in matrix form as (Zhu and Law, 1999):

$$M\ddot{q}(t) + C\dot{q}(t) + Kq(t) = F(t) \tag{2.19}$$

where

$$
\begin{aligned}
M &= \{m_{ij}, i = 1, 2, \ldots, n; j = 1, 2, \ldots, n\}; \\
K &= \{k_{ij}, i = 1, 2, \ldots, n; j = 1, 2, \ldots, n\}; \\
C &= \{c_{ij}, i = 1, 2, \ldots, n; j = 1, 2, \ldots, n\}; \\
q(t) &= \{q_1(t), q_2(t), \ldots, q_n(t)\}^T; \\
F(t) &= \{f_1(t), f_2(t), \ldots, f_n(t)\}^T.
\end{aligned}
\tag{2.20}
$$

The dynamic responses of the structure in Equations (2.14) and (2.19) can be obtained using the Newmark-Beta or the precise time step integration scheme described in Section 2.2.3.

2.2.2.1 Assumed Modes for a Uniform Beam

The general form of the vibration mode for a uniform Timoshenko beam can be written as follows:

$$
\begin{aligned}
W(x) &= D_1 \cos(\alpha x) + D_2 \sin(\alpha x) + D_3 \cosh(\beta x) + D_4 \sinh(\beta x) \\
\psi(x) &= D_1' \cos(\alpha x) + D_2' \sin(\alpha x) + D_3' \cosh(\beta x) + D_4' \sinh(\beta x)
\end{aligned}
\tag{2.21}
$$

where $D_1, D_2, D_3, D_4, D_1', D_2', D_3'$ and D_4' are constants, and α, β are frequency parameters. The vibration modes of a Timoshenko beam with simply supported ends are obtained as follows according to Huang (1961):

$$
\begin{cases}
W_i(x) = B_i \sin\left(\dfrac{i\pi x}{L}\right) \\
\phi_i(x) = \cos\left(\dfrac{i\pi x}{L}\right)
\end{cases}
\tag{2.22}
$$

where

$$
B_i = \frac{i\pi L}{(i\pi)^2 - b_i^2 s^2}; \qquad s^2 = \frac{EI}{\kappa A G L^2}; \qquad r^2 = \frac{1}{AL}
$$

$$
b_i^2 = \frac{1 + (i\pi)^2(r^2 + s^2) - \sqrt{(1 + (i\pi)^2(r^2 + s^2)^2)^2 - 4(i\pi)^4 r^2 s^2}}{2 r^2 s^2}
\tag{2.23}
$$

2.2.2.2 Assumed Modes for a Non-uniform Beam

The mode shapes of a uniform single span simply supported beam are expressed as:

$$W_{U_i}(x) = \sin\left(\frac{i\pi x}{L}\right), \quad (i = 1, 2, \ldots, n) \tag{2.24}$$

where n is the number of vibration modes. The vertical displacement of a non-uniform beam can be assumed as a combination of these mode shapes satisfying the boundary conditions as:

$$W(x) = \sum_{i=1}^{n} b_i W_{U_i(x)} \qquad (2.25)$$

where $\{b_i, i = 1, 2, \ldots, n\}$ is a set of coefficients that can be found by minimizing the following integral in Ritz's method:

$$Z = \int_{0}^{L} [EI(x)(W''(x))^2 - \omega^2 \rho A(x)(W(x))^2] dx \qquad (2.26)$$

to have

$$(K' - \omega^2 M)b = 0 \qquad (2.27)$$

where K' and M are $n \times n$ matrices with their components k'_{ij} and m_{ij} given in Equation (2.28), and $b = \{b_1, b_2, \ldots, b_n\}^T$ is a $n \times 1$ vector.

$$\begin{cases} k'_{ij} = \int_{0}^{L} EI(x) W''_{Ui}(x) W''_{Uj}(x) dx \\ m_{ij} = \int_{0}^{L} \rho A(x) W_{Ui}(x) W_{Uj}(x) dx \end{cases} \qquad (i = 1, 2, \ldots, n; \; j = 1, 2 \ldots n) \qquad (2.28)$$

Rewrite Equation (2.27) into

$$(D - \omega^2 I)b' = 0 \qquad (2.29)$$

and solving, we have

$$D = K'M^{-1}; \quad b' = Mb. \qquad (2.30)$$

where ω^2, b' are the eigenvalues and the eigenvectors of the matrix D. Coefficients b_i and hence $W(x)$ can be obtained from Equation (2.25), from which the natural frequencies and mode shapes are determined.

2.2.3   Precise Time Step Integration versus Newmark-Beta Method

2.2.3.1   Newmark-Beta Method

The central difference of Newmark-Beta and Wilson-θ methods can be used in the computation of Equation (2.19). This section introduces the Newmark-Beta method (Huebner et al., 2001) which is one of the implicit time integration methods. The Newmark-Beta equations for the displacement and velocity at time $t + \Delta t$ are as follows:

$$\{q\}_{t+\Delta t} = \{q\}_t + \{\dot{q}\}_t \Delta t + [(1 - \beta)\{\ddot{q}\}_t + \beta\{\ddot{q}\}_{t+\Delta t}] \Delta t^2 \qquad (2.31)$$

$$\{\dot{q}\}_{t+\Delta t} = \{\dot{q}\}_t + [(1 - \alpha)\{\ddot{q}\}_t + \alpha\{\ddot{q}\}_{t+\Delta t}] \Delta t \qquad (2.32)$$

where α and β are parameters that control the stability and accuracy of the method. The common choice is $\alpha = 1/2, \beta = 1/4$. We obtain the recurrence formula by writing Equation (2.19) at time $t + \Delta t$ as:

$$M\{\ddot{q}\}_{t+\Delta t} + C\{\dot{q}\}_{t+\Delta t} + K\{q\}_{t+\Delta t} = F(t + \Delta t) \tag{2.33}$$

Vector $\{\ddot{q}\}_{t+\Delta t}$ is solved in term of $\{q\}_{t+\Delta t}$ from Equation (2.31). Then the result is substituted into Equation (2.32) to have $\{\dot{q}\}_{t+\Delta t}$ in term of $\{q\}_{t+\Delta t}$. Finally, the recurrence formula in terms of an effective stiffness and load vector can be written as:

$$\overline{K}\{q\}_{t+\Delta t} = \overline{F}(t + \Delta t) \tag{2.34}$$

where the effective stiffness matrix can be expressed as:

$$\overline{K} = K + \frac{\alpha}{\beta \Delta t}C + \frac{1}{\beta \Delta t^2}M$$

and the effective load vector is:

$$\overline{F} = F_{t+\Delta t} + C\left(\frac{1}{\beta \Delta t}\{q\}_t + \left(\frac{\alpha}{\beta} - 1\right)\{\dot{q}\}_t + \frac{\Delta t}{2}\left(\frac{\alpha}{\beta} - 2\right)\{\ddot{q}\}_t\right)$$
$$+ M\left(\frac{1}{\beta \Delta t^2}\{q\}_t + \frac{1}{\beta \Delta t}\{\dot{q}\}_t + \left(\frac{1}{2\beta} - 1\right)\{\ddot{q}\}_t\right)$$

2.2.3.2　Precise Time Step Integration Method

The dynamic responses of structures are traditionally analyzed with the second-order differential equations by means of direct integration or step-by-step schemes, such as the Newmark-Beta, Wilson-θ and central difference methods. A precise time-step integration technique for structural dynamics has been proposed by Zhong and Williams (1994). This technique is unconditionally stable and exhibits high precision and efficiency in computation when compared with traditional methods. This section introduces this technique for the dynamic responses of a non-uniform continuous beam under a system of moving loads.

According to the precise time step integration method, the equation of motion of the beam in Equation (2.19) can be written as:

$$\dot{u} = Hu + f \tag{2.35}$$

where u is the response vector of size $2n \times 1$; H is a $2n \times 2n$ matrix; and f is the force vector of size $2n \times 1$, with:

$$u = \begin{bmatrix} q(t) \\ p(t) \end{bmatrix};$$

$$H = \begin{bmatrix} -\dfrac{M^{-1}C}{2} & M^{-1} \\ -\left(K - \dfrac{CM^{-1}C}{4}\right) & -\dfrac{CM^{-1}}{2} \end{bmatrix}; \tag{2.36}$$

$$f = \begin{bmatrix} 0 \\ F(t) \end{bmatrix} = \begin{bmatrix} 0 \\ A(t) \end{bmatrix} p(t);$$

$$A(t) = \begin{bmatrix} W_1(\hat{x}_1(t)) & W_1(\hat{x}_2(t)) & \dots & W_1(\hat{x}_{N_p}(t)) \\ W_2(\hat{x}_1(t)) & W_2(\hat{x}_2(t)) & \dots & W_2(\hat{x}_{N_p}(t)) \\ \dots & \dots & \dots & \dots \\ W_n(\hat{x}_1(t)) & W_n(\hat{x}_2(t)) & \dots & W_n(\hat{x}_{N_p}(t)) \end{bmatrix}$$

$$p(t) = M\dot{q}(t) + \frac{Cq(t)}{2}$$

where $W_i(\hat{x}_i(t))$ is defined in Equation (2.25) and matrix $A(t)$ is obtained from Equation (2.25). Integrating Equation (2.35), we have Equation (2.35) written into discrete equations using the exponential matrix representation as:

$$u(t) = e^{H(t-to)}u(t_o) + \int_{to}^{t} e^{H(t-\tau)}f(\tau)d\tau \tag{2.37}$$

Expressing Equation (2.33) in discrete form

$$u((j+1)h) = e^{Hh}u(jh) + \int_{jh}^{(j+1)h} e^{H((j+1)h-\tau)}f(\tau)d\tau \tag{2.38}$$

where h is the time step of integration. The force $f(\tau)$ is assumed constant within the time interval from jh to $(j+1)h$. We have:

$$u((j+1)h) = e^{Hh}u(jh) + \left[\int_{o}^{h} e^{H\tau'}d\tau' \right] f(jh) \tag{2.39}$$

$$= e^{Hh}u(jh) + H^{-1}[e^{Hh} - I]f(jh)$$

and the final discrete model for the $(j+1)$th step is rewritten as:

$$u_{j+1} = exp(H^*h)u_j + H^{-1}(exp(H^*h) - I)f_j \quad (j = 0,1,2,\dots) \tag{2.40}$$

The precision of integration depends on the accuracy of $exp(H^*h)$. The 2N algorithm presented by Zhong and Williams (1994) is used and $exp(H^*h)$ has the form:

$$exp(H^*h) = \left[exp\left(H^* \frac{h}{N_t} \right) \right]^{N_t} \tag{2.41}$$

$$= [exp(H^* \Delta t)]^{N_t}$$

where $\Delta t = h/N_t, N_t = 2^N$ and N can be any positive integer. Since h is not large such that Δt would be extremely small. The following truncated Taylor expansion of $exp(H^*\Delta t)$ may be used:

$$exp(H^* \Delta t) \approx I + H\Delta t + \frac{(H\Delta t)^2}{2!} + \frac{(H\Delta t)^3}{3!} + \frac{(H\Delta t)^4}{4!}$$

$$= I + R_0 \tag{2.42}$$

where

$$R_0 = H\Delta t + \frac{(H\Delta t)^2}{2!} + \frac{(H\Delta t)^3}{3!} + \frac{(H\Delta t)^4}{4!}$$

and R_i at the ith step of computation can be proved to take up the form of:

$$R_i = 2R_{i-1} + R_{i-1}R_{i-1}, \quad (i = 1, 2, \ldots, N),$$

Then

$$
\begin{aligned}
exp(H^*h) &= [exp(H\Delta t)]^{N_t} \\
&\approx I + R_N
\end{aligned}
\tag{2.43}
$$

The term $exp(H^*h)$ can be computed from Equations (2.42) and (2.43), and the vibration response can be calculated from Equation (2.40). The computed results are compared with those from the Newmark method using the same time step h in Equation (2.41), and the number of data points in the computation should be a multiple of two. It is noted that the accuracy of $exp(H^*h)$ and the vibration response u depends on the size of time step $\Delta t = h/N_t$ adopted, and there is no convergence error involved in the final results.

Example of Application

A simply support uniform beam subject to the excitation of a moving load is considered. The cross-sectional area and the material density of the beam are, respectively, $1.146 \times 10^{-3}\,\mathrm{m}^2$ and $7700\,\mathrm{kg/m}^3$. The overall length is one metre and the Young's modulus of material is 2.07×10^{-5} Mpa. The speed of the moving load is 17.3 m/s. Computation of the responses was done using the first 12 vibration modes with the integration time step h equals 9.0315×10^{-4} s. The number of data used is 64 and $N = 9$. Δt equals $h/2^9 = 1.76396 \times 10^{-6}$ s, and the Taylor series expansion for $exp(H\Delta t)$ contains infinitesimal approximation errors. Figure 2.3 shows the results obtained from using the method described above (Zhu and Law, 2001c), the Newmark method and the exact solution (Fryba, 1972). The deflection under the moving load has been normalized with the static deflection when the load is at midspan. Comparison on the computation error and computer time using PII-300 personal computer from the precise method and the Newmark method is also presented in Table 2.1. The computation error is defined as:

$$Error = \frac{\|x - x_{exact}\|}{\|x_{exact}\|} \times 100\%$$

where x and x_{exact} are the computed result and the exact solution respectively.

Both the curves from the precise integration method and the Newmark method match with the exact solution closely. The computation errors are almost the same for both methods as seen in Table 2.1. Table 2.1 also shows that the computation errors from both methods are comparable for different time step of integration, but the computer time required in the precise method is only one-quarter of that in the Newmark method. The precise method also gives a larger error than the Newmark method when a very large time step is used as seen in the last row of Table 2.1. This would indicate that a large time step should go together with a larger N value in the computation.

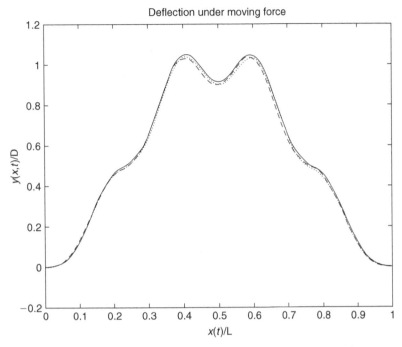

Figure 2.3 Deflection under the moving load at $v = 17.3\,\text{m/s}$ (— exact solution; --- precise integration; $\cdots$ Newmark method)

Table 2.1 Comparison between two methods

No. of data	N	Time Step (s)	Precise method		Newmark	
			Error (%)	Time (s)	Error (%)	Time (s)
500	9	1.1561×10^{-4}	1.4910	2.25	1.5250	10.65
200	9	2.8902×10^{-4}	1.5411	0.99	1.6806	4.01
128	9	4.5159×10^{-4}	1.6078	0.60	1.9091	2.52
64	9	9.0318×10^{-4}	2.5074	0.25	2.8021	1.31
32	9	0.0018	7.2589	0.16	4.9796	0.66

2.3 Multi-span Continuous Beam with Elastic Bearings

Elastic bearings often exist at the supports of bridge girders for load transference to the foundation. Li (2000) presented a unified approach on the vibration analysis of a beam with a general type of support. The boundary conditions are expressed by translational and rotational spring constraints, and the beam displacement is taken as the linear superposition of a Fourier cosine series and an auxiliary polynomial. Naguleswaran (2003) analyzed the transverse vibration of a uniform Euler-Bernoulli beam on five resilient supports (including the end supports). The closed-form solutions

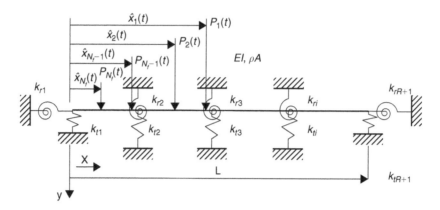

Figure 2.4 A continuous beam elastically restrained at both end subject to moving loads

for the responses of beams with different boundary conditions subject to a single deterministic moving force were obtained by Abu-Hilal and Zibdeh (2000). Yau et al. (2001) studied the impact response of bridges on elastic bearings from a series of moving loads simulating the action of a high-speed train. The bearings are modeled by linear transverse springs. Results show that the elastic bearings may increase the response of the beam when in the resonance condition. Chen et al. (2002) also studied the problem of an oscillator traversing an elastically supported continuum. The boundary flexibility amplifies the response of the beam, the dynamic interaction force and the shear force in general. This would be important to the understanding of the fatigue failure of the continuum in the moving load problem.

The bridge–vehicle system is modeled as a continuous multi-span Euler-Bernoulli beam elastically restrained at the supports and subject to a system of moving loads $P_l(t)(l = 1, 2, \ldots, N_p)$ as shown in Figure 2.4. The loads are assumed moving as a group at a prescribed velocity $v(t)$, along the axial direction of the beam from left to right. The equation of motion can be written as (Zhu and Law, 2006):

$$\rho A \frac{\partial^2 w(x,t)}{\partial t^2} + C \frac{\partial w(x,t)}{\partial t} + EI \frac{\partial^4 w(x,t)}{\partial x^4} = \sum_{l=1}^{N_p} P_l(t) \delta(x - \hat{x}_l(t)) \qquad (2.44)$$

where L is the total length of the beam; A is the cross-sectional area; E is the Young's modulus; I is the second moment of inertia of the beam cross-section; ρ, C and $w(x,t)$ are the mass per unit length, the damping and the displacement function of the beam respectively; $\hat{x}_l(t)$ is the location of moving load $P_l(t)$ at time t; $\delta(t)$ is the Dirac delta function and N_p is the number of loads. Express the transverse displacement $w(x,t)$ in modal coordinates:

$$w(x,t) = \sum_{i=1}^{\infty} \phi_i(x) q_i(t) \qquad (2.45)$$

where $\phi_i(x)$ is the mode shape function of the ith mode; $q_i(t)$ is the ith modal amplitude, which is determined from the eigenvalue and eigenfunction analysis described below.

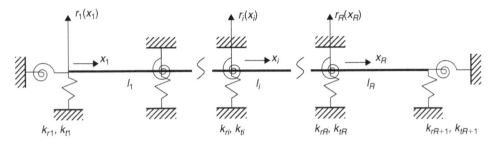

Figure 2.5 The R-span continuous beam

2.3.1 *Free Vibration*

The eigen-function of the R-span continuous Euler-Bernoulli beam shown in Figure 2.5 can be written in the following form:

$$r_i(x_i) = A_i \sin \beta x_i + B_i \cos \beta x_i + C_i \sinh \beta x_i + D_i \cosh \beta x_i; \quad (0 \le x_i \le l_i) \quad (2.46)$$

where $r_i(x_i)$ is the eigen-function for the ith span, l_i is the length of the ith span and β is the eigenvalue. The boundary conditions, the continuity of deflection and slope and the compatibility of the forces and moments in the R-span continuous Euler-Bernoulli beam shown in Figure 2.5 are listed as follows:

$$\left\{ \begin{aligned} & EI \frac{d^3 r_1(x_1)}{dx_1^3}\bigg|_{x_1=0} = -k_{t1} \, r_1(x_1)|_{x_1=0} \\ & EI \frac{d^2 r_1(x_1)}{dx_1^2}\bigg|_{x_1=0} = k_{r1} \frac{d r_1(x_1)}{dx_1}\bigg|_{x_1=0} \end{aligned} \right.$$

$$\left\{ \begin{aligned} & EI \frac{d^3 r_{i-1}(x_{i-1})}{dx_{i-1}^3}\bigg|_{x_{i-1}=l_{i-1}} = (-1)^i k_{ti} \, r_{i-1}(x_{i-1})|_{x_{i-1}=l_{i-1}} + EI \frac{d^3 r_i(x_i)}{dx_i^3}\bigg|_{x_i=0} \\ & EI \frac{d^2 r_{i-1}(x_{i-1})}{dx_{i-1}^2}\bigg|_{x_{i-1}=l_{i-1}} = (-1)^{i+1} k_{ri} \frac{d r_{i-1}(x_{i-1})}{dx_{i-1}}\bigg|_{x_{i-1}=l_{i-1}} + EI \frac{d^2 r_i(x_i)}{dx_i^2}\bigg|_{x_i=0} \quad (i=2,3,\dots,R) \\ & r_{i-1}(x_{i-1})|_{x_{i-1}=l_{i-1}} = r_i(x_i)|_{x_i=0} \\ & \frac{d r_{i-1}(x_{i-1})}{dx_{i-1}}\bigg|_{x_{i-1}=l_{i-1}} = \frac{d r_i(x_i)}{dx_i}\bigg|_{x_i=0} \end{aligned} \right.$$

$$\left\{ \begin{aligned} & EI \frac{d^3 r_R(x_R)}{dx_R^3}\bigg|_{x_R=l_R} = (-1)^{R+1} k_{tR+1} \, r_R(x_R)|_{x_R=l_R} \\ & EI \frac{d^2 r_R(x_R)}{dx_R^2}\bigg|_{x_R=l_R} = (-1)^{R+2} k_{rR+1} \frac{d r_R(x_R)}{dx_R}\bigg|_{x_R=l_R} \end{aligned} \right. \quad (2.47)$$

where $\{k_{ri}, k_{ti}, i=1,2,\dots,R+1\}$ are the stiffnesses of the rotational springs and the linear springs at $R+1$ supports respectively. Substituting Equation (2.46) into

Equation (2.47), the mode shape of the beam can be written as:

$$\phi(x) = \{A_i \sin \beta x_i + B_i \cos \beta x_i + C_i \sinh \beta x_i + D_i \cosh \beta x_i\}$$

$$\begin{cases} x_i = x - \sum_{j=1}^{i-1} l_j, & (i = 1, 2, \ldots, R) \\ 0 \le x_i \le l_i \end{cases} \tag{2.48}$$

where A_i, B_i, C_i, D_i are constants, and they are determined from the following equations:

$$[F]\{A\} = 0 \tag{2.49}$$

where $\{A\} = \{A_1, B_1, C_1, D_1, \ldots, A_R, B_R, C_R, D_R\}^T$ and the elements in matrix $[F]$ are given by:

$$f_{11} = EI\beta^3; \quad f_{12} = -k_{t1}; \quad f_{13} = -EI\beta^3; \quad f_{14} = -k_{t1};$$

$$f_{21} = k_{r1}\beta; \quad f_{22} = EI\beta^2; \quad f_{23} = k_{r1}\beta; \quad f_{24} = -EI\beta^2;$$

$$f_{4(i-2)+3,4(i-2)+1} = EI\beta^3 \cos \beta l_{i-1} + (-1)^i k_{ti} \sin \beta l_{i-1};$$
$$f_{4(i-2)+3,4(i-2)+2} = -EI\beta^3 \sin \beta l_{i-1} + (-1)^i k_{ti} \cos \beta l_{i-1};$$
$$f_{4(i-2)+3,4(i-2)+3} = -EI\beta^3 \cosh \beta l_{i-1} + (-1)^i k_{ti} \sinh \beta l_{i-1};$$
$$f_{4(i-2)+3,4(i-2)+4} = -EI\beta^3 \sinh \beta l_{i-1} + (-1)^i k_{ti} \cosh \beta l_{i-1};$$
$$f_{4(i-2)+3,4(i-2)+5} = -EI\beta^3; \quad f_{4(i-2)+3,4(i-2)+7} = EI\beta^3;$$

$$f_{4(i-2)+4,4(i-2)+1} = EI\beta \sin \beta l_{i-1} + (-1)^{i+1} k_{ri} \cos \beta l_{i-1};$$
$$f_{4(i-2)+4,4(i-2)+2} = EI\beta \cos \beta l_{i-1} - (-1)^{i+1} k_{ri} \sin \beta l_{i-1};$$
$$f_{4(i-2)+4,4(i-2)+3} = -EI\beta \sinh \beta l_{i-1} + (-1)^{i+1} k_{ri} \cosh \beta l_{i-1};$$
$$f_{4(i-2)+4,4(i-2)+4} = -EI\beta \cosh \beta l_{i-1} + (-1)^{i+1} k_{ri} \sinh \beta l_{i-1};$$
$$f_{4(i-2)+4,4(i-2)+6} = -EI\beta; \quad f_{4(i-2)+4,4(i-2)+8} = EI\beta;$$

$$f_{4(i-2)+5,4(i-2)+1} = \sin \beta l_{i-1}; \quad f_{4(i-2)+5,4(i-2)+2} = \cos \beta l_{i-1};$$
$$f_{4(i-2)+5,4(i-2)+3} = \sinh \beta l_{i-1}; \quad f_{4(i-2)+5,4(i-2)+4} = \cosh \beta l_{i-1};$$
$$f_{4(i-2)+5,4(i-2)+6} = -1; \quad f_{4(i-2)+5,4(i-2)+8} = -1;$$

$$f_{4(i-2)+6,4(i-2)+1} = \cos \beta l_{i-1}; \quad f_{4(i-2)+6,4(i-2)+2} = -\sin \beta l_{i-1};$$
$$f_{4(i-2)+6,4(i-2)+3} = \cosh \beta l_{i-1}; \quad f_{4(i-2)+6,4(i-2)+4} = \sinh \beta l_{i-1};$$
$$f_{4(i-2)+6,4(i-2)+5} = -1; \quad f_{4(i-2)+6,4(i-2)+7} = -1;$$

$$f_{4(R-1)+3,4(R-1)+1} = EI\beta^3 \cos \beta l_R + (-1)^{R+1} k_{tR+1} \sin \beta l_R;$$
$$f_{4(R-1)+3,4(R-1)+2} = -EI\beta^3 \sin \beta l_R + (-1)^{R+1} k_{tR+1} \cos \beta l_R;$$
$$f_{4(R-1)+3,4(R-1)+3} = -EI\beta^3 \cosh \beta l_R + (-1)^{R+1} k_{tR+1} \sinh \beta l_R;$$
$$f_{4(R-1)+3,4R} = -EI\beta^3 \sinh \beta l_R + (-1)^{R+1} k_{tR+1} \cosh \beta l_R;$$

$$f_{4R,4(R-1)+1} = EI\beta \sin \beta l_R + (-1)^{R+2} k_{rR+1} \cos \beta l_R;$$
$$f_{4R,4(R-1)+2} = EI\beta \cos \beta l_R - (-1)^{R+2} k_{rR+1} \sin \beta l_R;$$
$$f_{4R,4(R-1)+3} = -EI\beta \sinh \beta l_R + (-1)^{R+2} k_{rR+1} \cosh \beta l_R; \quad (i = 2, 3, \ldots, R) \tag{2.50}$$
$$f_{4R,4R} = -EI\beta \cosh \beta l_R + (-1)^{R+2} k_{rR+1} \sinh \beta l_R;$$

while other coefficients f_{ij} all equal to zero.

Table 2.2 Undamped Frequency parameters (Gorman, 1975)

Support Conditions	Order of frequency parameter									
	1	2	3	4	5	6	7	8	9	10
Simple-Simple	**3.142**	**6.283**	**9.425**	**12.566**	**15.708**	**18.850**	**21.991**	**25.133**	**28.274**	**31.416**
	3.142	6.283	9.425	12.566	15.708	18.850	21.991	25.133	28.274	31.416
Free-Free	**0**	**4.730**	**4.853**	**10.996**	**14.137**	**17.274**	**20.420**	**23.562**	**26.703**	**29.845**
	0	4.730	7.853	10.996	14.137	17.279	20.420	23.562	26.704	29.845
Simple-Free	**0**	**3.927**	**7.069**	**10.210**	**13.352**	**16.493**	**19.635**	**22.777**	**25.918**	**29.060**
	0	3.927	7.069	10.210	13.352	16.493	19.635	22.777	25.918	29.060
Clamped-Free	**1.875**	**4.694**	**7.855**	**10.996**	**14.137**	**17.279**	**20.420**	**23.562**	**26.704**	**29.845**
	1.875	4.694	7.855	10.996	14.137	17.279	20.420	23.562	26.704	29.845
Clamped-Simple	**3.927**	**7.069**	**10.210**	**13.352**	**16.493**	**19.635**	**22.777**	**25.918**	**29.060**	**32.201**
	3.927	7.069	10.210	13.352	16.493	19.635	22.777	25.918	29.060	32.200
Clamped-Clamped	**4.730**	**7.853**	**10.996**	**14.137**	**17.274**	**20.420**	**23.562**	**26.703**	**29.845**	**32.987**
	5.320	8.522	11.696	14.855	18.009	21.159	24.306	27.453	30.598	33.742

Note: Results shown **bolded** are close-form solutions from Gorman (1975).

Clamped – $k_t = 10^{20}$ N/m; $k_r = 10^{20}$ Nm/rad; Simple support – $k_t = 10^{20}$ N/m; $k_r = 0$; Free – $k_t = k_r = 0$

Many familiar classical boundary conditions can be obtained by assuming the translational and rotational spring constants to be extremely large or small. Table 2.2 shows the first ten dimensionless undamped frequency parameters, $\beta = \sqrt{\omega L^2 \sqrt{\rho A/EI}}$, for different combinations of boundary conditions with comparison to the close-form solution from Gorman (1975). The results are very similar indicating the accuracy of the computational algorithm for vibration analysis except for the case with the clamped-clamped boundary conditions. This difference is the natural result arising from the over-simplified modeling of an infinitely rigid support with large finite support stiffness.

$$\alpha_{ti} = k_{ti}L^3/(EI), \quad \alpha_{ri} = k_{ri}L/(EI), \quad i = 1, 2$$

Further comparison with results from Rao and Mirza (1989) is made for different values of the translational and rotational stiffnesses, and the results are found to be almost identical, as shown in Table 2.3. This is because the formulation by Rao and Mirza (1989) is a special solution of the present formulation using elastic supports.

Previous study by Law (1988) on the effect of support restraints on the eigenvalues shows that the lower frequencies are significantly affected when the stiffness of the translational springs at both ends is small compared with the stiffness of the beam. When the dimensionless vertical stiffness ratio, $\alpha_t = k_t L^3/EI \geq 10^7$, the first ten natural frequencies are close to those from the simply-supported beam model.

2.3.2 *Dynamic Behavior under Moving Loads*

The present formulation can be used for the dynamic analysis of the beam under moving loads. Substituting Equation (2.45) into Equation (2.44), and multiplying by

Table 2.3 Undamped Frequency parameters (Rao and Mirza, 1989)

Support parameters	Order of frequency parameter				
	1	2	3	4	5
$\alpha_{r1} = \alpha_{r2} = 0.01$	**0.668464**	**0.956995**	**4.735993**	**7.856163**	**10.997575**
$\alpha_{t1} = \alpha_{t2} = 0.10$	0.6684630	0.9569953	4.7359936	7.85616370	10.997575
$\alpha_{r1} = \alpha_{r2} = 1000$	**0.668694**	**3.141694**	**6.277334**	**9.415527**	**12.553946**
$\alpha_{t1} = \alpha_{t2} = 0.10$	0.6686934	3.1416944	6.2773346	9.4155275	12.5539463
$\alpha_{r1} = \alpha_{r2} = 1000$	**3.515830**	**4.664770**	**6.677590**	**9.538559**	**12.605675**
$\alpha_{t1} = \alpha_{t2} = 100$	3.5158306	4.6647769	6.6775907	9.5385589	12.6056751
$\alpha_{r1} = 1.0, \alpha_{r2} = 100$	**1.634071**	**2.847163**	**5.650639**	**8.708708**	**11.806661**
$\alpha_{t1} = 0.10, \alpha_{t2} = 10$	1.6340719	2.8471633	5.6506390	8.70870851	11.8066612
$\alpha_{r1} = 10^3, \alpha_{r2} = 10^5$	**3.844697**	**5.804663**	**8.650176**	**11.624110**	**14.516931**
$\alpha_{t1} = 10^2, \alpha_{t2} = 10^4$	3.8446974	5.8046630	8.6501770	11.6241107	14.5169302

Note: Results shown **bolded** are close-form solutions from Rao and Mirza (1989).

$\phi_i(x)$, integrating with respect to x between 0 and L, and applying the orthogonality conditions, we obtain:

$$\frac{d^2 q_i(t)}{dt^2} + 2\xi_i \omega_i \frac{dq_i(t)}{dt} + \omega_i^2 q_i(t) = \frac{1}{M_i} \sum_{l=1}^{N_p} P_l(t)\phi_i(\hat{x}_l(t)) \tag{2.51}$$

where ω_i, ξ_i, M_i are the modal frequency, the damping ratio and the modal mass of the ith mode, with:

$$M_i = \int_0^L \rho A \phi_i^2(x)dx \tag{2.52}$$

The displacement of the beam at point x and time t can be found from Equations (2.47) and (2.52) as:

$$w(x,t) = \sum_{i=1}^{\infty} \frac{\phi_i(x)}{M_i} \int_0^t h_i(t-\tau) \sum_{l=1}^{N_p} P_l(\tau)\phi_i(\hat{x}_l(\tau))d\tau \tag{2.53}$$

where

$$h_i(t) = \frac{1}{\omega_i'} e^{-\xi_i \omega_i t} \sin \omega_i' t; \quad \omega_i' = \omega_i \sqrt{1 - \xi_i^2}$$

2.4 Summary

This chapter presents analysis methods for the dynamic behavior of a multi-span continuous beam subject to the effects of moving loads. The Newmark-Beta and precise integration methods are illustrated for the calculation of the dynamic response of a continuous beam under the action of moving forces. The precise integration method is shown to exhibit high precision and efficiency in computation when compared with the Newmark-Beta method.

Chapter 3

Dynamic Response of Orthotropic Plates under Moving Loads

3.1 Introduction

The problem of a plate under the action of moving forces has attracted much research attention in the last two decades. Results of these investigations can be employed in many branches of modern transportation engineering, such as the design of track, the track/road beds and bridges for high-speed trains, cars, trucks, parking garages, military ballistic systems, aircraft runways, high-speed precision machining, magnetic disk drivers, and so forth. Most previous studies of a plate subject to moving loads are based on the two-dimensional theories such as the classical plate theory and the first-order shear deformation theory. Fryba (1972) has solved analytically the dynamic responses of a uniform flat plate under a moving load along a specified path. Wu et al. (1987) analyzed the dynamic responses of a flat plate subject to various types of moving loads by the finite element method. Later Wang and Lin (1996) analyzed the dynamic behavior of a multi-span continuous Mindlin plate subject to a moving load. Transfer matrix is used to determine the natural frequency and vibration modes of the plate. Marchesiello et al. (1999) analyzed the dynamics of multi-span continuous straight bridges subject to multi-degrees-of-freedom moving vehicle excitation by applying the modal superposition principle.

In this chapter, the bridge deck is modeled as an orthotropic plate, and the dynamic behavior of the bridge deck under moving loads is analyzed basing on the modal superposition principle. The single-span plate under moving loads is introduced firstly in Section 3.2 and the multi-span continuous plate under moving loads is discussed in Section 3.3. These models will be applied to the vehicle–bridge interaction analysis in Chapter 4.

3.2 Orthotropic Plates under Moving Loads

3.2.1 Free Vibration

According to Huffington and Hoppmann (1958), the governing equations of motion of an orthotropic plate shown in Figure 3.1 can be written as follows:

$$D_x \frac{\partial^4 w}{\partial x^4} + 2D_{xy} \frac{\partial^4 w}{\partial x^2 \partial y^2} + D_y \frac{\partial^4 w}{\partial y^4} + C\frac{\partial w}{\partial t} + \rho h \frac{\partial^2 w}{\partial t^2} = p \tag{3.1}$$

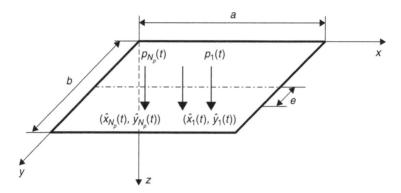

Figure 3.1 An orthotropic plate subject to action of moving loads

where $D_{xy} = (D_x v_{yx} + 2D_k)$, $D_x = E_x h^3/12(1 - v_{xy}v_{yx})$; $D_y = E_y h^3/12(1 - v_{xy}v_{yx})$; $D_k = G_{xy}h^3/12$. D_x, D_y are the flexural rigidities of the orthotropic plate in the x- and y-directions, respectively. D_k is the twisting rigidity of plate. C is the damping coefficient. h is the thickness of the plate. E_x, E_y are the modulus of the plate in the x- and y-directions, respectively; v_{xy} is the Poisson's ratio associated with a strain in the y-direction for a load in the x-direction. G_{xy} is the shear modulus. $p(x, y, t)$ is the external moving load. $w(x, y, t)$ is the displacement of plate in the z-direction. ρ is the mass density of plate material.

The free vibration of the plate without damping is firstly analyzed. Assuming the plate is simply supported along $x = 0$ and $x = a$ with the other two sides free to vibrate, the displacement of the plate can be written as:

$$w(x, y, t) = \sum_{m,n} Y_{mn}(y) \sin\left(\frac{m\pi x}{a}\right) \sin(\omega_{mn} t + \theta) \tag{3.2}$$

where ω_{mn} is the natural frequency corresponds to the mth mode in the x-direction and nth node in the y-direction; θ is the initial phase angle. $Y_{mn}(y) \sin(m\pi x/a)$ is the mode shape. Substituting Equation (3.2), Equation (3.1) becomes:

$$D_y Y_{mn}^{(4)}(y) - 2D_{xy}\left(\frac{m\pi}{a}\right)^2 Y_{mn}''(y) + \left[D_x\left(\frac{m\pi}{a}\right)^4 - \rho h\omega_{mn}^2\right] Y_{mn}(y) = 0 \tag{3.3}$$

The solution on $Y_{mn}(y)$ can be obtained and classified according to the following properties of the plate (Zhu and Law, 2000; Zhu and Law, 2003b).

1) When $D_x < \rho h\omega_{mn}^2\left(\dfrac{a}{m\pi}\right)^4$,

$$Y_{mn}(y) = A_{mn} \sin(r_{2mn}y) + B_{mn} \cos(r_{2mn}y)$$
$$+ C_{mn} \sinh(r_{1mn}y) + D_{mn} \cosh(r_{1mn}y) \tag{3.4}$$

where

$$r_{1mn} = \frac{m\pi}{a} \sqrt{\frac{D_{xy} + \sqrt{D_{xy}^2 + D_y \rho h \omega_{mn}^2 \left(\frac{a}{m\pi}\right)^4 - D_x D_y}}{D_y}}$$

$$r_{2mn} = \frac{m\pi}{a} \sqrt{\frac{-D_{xy} + \sqrt{D_{xy}^2 + D_y \rho h \omega_{mn}^2 \left(\frac{a}{m\pi}\right)^4 - D_x D_y}}{D_y}}$$

(3.5)

2) When $\dfrac{D_{xy}^2}{D_y} + \rho h \omega_{mn}^2 \left(\dfrac{a}{m\pi}\right)^4 > D_x > \rho h \omega_{mn}^2 \left(\dfrac{a}{m\pi}\right)^4$,

$$Y_{mn}(y) = A_{mn} \sinh(r_{1mn}y) + B_{mn} \cosh(r_{1mn}y)$$
$$+ C_{mn} sh(r_{3mn}y) + D_{mn} ch(r_{3mn}y)$$

(3.6)

where

$$r_{3mn} = \frac{m\pi}{a} \sqrt{\frac{D_{xy} - \sqrt{D_{xy}^2 + D_y \rho h \omega_{mn}^2 \left(\frac{a}{m\pi}\right)^4 - D_x D_y}}{D_y}}$$

(3.7)

3) When $D_x > \dfrac{D_{xy}^2}{D_y} + \rho h \omega_{mn}^2 \left(\dfrac{a}{m\pi}\right)^4$,

$$Y_{mn}(y) = \cosh(r_{4mn}y)(A_{mn} \cos(r_{5mn}y) + B_{mn} \sin(r_{5mn}y))$$
$$+ \sinh(r_{4mn}y)(C_{mn} \cos(r_{5mn}y) + D_{mn} \sin(r_{5mn}y))$$

(3.8)

where

$$r_{4mn} = \frac{m\pi}{a} \sqrt{\frac{1}{2}\left(\frac{D_{xy}}{D_y} + \sqrt{\frac{D_x}{D_y} - \frac{\rho h \omega_{mn}^2}{D_y}\left(\frac{a}{m\pi}\right)^4}\right)}$$

$$r_{5mn} = \frac{m\pi}{a} \sqrt{\frac{1}{2}\left(-\frac{D_{xy}}{D_y} + \sqrt{\frac{D_x}{D_y} - \frac{\rho h \omega_{mn}^2}{D_y}\left(\frac{a}{m\pi}\right)^4}\right)}$$

(3.9)

The parameters $A_{mn}, B_{mn}, C_{mn}, D_{mn}$ and the natural frequencies are determined with the free boundary conditions at $y = 0$ and $y = b$. The moment, the transverse shear and the torsional moment are zero at these edges, giving:

$$\begin{cases} \dfrac{\partial^2 w}{\partial y^2} + v_{xy}\dfrac{\partial^2 w}{\partial x^2} = 0 \\[2mm] -D_{xy}\dfrac{\partial^3 w}{\partial x^2 \partial y} - D_y\dfrac{\partial^3 w}{\partial y^3} = 0 \\[2mm] 2D_k\dfrac{\partial^3 w}{\partial x^2 \partial y} = 0 \\[2mm] -D_{xy}\dfrac{\partial^3 w}{\partial x^2 \partial y} - D_y\dfrac{\partial^3 w}{\partial y^3} - 2D_k\dfrac{\partial^3 w}{\partial x^2 \partial y} = 0 \end{cases}, \quad (y = 0 \text{ or } y = b) \qquad (3.10)$$

Substitute Equation (3.2) and Equations (3.4) to (3.9) into Equation (3.10). The following equation can be obtained:

$$A\{C\} = 0 \tag{3.11}$$

where $A = \{a_{ij}\}$ and the detail formulation of each coefficient for the different classifications presented above are listed as follows:

When $D_x < \rho h \omega_{mn}^2 \left(\dfrac{a}{m\pi}\right)^4$

$$a_{11} = 0; \quad a_{12} = -r_{2mn}^2 - v\left(\frac{m\pi}{a}\right)^2; \quad a_{13} = 0; \quad a_{14} = -r_{1mn}^2 - v\left(\frac{m\pi}{a}\right)^2;$$

$$a_{21} = a_{12}\sin(r_{2mn}b); \quad a_{22} = a_{14}\sinh(r_{1mn}b);$$

$$a_{23} = a_{12}\cos(r_{2mn}b); \quad a_{24} = a_{14}\cosh(r_{1mn}b);$$

$$a_{31} = D_y r_{2mn}^3 + (D_{xy} + 2D_k)\left(\frac{m\pi}{a}\right)^2 r_{2mn}; \quad a_{32} = 0; \tag{3.12}$$

$$a_{33} = -D_y r_{1mn}^3 + (D_{xy} + 2D_k)\left(\frac{m\pi}{a}\right)^2 r_{1mn}; \quad a_{34} = 0;$$

$$a_{41} = a_{31}\cos(r_{2mn}b); \quad a_{42} = -a_{31}\sin(r_{2mn}b);$$

$$a_{43} = a_{33}\cosh(r_{1mn}b); \quad a_{44} = a_{33}\sinh(r_{1mn}b);$$

When $\dfrac{D_{xy}^2}{D_y} + \rho h \omega_{mn}^2 \left(\dfrac{a}{m\pi}\right)^4 > D_x > \rho h \omega_{mn}^2 \left(\dfrac{a}{m\pi}\right)^4$

$$a_{11} = 0; \quad a_{12} = r_{1mn}^2 - v\left(\frac{m\pi}{a}\right)^2; a_{13} = 0; \quad a_{14} = r_{3mn}^2 - v\left(\frac{m\pi}{a}\right)^2;$$

$$a_{21} = a_{12}\sinh(r_{1mn}b); \quad a_{22} = a_{12}\cosh(r_{1mn}b);$$

$$a_{23} = a_{14}\sinh(r_{3mn}b); \quad a_{24} = a_{14}\cosh(r_{3mn}b);$$

$$a_{31} = -D_y r_{1mn}^3 + (D_{xy} + 2D_k)\left(\frac{m\pi}{a}\right)^2 r_{1mn}; \quad a_{32} = 0; \tag{3.13}$$

$$a_{33} = -D_y r_{3mn}^3 + (D_{xy} + 2D_k)\left(\frac{m\pi}{a}\right)^2 r_{3mn}; \quad a_{34} = 0;$$

$$a_{41} = a_{31}\cosh(r_{1mn}b); \quad a_{42} = a_{31}\sinh(r_{1mn}b);$$

$$a_{43} = a_{33}\cosh(r_{3mn}b); \quad a_{44} = a_{33}\sinh(r_{3mn}b);$$

When $D_x > \dfrac{D_{xy}^2}{D_y} + \rho h \omega_{mn}^2 \left(\dfrac{a}{m\pi}\right)^4$

$$a_{11} = r_{4mn}^2 - r_{5mn}^2 - v\left(\frac{m\pi}{a}\right)^2; \quad a_{12} = 0; \quad a_{13} = 0; \quad a_{14} = 2r_{4mn}r_{5mn};$$

$$a_{21} = a_{11}\cosh(r_{4mn}b)\cos(r_{5mn}b) - 2r_{4mn}r_{5mn}\sinh(r_{4mn}b)\sin(r_{5mn}b);$$

$$a_{22} = a_{11}\cosh(r_{4mn}b)\sin(r_{5mn}b) + 2r_{4mn}r_{5mn}\sinh(r_{4mn}b)\cos(r_{5mn}b);$$

$$a_{23} = a_{11}\sinh(r_{4mn}b)\cos(r_{5mn}b) - 2r_{4mn}r_{5mn}\cosh(r_{4mn}b)\sin(r_{5mn}b);$$

$$a_{24} = a_{11}\sinh(r_{4mn}b)\sin(r_{5mn}b) + 2r_{4mn}r_{5mn}\cosh(r_{4mn}b)\cos(r_{5mn}b); \tag{3.14}$$

$$a_{31} = 0; \quad a_{32} = (D_{xy} + 2D_k)\left(\frac{m\pi}{a}\right)^2 r_{4mn} - 3D_y r_{4mn}^2 r_{5mn} + D_y r_{5mn}^3;$$

$$a_{33} = -D_y r_{4mn}^3 + (D_{xy} + 2D_k)\left(\frac{m\pi}{a}\right)^2 r_{4mn} + 3D_y r_{5mn}^2 r_{4mn}; \quad a_{34} = 0;$$

$$a_{41} = a_{33}\sinh(r_{4mn}b)\cos(r_{5mn}b) - a_{32}\cosh(r_{4mn}b)\sin(r_{5mn}b);$$

$$a_{42} = a_{33}\sinh(r_{4mn}b)\sin(r_{5mn}b) + a_{32}\cosh(r_{4mn}b)\cos(r_{5mn}b);$$

$$a_{43} = a_{33}\cosh(r_{4mn}b)\cos(r_{5mn}b) - a_{32}\sinh(r_{4mn}b)\sin(r_{5mn}b);$$

$$a_{44} = a_{33}\cosh(r_{4mn}b)\sin(r_{5mn}b) + a_{32}\sinh(r_{4mn}b)\cos(r_{5mn}b);$$

and $\{C\} = \{A_{mn} \quad B_{mn} \quad C_{mn} \quad D_{mn}\}^T$.

Since only the non-zero solution of Equation (3.11) is of interest, the determinant of the coefficient matrix is set equal to zero from which the natural frequencies ω_{mn} $(m = 1, 2, \ldots; n = 1, 2, \ldots)$ are obtained. Vector $\{C\}$ can then be obtained for each natural frequency, and subsequently $Y_{mn}(y)$ can be found.

3.2.2 Dynamic Behavior under Moving Loads

The equations of motion of a damped orthotropic plate under moving loads expressed in Equation (3.1) can be written as follows by expressing the force p as a time step function:

$$D_x\frac{\partial^4 w}{\partial x^4} + 2D_{xy}\frac{\partial^4 w}{\partial x^2\partial y^2} + D_y\frac{\partial^4 w}{\partial y^4} + C\frac{\partial w}{\partial t} + \rho h\frac{\partial^2 w}{\partial t^2}$$

$$= \sum_{l=1}^{N_p} p_l(t)\delta(x - \hat{x}_l(t))\delta(y - \hat{y}_l(t)) \tag{3.15}$$

where $\{p_l(t), l = 1, 2, \ldots, N_p\}$ are the moving loads moving as a group at a fixed spacing. $(\hat{x}_l(t), \hat{y}_l(t))$ is the position of the moving load $p_l(t)$. $\delta(x), \delta(y)$ are the Dirac functions. By modal superposition, the displacement of the orthotropic plate can be written as:

$$w(x, y, t) = \sum_{m,n} W_{mn}(x, y)q_{mn}(t) \tag{3.16}$$

where $W_{mn}(x, y) = Y_{mn}(y)\sin(m\pi x/a)$ is the mode shape of the orthotropic plate which can be obtained from the formulation in Chapter 2, and $q_{mn}(t)$ is the corresponding modal coordinate.

Substituting Equation (3.16) into Equation (3.15) results in:

$$\ddot{q}_{mn}(t) + 2\zeta_{mn}\omega_{mn}\dot{q}_{mn}(t) + \omega_{ij}^2 q_{mn}(t)$$

$$= \frac{2}{\rho h a \int_0^b Y_{mn}^2(y)dy}\sum_{l=1}^{N_p} p_l(t)Y_{mn}(\hat{y}_l(t))\sin\left(\frac{m\pi}{a}\hat{x}_l(t)\right) \quad (m, n = 1, 2, \ldots) \tag{3.17}$$

where $\zeta_{mn} = C/2\rho h\omega_{mn}$. a, b are the dimensions of the orthotropic plate in x- and y-directions respectively. Equation (3.17) can be solved in the time domain by the convolution integral with the plate initially at rest, yielding:

$$q_{mn}(t) = \frac{1}{M_{mn}}\int_0^t H_{mn}(t - \tau)f_{mn}(\tau)d\tau \tag{3.18}$$

where

$$M_{mn} = \frac{\rho h a}{2} \int_0^b Y_{mn}^2(y)dy$$

$$H_{mn}(t) = \frac{1}{\omega'_{mn}} e^{-\zeta_{mn}\omega_{mn}t} \sin(\omega'_{mn}t), \quad t \geq 0 \tag{3.19}$$

$$f_{mn}(t) = \sum_{l=1}^{N_p} p_l(t) Y_{mn}(\hat{y}_l(t)) \sin\left(\frac{m\pi}{a}\hat{x}_l(t)\right)$$

$$\omega'_{mn} = \omega_{mn}\sqrt{1 - \zeta_{mn}^2}$$

Substituting Equation (3.18) into Equation (3.16), the displacement of the orthotropic plate at point (x, y) and time t can be found as:

$$w(x, y, t) = \sum_{m=1}^{\infty}\sum_{n=1}^{\infty} Y_{mn}(y) \sin\left(\frac{m\pi}{a}x\right) \frac{1}{M_{mn}} \int_0^t H_{mn}(t - \tau) f_{mn}(\tau) d\tau \tag{3.20}$$

3.2.3 Numerical Simulation

3.2.3.1 Natural Frequency of Orthotropic Plates

The example of a simply-supported orthotropic plate by Jayaraman, et al. (1990), is adopted using the formulation in Equations (3.1) to (3.14) for the free vibration analysis. The plate is simply supported along edges $x = 0$, $x = a$, and is free along edges $y = 0$, and $y = b$. $a = 10$ m and the aspect ratio is $a/b = 1.0$. $v_x = v_y = 0.3$, and parameters E, G and ρ are taken as unity because they will be cancelled out in the calculation. Thickness of the plate is 0.2 m. The dimensionless frequency parameter $\lambda^* = \omega a^2 \sqrt{\rho h / D_y}$ for the first 16 flexural modes of free vibrations is presented in Table 3.1 for different ratios of D_x/D_y and D_{xy}/D_y. Excellent agreement is observed with results from Jayaraman et al. (1990). It may be concluded that the formulation presented in this chapter and the associated classification of plates to obtain the natural frequency and mode shapes is correct.

3.2.3.2 Simply Supported Beam-Slab Type Bridge Deck under Moving Loads

A simply supported beam-slab type bridge deck is shown in Figure 3.2. It is of sufficient width for three-lane traffic in the study. The Physical parameters of the structure are:

For the deck slab: $a = 20$ m, $b = 11$ m, $E = 2.1 \times 10^9$ N/m^2; $\rho = 2300$ kg/m^3.
For the I-beams: $A_I = 0.299$ m^2; $I = 0.118$ m^4; $J = 0.04385$ m^2; $m_1 = 0.175$ m; $n_1 = 1.13$ m; $\alpha = 0.3$; $b_1 = 2.25$ m; and $v = 0.33$.

The rigidities in the x- and y-directions of the orthotropic bridge deck can be calculated as:

$$D_x = \frac{Eh^3}{12(1 - Sv^2)} + \frac{EI}{b_1}$$

$$D_y = \frac{Eh^3}{12(1 - Sv^2)}, \quad D_{xy} = vD_y + \frac{Gh^3}{6} + \frac{Gm_1^3 n_1\alpha}{b_1} \tag{3.21}$$

Table 3.1 Frequency parameters for orthotropic plates

D_x/D_y	1/2			1			2		
D_{xy}/D_y	1/2	1	2	1/2	1	2	1/2	1	2
(1,1)	6.4705	6.6377	6.7529	9.5170	9.6314	9.7111	13.7106	13.7903	13.8461
	(6.4705)	(6.6377)	(6.7529)	(9.5169)	(9.6314)	(9.7111)	(13.7106)	(13.7903)	(13.8461)
(1,2)	9.5537	14.5474	20.8297	11.8312	16.1348	21.9677	15.4074	18.9140	24.0830
	(9.5537)	(14.5474)	(20.8297)	(11.8312)	(16.1348)	(21.9677)	(15.4074)	(18.9140)	(24.0830)
(1,3)	28.6169	36.0565	47.1237	29.4556	36.7256	47.6376	31.0651	38.0287	48.6493
	(28.6169)	(36.0564)	(47.1236)	(29.4556)	(36.7256)	(47.6376)	(31.0651)	(38.0287)	(48.6493)
(1,4)	67.4853	74.9592	87.7702	67.8452	75.2834	88.0472	68.5593	75.9276	88.5987
(2,1)	26.4882	27.1558	27.4506	38.4824	38.9450	39.1511	55.1311	55.4550	55.5999
	(26.4882)	(27.1557)	(27.4505)	(38.4824)	(38.9449)	(39.1510)	(55.1311)	(55.4550)	(55.5998)
(2,2)	29.9614	37.4858	47.6179	40.9507	46.7381	55.1972	56.8815	61.1801	67.8622
	(29.9614)	(37.4857)	(47.6178)	(40.9506)	(46.7381)	(55.1972)	(56.8814)	(61.1800)	(67.8621)
(2,3)	48.4665	64.9992	87.5872	55.9310	70.7401	91.9282	68.4603	81.0105	100.0466
	(48.4665)	(64.9992)	(87.5872)	(55.9310)	(70.7401)	(91.9282)	(68.4603)	(81.0106)	(100.0467)
(2,4)	85.9053	107.4587	139.3780	90.3271	111.0254	142.1461	98.5776	117.8354	147.5265
(3,1)	60.2723	61.6165	62.1033	87.0507	87.9867	88.3283	124.3702	125.0272	125.2678
	(60.2722)	(61.6164)	(62.1032)	(87.0505)	(87.9866)	(88.3282)	(124.3700)	(125.0270)	(125.2676)
(3,2)	63.5631	72.6547	84.9931	89.3607	96.0405	105.6380	125.9979	130.8202	138.0544
	(63.5631)	(72.6547)	(84.9931)	(89.3607)	(96.0405)	(105.6830)	(125.9979)	(130.8202)	(138.0545)
(3,3)	82.3857	104.6360	135.9387	103.5977	122.0400	149.7478	136.4647	150.9433	174.1107
	(82.3857)	(104.6360)	(135.9387)	(103.5977)	(122.0400)	(149.7478)	(136.4647)	(150.9434)	(174.1107)
(3,4)	118.2938	152.2487	200.6094	133.9347	164.6959	210.2122	160.7129	187.1226	228.2089
(4,1)	107.7851	110.0135	110.7104	155.1966	156.7525	157.2423	221.4108	222.5041	222.8494
(4,2)	110.7019	121.1869	134.8802	157.2364	164.7866	175.1029	222.8452	228.2353	235.7917
(4,3)	130.2873	156.0280	193.8604	171.5900	191.8672	223.7191	233.1949	248.4951	273.8374
(4,4)	165.0860	208.2100	270.3388	199.3031	236.2620	292.4918	254.2803	284.1768	332.3976

Note: The values in ( ) are from Jayaraman et al. (1990).
(m,n) denotes mode shapes with m half sine waves in the x-direction and n half sine waves in the y-direction respectively.

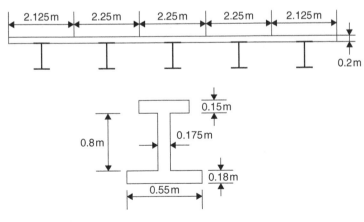

Figure 3.2 A simply supported beam-slab type bridge deck

Table 3.2 Effect of eccentricity of moving loads

Eccentricity (m)	Maximum displacement (metres) when travelling at					
	10 m/s			20 m/s		
	a/2,b/4	a/2,b/2	a/2,3b/4	a/2,b/4	a/2,b/2	a/2,3b/4
0	0.0043	0.0067	0.0031	0.0045	0.0069	0.0039
0.5	0.0036	0.0066	0.0038	0.0037	0.0068	0.0040
1.5	0.0021	0.0060	0.0055	0.0022	0.0062	0.0058
2.5	0.0009	0.0049	0.0072	0.0010	0.0051	0.0076
3.5	0.0002	0.0033	0.0090	0.0004	0.0035	0.0095
4.5	0.0002	0.0016	0.0106	0.0003	0.0019	0.0112

where G is the torsional modulus of rigidity of material; α is a coefficient on the equivalent torsional moment of inertia of the I-section; and S is the ratio D_y/D_x. The damping ratio of the bridge deck is 0.002. The calculated rigidities are $D_x = 1.1153 \times 10^9$ kN-m; $D_y = 1.4 \times 10^7$ kN-m and $D_{xy} = 2.1665 \times 10^7$ kN-m. The variations of D_y, D_{xy} and single and multi-lane loads are studied in the following Sections for their effect on the impact factor of the bridge deck.

Effect of Eccentricity of Moving Loads

Table 3.2 shows the effect of eccentricity of moving load on the maximum displacements along the midspan cross-section. The moving load is $P = 300$ kN and is moving at a speed of 10 m/s or 20 m/s. The first 12 modes are used in the calculation. Results show that when the eccentricity increases, the maximum displacement of the plate on the loaded side increases, but the maximum central displacement at $(a/2, b/2)$ decreases. This is because the torsional rigidity D_{xy} of the bridge deck is larger than the flexural rigidity D_y in the y-direction and the eccentric loads have less effect on the central displacement.

Figure 3.3 shows the responses $w(x, y_1, t)$ at mid-span normalized with $max(w_{st}(y_1, y_0))$ when the moving load moves at 10 m/s on top of the third, fourth and the fifth beams. $w_{st}(y_1, y_0)$ is the static displacement along the line $y = y_1$ due to the load on $(a/2, y_0)$ and $w(x, y_1, t)$ is the dynamic displacement at $(a/2, y_1)$ due to the load moving along $y = y_0$.

Effect of Multi-Lane Loads

Humar and Kashif (1995) have studied the deflection of a slab-type bridge deck under the action of two vehicles moving in series along the bridge centerline. Nowak et al. (1993) studied the effect of multi-lane vehicles on the lateral load distribution of highway bridges. Yang et al. (1995) studied the impact effect of vehicles on a bridge modeled as a continuous beam. The following gives the analytical study of impact factors in an orthotropic bridge deck with multi-lanes loading.

The bridge deck in Figure 3.2 is wide enough to accommodate three-lanes of traffic. Constant loads of 150 kN each are considered moving at 10 m/s on top of the deck.

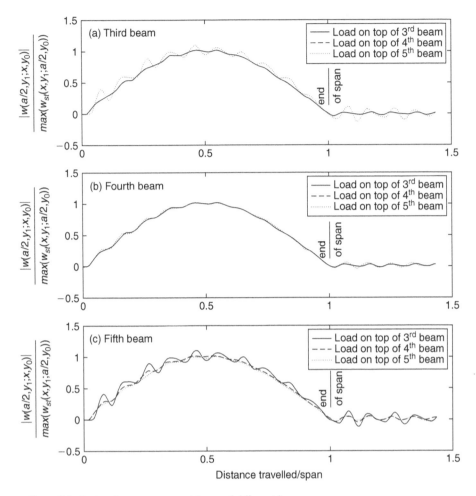

Figure 3.3 Ratio of responses at midspan of different beams

The displacement at midspan of the third beam is computed for the following cases for comparison:

(a) two loads moving along the centerline of the third beam at a spacing of 4 metres;
(b) two loads moving along the centerline of the first and third lanes at $y = 2.75$ m and 8.25 m simultaneously and at the same speed but in different directions; and
(c) two loads moving along the centerline of the first and second lanes at $y = 2.75$ m and 5.5 m simultaneously and at the same speed but in different directions.

The displacements at midspan of the third beam are plotted in Figure 3.4. Comparison between cases (b) and (c) shows that any loading in a side lane would reduce the displacement at the centre of the bridge deck when the loads are travelling in opposite direction. This is also expected to happen in a side lane when the loads are travelling

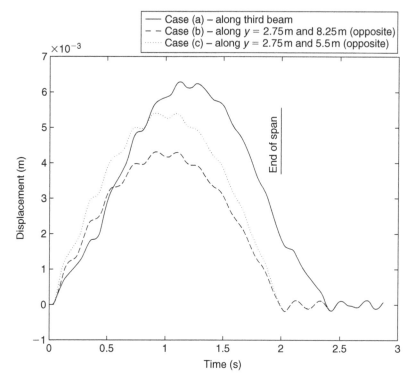

Figure 3.4 Displacement at (a/2, b/2) under single and multilane loads

in multi-lanes in the same direction. The most critical case would be having the loads moving in series in a single lane along the centerline of the deck.

3.3 Multi-span Continuous Orthotropic Plate under Moving Loads

The following assumptions (Zhu and Law, 2002a) are made for the formulation of the problem in this Section:

1. The bridge is treated as a continuous rectangular orthotropic plate with simple supports at its two ends ($x = 0, x = a$), and the other two opposite edges are free ($y = 0, y = b$) as shown in Figure 3.5. A linear elastic behavior is assumed, and the effects of shear deformation and rotary inertia are neglected.
2. The intermediate line supports of the bridge are assumed as linear rigid and they are orthogonal to the free edges of the plate.
3. Since the horizontal dimensions of the bridge deck are much larger than its thickness, the thin plate assumption is made.
4. The wheel loads are assumed to be in contact with the bridge deck all the time.

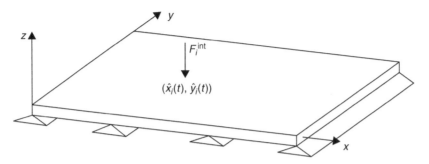

Figure 3.5 Model of the continuous bridge deck

3.3.1 *Dynamic Behavior under Moving Loads*

From the vibrational theory of thin plate, the strain energy of the continuous orthotropic plate in Cartesian co-ordinates is:

$$
\begin{aligned}
U_e &= \iiint_V \frac{1}{2} \sigma_i \varepsilon_i dV \\
&= \frac{1}{2} \iint_S \left[D_x \left(\frac{\partial^2 w}{\partial x^2} \right)^2 + \left(D_x \nu_{yx} + D_y \nu_{xy} \right) \frac{\partial^2 w}{\partial x^2} \frac{\partial^2 w}{\partial y^2} \right. \\
&\quad \left. + D_y \left(\frac{\partial^2 w}{\partial y^2} \right)^2 + 4 D_k \left(\frac{\partial^2 w}{\partial x \partial y} \right)^2 \right] dS
\end{aligned}
\tag{3.22}
$$

where D_x, D_y, D_k are the rigidity constants of the orthotropic plate; ν_{yx}, ν_{xy} are Poisson's ratio of the orthotropic material. For the bridge deck with material orthotropy and an equivalent uniform plate thickness h, $D_x = E_x h^3 / 12(1 - \nu_{xy} \nu_{yx})$, $D_y = E_y h^3 / 12(1 - \nu_{xy} \nu_{yx})$, $D_k = G_{xy} h^3 / 12$, in which E_x, E_y are Young's moduli in the x- and y-directions respectively, G_{xy} is the shear modulus. These rigidities for the bridge deck with shape orthotropy can be determined by the method of Bakht and Jaeger (1985).

The kinetic energy of the system is expressed as:

$$
T = \frac{1}{2} \iint_S \rho h \left(\frac{\partial w}{\partial t} \right)^2 dS
\tag{3.23}
$$

where ρ is the mass density of plate material, and w is the vertical deflection.

The work done by the damping of the plate and moving loads are as follows:

$$
W_c = -\iint_S c_b w \frac{\partial w}{\partial t} dS
$$

$$
W_e = \iint_S \sum_{N_f} F_i^{\text{int}} \delta(x - \hat{x}_i(t)) \delta(y - \hat{y}_i(t)) w \, dS
\tag{3.24}
$$

where c_b is the damping coefficient of the plate; F_i^{int} is the ith interaction force between the vehicular wheel and the bridge. $(\hat{x}_i(t), \hat{y}_i(t))$ is the location of the interaction force

F_i^{int}. When the vehicle is moving along one lane, $\hat{y}_i(t)$ is a constant. $\delta(x), \delta(y)$ are Dirac functions.

Based on modal superposition, the dynamic deflection $w(x, y, t)$ can be described as:

$$w(x, y, t) = \sum_{i=0}^{\infty} W_i(x, y) q_i(t) \tag{3.25}$$

where $W_i(x, y)$ is the vibration mode shape of the plate and $q_i(t)$ is the corresponding modal amplitude. When Equation (3.25) is substituted into Equations (3.22) to (3.24), the equations of motion for the bridge deck are:

$$M_b \ddot{Q} + C_b \dot{Q} + K_b Q = W_b F_b^{\text{int}} \tag{3.26}$$

where M_b, C_b, K_b are the mass, damping and stiffness matrices of the bridge, respectively; F_b^{int} is the vector of interaction force under the wheels of the moving vehicles. $\dot{Q}, \ddot{Q}$ are the first and second derivatives of Q, and Q is the vector of modal amplitudes. M_n is the number of mode shapes for the continuous orthotropic plate.

$$Q = \{q_1(t), q_2(t), \ldots, q_{M_n}(t)\}^T; \quad W = \{W_1(x, y), W_2(x, y), \ldots, W_{M_n}(x, y)\};$$

$$M_b = \iint_S \rho h W^T W \, dS; \quad C_b = \iint_S c_b W^T W \, dS;$$

$$W_b = \begin{bmatrix} W_1(\hat{x}_1(t), \hat{y}_1(t)) & W_1(\hat{x}_2(t), \hat{y}_2(t)) & \cdots & W_1(\hat{x}_{N_p}(t), \hat{y}_{N_p}(t)) \\ W_2(\hat{x}_1(t), \hat{y}_1(t)) & W_2(\hat{x}_2(t), \hat{y}_2(t)) & \cdots & W_2(\hat{x}_{N_p}(t), \hat{y}_{N_p}(t)) \\ \vdots & \vdots & \cdots & \vdots \\ W_{M_n}(\hat{x}_1(t), \hat{y}_1(t)) & W_{M_n}(\hat{x}_2(t), \hat{y}_2(t)) & \cdots & W_{M_n}(\hat{x}_{N_p}(t), \hat{y}_{N_p}(t)) \end{bmatrix}$$

$$K_b = \iint_S \left[D_x W_x''^T W_x'' + \frac{1}{2}(D_x \nu_{yx} + D_y \nu_{xy})(W_x''^T W_y'' + W_y''^T W_x'') \right. $$
$$\left. + D_y W_y''^T W_y'' + 4 D_k W_{xy}''^T W_{xy}'' \right] dS$$

$$W_{xi}'' = \sum_m \sum_n A_{mn} \varphi_m''(x) \psi_n(y); \quad W_{xyi}'' = \sum_m \sum_n A_{mn} \varphi_m'(x) \psi_n'(y);$$
$$W_{yi}'' = \sum_m \sum_n A_{mn} \varphi_m(x) \psi_n''(y) \quad (i = 1, 2, \ldots, M_n)$$
$$F_b^{\text{int}} = \{F_{t1}, F_{t2}, F_{t3}, F_{t4}\}^T$$

3.3.2 Modal Analysis of Multi-span Continuous Plates

For free vibration of the plate, the vertical displacement may be expressed as:

$$w(x, y, t) = W(x, y) e^{j\omega t} \tag{3.27}$$

where ω is the natural frequency of vibration and $j = \sqrt{-1}$. Assuming the variables in $W(x,y)$ are separable, the mode shape function $W(x,y)$ can be expressed in terms of a series as:

$$W(x,y) = \sum_m \sum_n A_{mn}\varphi_m(x)\psi_n(y) \tag{3.28}$$

where $\varphi_m(x)$ and $\psi_n(y)$ are the assumed admissible functions along the x- and y-directions respectively, while A_{mn} are the unknown coefficients. A set of series consisting of a combination of beam eigenfunctions and polynomials has been selected as the admissible functions in the line-supported plates by Zhou (1994). Here $\varphi_m(x)$ are taken as the eigen-functions of the continuous multi-span Euler-Bernoulli beam, and $\psi_n(y)$ are the eigen-functions of the single-span Euler-Bernoulli beam satisfying the free boundary conditions. Substituting Equation (3.28) into Equations (3.22), (3.23) and (3.27), and minimizing the Rayleigh's quotient with respect to each coefficient A_{mn} leads to the eigenvalue equations in matrix form as follows:

$$(K_b - \omega^2 M_b)A = 0 \tag{3.29}$$

where

$$A = \{A_{11}, A_{12}, \ldots, A_{1N}, A_{21}, \ldots, A_{MN}\}^T;$$

$$m_{b_{ij}} = \rho h \int_0^a \varphi_{m1}(x)\varphi_{m2}(x)dx \int_0^b \psi_{n1}(y)\psi_{n2}(y)dy$$

$$k_{b_{ij}} = D_x \int_0^a \varphi_{m1}''(x)\varphi_{m2}''(x)dx \int_0^b \psi_{n1}(y)\psi_{n2}(y)dy$$

$$+ D_y \int_0^a \varphi_{m1}(x)\varphi_{m2}(x)dx \int_0^b \psi_{n1}''(y)\psi_{n2}''(y)dy$$

$$+ v_{xy}D_y \left(\int_0^a \varphi_{m1}''(x)\varphi_{m2}(x)dx \int_0^b \psi_{n1}(y)\psi_{n2}''(y)dy \right.$$

$$\left. + \int_0^a \varphi_{m1}(x)\varphi_{m2}''(x)dx \int_0^b \psi_{n1}''(y)\psi_{n2}(y)dy \right)$$

$$+ 2(1 - v_{xy})D_{xy} \int_0^a \varphi_{m1}'(x)\varphi_{m2}'(x)dx \int_0^b \psi_{n1}'(y)\psi_{n2}'(y)dy$$

$$(m1 = 1, 2, \ldots, M; \quad m2 = 1, 2, \ldots, M; \quad n1 = 1, 2, \ldots, N; \quad n2 = 1, 2, \ldots, N)$$

$$i = (m1 - 1)N + n1; \quad j = (m2 - 1)N + n2.$$

and M, N are the number of admissible functions in x- and y-directions respectively. $\varphi_m''(x)$, $\varphi_m'(x)$ are the second and first derivatives of $\varphi_m(x)$; $\psi_n''(y)$, $\psi_n'(y)$ are the second and first derivatives of $\psi_n(y)$.

The natural frequencies ω and coefficients A_{mn} can be determined from Equation (3.29). Then the mode shape functions of the continuous orthotropic plate are determined from Equation (3.28). Since the admissible functions are eigen-functions of the

Table 3.3 Natural frequencies for the three-span continuous bridge (Hz)

Mode	Beam model	Marchesiello et al. (1999)		Zhou's method (1994)		Proposed method	
		$9 \times 5^*$	$17 \times 9^*$	$9 \times 5^*$	$17 \times 9^*$	$9 \times 5^*$	$17 \times 9^*$
1	4.71	4.79	4.77	4.90	4.88	4.90	4.88
2	6.04	6.19	6.17	6.30	6.28	6.29	6.27
3	8.82	9.11	9.09	9.26	9.20	9.21	9.18
4	–	16.65	16.65	15.05	15.04	15.04	15.03
5	–	17.55	17.53	16.00	15.95	15.95	15.92
6	18.85	19.37	19.31	18.21	17.97	17.98	17.91
7	–	19.57	19.51	19.68	19.61	19.67	19.60
8	21.49	22.16	22.10	22.59	22.40	22.44	22.37

Note: $9 \times 5^*$ denotes 9 number of eigen-functions in $\varphi_m(x)$ and 5 number of eigen-functions in $\psi_n(y)$.

Euler-Bernoulli beam, the mode shape functions of the continuous orthotropic plate satisfy the orthogonality relationships. It should be noted that this approach is much more simple and direct then existing methods by Zhou (1994) and Marchesiello et al. (1999).

3.3.3 *Numerical Examples*

The natural frequencies of the continuous three-span isotropic plate in Figure 3.5 are shown in Table 3.3. Since the aspect ratio b/L is small, the number of series in the assumed beam functions in x-direction is selected larger than that in y-direction as shown. The results obtained by the proposed method are very close to those obtained using Zhou's method (Zhou, 1994), and are approximately equal to the results in Marchesiello et al. (1999). This shows that the proposed method and algorithm to obtain the natural frequencies of the continuous plate are correct.

3.4 Summary

This chapter presents the dynamic analysis methods for the dynamic behavior of single span and multi-span continuous orthotropic plates under a system of moving loads. The natural frequencies and mode shapes of the single span plate are determined according to a new classification of plates. The effect of multiple moving loads on the dynamic response is studied. The natural frequencies and mode shapes of a multi-span plate can also be obtained via the modal analysis with a new and more simplistic approach than existing methods.

Chapter 4

Application of Vehicle–Bridge Interaction Dynamics

4.1 Introduction

This chapter presents the moving load problem in the context of the vehicle–bridge interaction analysis. Section 4.2 presents the effect of vehicle braking on a single span and multi-span continuous bridge deck, together with the influence of parameters such as initial vehicle velocity, road surface roughness (ISO 8608, 1995), braking rise time, amplitude of braking force and vehicle braking location (Law and Zhu, 2005). The analytical vehicle is simulated as a 7 degrees-of-freedom three-axle tractor-trailer vehicle and as a two-axle three-dimensional vehicle model with 7 degrees-of-freedom according to the H20-44 vehicle design loading in AASHTO (2007). Section 4.3 investigates the dynamic loading on a multi-lane continuous bridge due to multiple vehicles moving on top (Zhu and Law, 2002a). The vehicle–bridge interaction is discussed in terms of the impact factor in both set of studies.

4.2 Bridge Dynamic Response

The design for the dynamic effect of a moving vehicle on a bridge deck is usually realized through a specified impact factor on the static load, and this approach is considered satisfactory to allow for the braking of vehicle on the bridge deck. But in real life this effect could be significantly different.

This section studies the effect on a single span and multi-span continuous bridge deck, together with the influence of parameters such as initial vehicle velocity, road surface roughness (ISO 8608, 1995), braking rise time, amplitude of braking force and vehicle braking location. A three-axle tractor-trailer vehicle is selected for the study as the design of the modern bridge deck is governed by heavy freight loading. Results from this simulation study indicate that the effect from vehicle braking is very significant in the single span bridge deck, and an impact factor larger than that allowed for in the recommendation (AASHTO, 2007) may be necessary for the bridge–vehicle combination under study. The situation is reversed in the case of the three-span bridge deck under study, in which the effect from vehicle braking is negligibly small in terms of the impact factor. The influence from other factors on the impact factor is also small. If a smaller impact factor is used there would be a saving in the cost of the bridge deck.

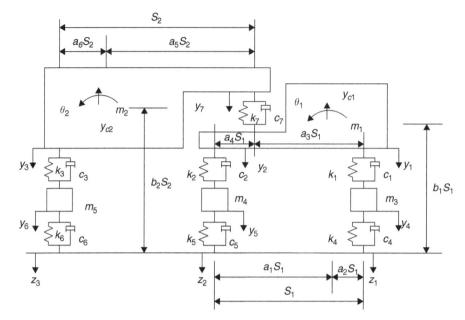

Figure 4.1   Model of a seven degrees-of-freedom vehicle

4.2.1 Vehicle and Bridge Models

A tractor-trailer vehicle is studied as an example of modern freight vehicle. It is modeled as a seven degrees-of-freedom (DOFs) system as shown in Figure 4.1. The seven vertical DOFs are denoted by y_i and those at the contact points between the wheels and the pavement are denoted by z_i. Each vehicle axle has a 'friction' stiffness and damping from the suspensions denoted with subscripts 1, 2 and 3, and a tyre stiffness and damping denoted with subscripts 4, 5 and 6. The mass of each axle assembly is denoted by m_3, m_4 and m_5 and that of the tractor and trailer are m_1 and m_2 respectively.

The bridge superstructure is modeled as a non-uniform continuous Euler-Bernoulli beam as shown in Figure 2.1 with $(Q - 1)$ intermediate vertical point supports. This model is representative of a modern bridge superstructure, which is usually non-uniform and continuous. $\{P_s(t), s = 1, 2, 3\}$ are the axle forces from the moving vehicle. The force locations are denoted by $\hat{x}_S(t), (s = 1, 2, 3)$ measured from the left support. The motion of the centroids of the tractor and trailer are expressed in terms of the seven DOFs of the vehicle model and the co-ordinates at the contact points between the bridge and the vehicle.

4.2.2 Vehicle–Bridge Interaction

The vertical displacements and rotations of the vehicle are assumed to be relative to the static equilibrium position, and the co-ordinates z_i are measured from the static equilibrium position without the vehicle. The tyres are assumed to remain in contact with the bridge surface at all times. The intermediate supports are modeled as vertical linear springs with large stiffness to simulate bridge piers which are practically not perfectly rigid. The kinetic energy T and the potential energy U of the vehicle–bridge system can be obtained as:

$$T = T_b + T_c$$

$$= \frac{1}{2} \int_0^L \rho A(x) \left(\frac{\partial w(x,t)}{\partial t} \right)^2 dx + \frac{1}{2} m_1 \dot{y}_{c1}^2 + \frac{1}{2} J_1 \dot{\theta}_1^2 + \frac{1}{2} m_2 \dot{y}_{c2}^2 + \frac{1}{2} J_2 \dot{\theta}_2^2$$

$$+ \frac{1}{2} m_3 \dot{y}_4^2 + \frac{1}{2} m_4 \dot{y}_5^2 + \frac{1}{2} m_5 \dot{y}_6^2 + \frac{1}{2} \sum_{i=3}^{5} m_i \dot{\hat{x}}_1(t)^2 + \frac{1}{2} m_1 \dot{x}_{c1}^2 + \frac{1}{2} m_2 \dot{x}_{c2}^2$$

$$U = U_b + U_c \tag{4.1}$$

$$= \frac{1}{2} \int_0^L EI(x) \left(\frac{\partial^2 w(x,t)}{\partial x^2} \right)^2 dx + \frac{1}{2} k_s \sum_{i=1}^{R-1} w(x_i, t)^2 + \frac{1}{2} k_1 (y_1 - y_4)^2$$

$$+ \frac{1}{2} k_2 (y_2 - y_5)^2 + \frac{1}{2} k_3 (y_3 - y_6)^2 + \frac{1}{2} k_7 (y_7 - a_3 y_2 - a_4 y_1)^2$$

$$- m_1 g z_1^c - m_2 g z_2^c - m_3 g z_1 - m_4 g z_2 - m_5 g z_3$$

where subscripts c and b denote the contributions from the vehicle and the bridge respectively; ρ is the density of material of the bridge; $A(x)$ is the cross-sectional area; E is the Young's modulus; $I(x)$ is the moment of inertia of the beam cross-section; $w(x,t)$ is the vertical displacement of the beam; $x_i (i = 0, 1, 2, \ldots, R)$ are the co-ordinates of intermediate point supports and end supports; k_s is the stiffness of the linear spring at the point constraints. y_{c1}, y_{c2} are the vertical displacements and x_{c1} and x_{c2} are the horizontal locations of the centroids of the tractor and the trailer respectively, with subscripts 1 and 2 denote the tractor and the trailer respectively. J_1 and J_2 are the rotational moments of inertia of the tractor and the trailer respectively, and g is the acceleration due to gravity. The bridge is assumed to be in equilibrium under its own weight before the vehicle enters the deck.

By separation of variables, the vertical displacement of the beam $w(x,t)$ can be expressed as:

$$w(x,t) = \sum_{i=1}^{n} q_i(t) W_i(x), \quad \{i = 1, 2, \ldots, n\} \tag{4.2}$$

where $\{W_i(x), i = 1, 2, \ldots, n\}$ are the assumed vibration modes that satisfy the boundary conditions and $\{q_i(t), i = 1, 2, \ldots, n\}$ are the modal coordinates of the bridge. Mode shapes of the bridge deck can be obtained following the procedures described in Section 2.2.2.2.

The motion of the vehicle is defined by the vehicle co-ordinates $\hat{x}_S(t)$ and y_i with the vehicle moving from left to right. The longitudinal position of the centroids x_{c1} and x_{c2} of the tractor and trailer respectively, and the axle locations $\hat{x}_S(t), (s = 1, 2, 3)$

are relative to the point of bridge entry. The longitudinal position and the vertical displacement of the centroids of the tractor and trailer are:

$$x_{c1} = \hat{x}_1(t) - a_1 S_1 - b_1 S_1 \theta_1$$
$$x_{c2} = \hat{x}_1(t) - a_3 S_1 - b_1 S_1 \theta_1 - a_5 S_2 - b_2 S_2 \theta_2$$
$$y_{c1} = a_2 y_1 + a_1 y_2$$
$$y_{c2} = a_5 y_3 + a_6 y_7$$

(4.3)

and the rotational deformations θ_1 and θ_2 of the centroids of the tractor and the trailer are:

$$\theta_1 = (y_1 - y_2)/S_1$$
$$\theta_2 = (y_7 - y_3)/S_2$$

(4.4)

The vertical displacements of the centroids of the tractor and trailer due to the motion of the vehicle relative to the contact points with the pavement are:

$$z_1^c = a_2 z_1 + a_1 z_2$$
$$z_2^c = a_6(a_4 z_1 + a_3 z_2) + a_5 z_3$$

(4.5)

The work done by the system of non-conservative forces of the bridge–vehicle system can then be written as:

$$
\begin{aligned}
W &= W_b + W_d + W_a + W_c \\
&= Q^T C_b \dot{Q} + Y^T C_c \dot{Y} + F_d \hat{x}_1(t) - P_1(t)(z_1 - y_4) - P_2(t)(z_2 - y_5) \\
&\quad - P_3(t)(z_3 - y_6)
\end{aligned}
$$

(4.6)

where W_b, W_d, W_a, W_c are the work done by the damping force of the bridge, the driving force of the vehicle, the interaction forces and the damping force of the vehicle respectively. $Q = \{q_1(t), q_2(t), \ldots, q_n(t)\}^T$ is the vector of modal coordinates of the bridge; F_d is the longitudinal drive force of the vehicle; C_c, C_b are the damping coefficient matrices of the vehicle and the bridge respectively. $\{P_1(t), P_2(t), P_3(t)\}$ are the interaction forces between the vehicle and the bridge written as:

$$P_1(t) = k_4(y_4 - z_1) + c_4(\dot{y}_4 - \dot{z}_1)$$
$$P_2(t) = k_5(y_5 - z_2) + c_5(\dot{y}_5 - \dot{z}_2)$$
$$P_3(t) = k_6(y_6 - z_3) + c_6(\dot{y}_6 - \dot{z}_3)$$

(4.7)

Sharp changes in the road surface cause local changes in the profile at the bottom of the tyre spring. Mulcahy (1983) developed a 'tyre enveloping' model on the road surface roughness in which the profile is approximated by a quadratic parabola. This tends to smooth out the excitation from changes in the road profile. In the present study, the vertical displacements, z_i, at the points of contact of the wheels and the bridge given below includes the road surface roughness function $d(x)$ along the horizontal direction. The relation of function $d(x)$ to the road roughness definition is referred to Section 4.2.3.

$$z_i = w(\hat{x}_i(t), t) + d(\hat{x}_i(t)), \qquad (i = 1, 2, 3)$$

$$\dot{z}_i = v \left.\frac{\partial w(x,t)}{\partial x}\right|_{x=\hat{x}_i(t)} + \left.\frac{\partial w(x,t)}{\partial t}\right|_{x=\hat{x}_i(t)} + v \left.\frac{\partial d(x)}{\partial x}\right|_{x=\hat{x}_i(t)}, \qquad (i = 1, 2, 3) \qquad (4.8)$$

where v is the horizontal velocity of the moving vehicle. Since $d(x)$ is independent of x, the last term in $\dot{z}_i$ can be put equal to zero. The longitudinal position of the second and the third axles are related to that of the first as:

$$\hat{x}_2(t) = \hat{x}_1(t) - S_1$$

$$\hat{x}_3(t) = \hat{x}_1(t) - a_3 S_1 - a_5 S_2 \qquad (4.9)$$

Mulcahy (1983) has presented the equation of motion for a three-axle vehicle on a single span bridge using the Lagrange approach. However, the equation of motion is presented in this section using the Hamilton principle as:

$$\int_{t_1}^{t_2} \delta(T - V + W)dt = 0 \qquad (4.10)$$

The equation of motion of the vertical motion of the vehicle–bridge system can be obtained in both y_i and q_i coordinates as:

$$M_c \ddot{Y} + C_c \dot{Y} + K_c Y = F_c$$

$$M_b \ddot{Q} + C_b \dot{Q} + K_b Q = F_b \qquad (4.11)$$

where $Y = \{y_1, y_2, \ldots, y_7\}^T$ is the vector of displacements at the seven DOFs of the vehicle; F_c and F_b are vectors of generalized forces acting on the vehicle and the bridge structure respectively. M_c, K_c, C_c and M_b, K_b, C_b are the mass, stiffness and damping matrices of the vehicle and the bridge deck modeled as a beam respectively. The parameters in Equation (4.11) are given as:

$$F_c = \left\{ -(m_1 + m_2)b_1\ddot{\hat{x}}_1(t), \ (m_1 + m_2)b_1\ddot{\hat{x}}_1(t), \ m_2 b_2\ddot{\hat{x}}_1(t), \ -P_1(t), \ -P_2(t), \right.$$
$$\left. -P_3(t), \ -m_2 b_2\ddot{\hat{x}}_1(t) \right\}^T$$

$$F_b = \left\{ \sum_{s=1}^{3} W_i(\hat{x}_s(t))P_s'(t), \quad i = 1, 2, \ldots, n \right\}$$

$$P_1'(t) = P_1(t) + (m_1 a_2 + m_2 a_4 a_6 + m_3)g$$

$$P_2'(t) = P_2(t) + (m_1 a_1 + m_2 a_3 a_6 + m_4)g$$

$$P_3'(t) = P_3(t) + (m_2 a_5 + m_5)g$$

$$Y = \{y_1, y_2, \ldots, y_7\}^T;$$

$$Q = \{q_1(t), q_2(t), \ldots, q_n(t)\}^T.$$

$$M_b = \left\{ \int_0^L \rho A(x) W_i(x) W_j(x) dx, \quad (i,j = 1, 2, \ldots, n) \right\}$$

$$K_b = \left\{ \int_0^L EI(x) W_i''(x) W_j''(x) dx + k_s \sum_{l=1}^{R-1} W_i(x_l) W_j(x_l), \quad (i,j = 1, 2, \ldots, n) \right\}$$

$$K_c = \begin{bmatrix} K_{c1} & -K_{c2} & K_{c3}^T \\ -K_{c2} & K_{c2} & 0 \\ K_{c3} & 0 & K_{c4} \end{bmatrix}; \quad M_c = \begin{bmatrix} M_{c1} & 0 & M_{c3}^T \\ 0 & M_{c2} & 0 \\ M_{c3} & 0 & M_{c4} \end{bmatrix}$$

$$M_{c1} = \begin{bmatrix} m_1 a_2^2 + \dfrac{J_1}{S_1^2} + (m_1 + m_2)b_1^2 & m_1 a_1 a_2 - \dfrac{J_1}{S_1^2} - (m_1 + m_2)b_1^2 & -m_2 b_1 b_2 \\ m_1 a_1 a_2 - \dfrac{J_1}{S_1^2} - (m_1 + m_2)b_1^2 & m_1 a_1^2 + \dfrac{J_1}{S_1^2} + (m_1 + m_2)b_1^2 & m_2 b_1 b_2 \\ -m_2 b_1 b_2 & m_2 b_1 b_2 & m_2(a_5^2 + b_2^2) + \dfrac{J_2}{S_2^2} \end{bmatrix}$$

$$M_{c2} = diag(m_3, m_4, m_5)$$

$$M_{c3} = \begin{bmatrix} m_2 b_1 b_2 & -m_2 b_1 b_2 & m_2 a_5 a_6 - \dfrac{J_2}{S_2^2} - m_2 b_2^2 \end{bmatrix}$$

$$M_{c4} = \begin{bmatrix} m_2(a_6^2 + b_2^2) + \dfrac{J_2}{S_2^2} \end{bmatrix}$$

$$K_{c1} = \begin{bmatrix} k_1 + k_7 a_4^2 & k_7 a_3 a_4 & 0 \\ k_7 a_3 a_4 & k_2 + k_7 a_3^2 & 0 \\ 0 & 0 & k_3 \end{bmatrix}$$

$$K_{c2} = diag(k_1, k_2, k_3, k_4)$$

$$K_{c3} = \{-k_7 a_4 \quad -k_7 a_3 \quad 0\}$$

$$K_{c4} = \{k_7\}.$$

The equation of motion of the horizontal motion of the vehicle in $\hat{x}_S(t)$ co-ordinate can also be written as:

$$\sum_{i=1}^5 m_i \ddot{\hat{x}}_1(t) - (m_1 + m_2)b_1(\ddot{y}_1 - \ddot{y}_2) - m_2 b_2(\ddot{y}_7 - \ddot{y}_3)$$

$$= F_d + \sum_{s=1}^3 \left[\sum_{i=1}^n W_i'(\hat{x}_s(t)) q_i(t) \right] P_s'(t) \tag{4.12}$$

with

$$W_i'(\hat{x}_s(t)) = \left. \frac{\partial W_i(x)}{\partial x} \right|_{x = \hat{x}_s(t)} \quad (i = 1, 2, \ldots, n; \quad s = 1, 2, 3) \tag{4.13}$$

4.2.3 Road Surface Roughness

The distribution of the road surface roughness can be represented with a periodic modulated random process. In ISO-8608 (1995) specifications, the road surface roughness is related to the vehicle speed by a formula between the velocity power spectral density (PSD) and the displacement PSD. The general form of the displacement PSD of the road surface roughness is given as:

$$S_d(f) = S_d(f_0) \cdot (f/f_0)^{-a} \tag{4.14}$$

where $f_0(=0.1$ cycles/m$)$ is the reference spatial frequency; a is an exponent of the PSD, and f is the spatial frequency (cycles/m). Equation (4.14) gives an estimate on the degree of roughness of the road from the value of $S_d(f_0)$. This surface roughness classification is based on a constant vehicle velocity PSD taking a equals to 2.

The road surface roughness function $d(x)$ in the time domain can be simulated by applying the Inverse Fast Fourier Transformation on $S_d(f_i)$ to give (Henchi et al., 1998):

$$d(x) = \sum_{i=1}^{N} \sqrt{4S_d(f_i)\Delta f} \cos(2\pi f_i x + \theta_i) \tag{4.15}$$

where $f_i = i\Delta f$ is the spatial frequency; $\Delta f = 1/N\Delta$; Δ is the distance interval between successive ordinates of the surface profile; N is the number of data points, and θ_i is a set of independent random phase angle uniformly distributed between 0 and 2π.

4.2.4 Braking of Vehicle

Initially both the vehicle and bridge are assumed to be at rest and the vehicle is travelling forward at a uniform velocity. A ramp function is assumed for the braking force (Kishan and Traill-Nash, 1977; Gupta and Traill-Nash, 1980). This is based on the test results of highway vehicles conducted by the Transport and Road Research Laboratory in England, in 1975. The braking force increases linearly to a maximum $F_{d\max}$ and then stays constant until the vehicle either comes to a stop or crosses the bridge span and is written as:

$$F_d = \begin{cases} -F_{d\max}t/t_b, & t < t_b \\ -F_{d\max}, & t \geq t_b \end{cases} \tag{4.16}$$

where t_b is the braking rise time.

The impact factors I_d and I_m calculated from the mid-span deflections and bending moments are defined respectively as:

$$I_d = \frac{w_{dynamic}}{w_{static}}$$

$$I_m = \frac{B_{dynamic}}{B_{static}} \tag{4.17}$$

where $w_{dynamic}$, w_{static}, $B_{dynamic}$ and B_{static} are the dynamic and static maximum deflections and bending moments at mid-span of the beam. w_{static} and B_{static} are obtained from an analysis with the vehicle crossing the bridge deck at a crawling speed.

4.2.5 Computational Algorithm

The coupled equations of motion of the bridge–vehicle system presented in Equations (4.11) and (4.12) are subject to the compatibility constraints on the interaction forces and the displacements of the two sub-systems. The procedure to solve the problem is implemented as follows:

Step 1: The mode shapes $W_i(x)$ of the non-uniform multi-span continuous bridge deck are calculated from Equations (2.27) to (2.30) in Chapter 2.

Step 2: Determine the mass, stiffness and damping matrices of both the vehicle and the bridge deck.

Step 3: Calculate the road surface roughness function $d(x)$ from Equation (4.15) according to the selected road class in ISO-8608 (1995).

Step 4: The responses of the bridge and vehicle are calculated by the Newmark Method. The time step, parameters of Newmark Method and the error for convergence are determined before the iteration. Set the initial values of Q and Y.

Step 5: Determine the initial vehicle position on the bridge deck.

Step 6: Calculate the excitation force on vehicle, F_c, and solve for the motion of the vehicle, Y, at time t from Equations (4.11) and (4.12).

Step 7: Calculate the excitation force on the bridge, F_b, and solve for the motion of the bridge, Q, at time t from Equation (4.11).

Step 8: Solve for the displacement of the bridge $w(x, t)$ from Equation (4.2).

Step 9: Repeat Steps 6 to 8 using the calculated Q and Y. Check the convergence of the difference between the two successively calculated $w(x, t)_i$ and $w(x, t)_{i+1}$. The Tolerance error has been set equal to 0.001.

$$\|w(x,t)_{i+1} - w(x,t)_i\| \leq Tolerance\ error \qquad (4.18)$$

Step 10: If convergence is not achieved, repeat Steps 6 to 9.

Step 11: Otherwise, repeat Steps 5 to 10 for the next time step.

4.2.6 Numerical Simulation

The three-axle tractor-trailer vehicle shown in Figure 4.1 was used in the study. The properties of the vehicle are physically measured (Mulcahy, 1983):

The body masses are:

$$m_1 = 3930\,\text{kg}, \; m_2 = 15700\,\text{kg}, \; m_3 = 220\,\text{kg}, \; m_4 = 1500\,\text{kg}, \; m_5 = 1000\,\text{kg}.$$

The vertical stiffness at each DOF is:

$$k_1 = 2.00 \times 10^6\,\text{N/m}, \quad k_2 = 4.60 \times 10^6\,\text{N/m}, \quad k_3 = 5.00 \times 10^6\,\text{N/m},$$
$$k_4 = 1.73 \times 10^6\,\text{N/m}, \quad k_5 = 3.74 \times 10^6\,\text{N/m}, \quad k_6 = 4.60 \times 10^6\,\text{N/m},$$
$$k_7 = 2.00 \times 10^6\,\text{N/m}$$

and their respective viscous damping constants are:

$$c_1 = 5000\,\text{N/m}^{-1}\text{s}, \quad c_2 = 30000\,\text{N/m}^{-1}\text{s}, \quad c_3 = 40000\,\text{N/m}^{-1}\text{s},$$
$$c_4 = 1200\,\text{N/m}^{-1}\text{s}, \quad c_5 = 3900\,\text{N/m}^{-1}\text{s}, \quad c_6 = 4300\,\text{N/m}^{-1}\text{s}, \quad c_7 = 5000\,\text{N/m}^{-1}\text{s}$$

The axle spacing and the pitching moment of inertia of the tractor and the trailer are:

$$S_1 = 3.66\,\text{m}, \quad S_2 = 6.20\,\text{m}, \quad J_1 = 1.05 \times 10^4\,\text{kgm}^2, \quad J_2 = 1.47 \times 10^5\,\text{kgm}^2$$

The parameters on the dimensions of the vehicle are:

$$a_1 = 0.5, \quad a_2 = 0.5, \quad a_3 = 1.0, \quad a_4 = 0.0, \quad a_5 = 0.58, \quad a_6 = 0.42,$$
$$b_1 = 0.25, \quad b_2 = 0.40.$$

Example 1: A Uniform Single Span Bridge Deck

No reference can be found in the literature on the use of a non-uniform beam model in simulation studies, so the simply supported bridge deck studied by Mulcahy (1983) is adopted for comparison. It is 32.6 m long with 16 m effective width, and the mass per unit area is 1240 kg/m². The flexural stiffness of the bridge superstructure is 4.592×10^{10} Nm². The vehicle–bridge mass ratio is 0.0346. It is modeled as a simply supported uniform beam. The first ten modes are used in the solution of the equation of motion in Equations (4.11) and (4.12). A time step of 0.008 second is used in the integration, and it is approximately one-tenth of the highest natural frequency of the bridge superstructure included in the analysis. This small time step is essential because braking produces an impulsive force that consists of frequency components over a wide spectrum.

The three-axle vehicle described previously is traveling at 17 m/s and it brakes at one-quarter span. The impact factors calculated from mid-span bending moments and deflections are studied with variations in the following parameters:

(a) amplitude of the braking force ($F_{d\,\text{max}} = 0.6\,\text{m}_c\text{g}, 0.4\,\text{m}_c\text{g}, 0.2\,\text{m}_c\text{g}$), where m_cg is the vehicle static weight;
(b) braking rise time, $t_b = 0.6\,\text{s}, 0.3\,\text{s}$ and $0.0\,\text{s}$;
(c) braking position of the vehicle;
(d) vehicle horizontal moving velocity; and
(e) different classes of road surface roughness as specified in the ISO-8608 (1995). Road Classes A to E are used in the study.

The maximum impact factors computed for different braking force and braking locations are plotted in Figures 4.2 and 4.3 with no surface roughness included in the analysis. These factors are the maximum values for the duration when the vehicle is on top of the bridge deck. The maximum impact factors for different classes of roads are presented in Table 4.1. The following observations are made:

(a) Figure 4.2 shows that the variation of braking force $F_{d\,\text{max}}$ has little effect on the maximum impact factor which is around 1.0 for all the magnitudes of braking force under study.

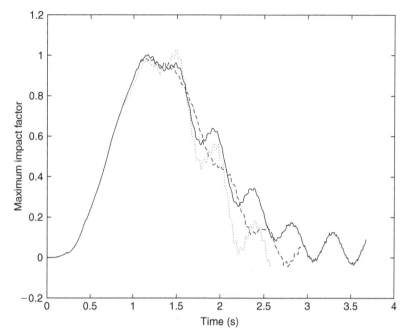

Figure 4.2 Mid-span impact factor from moment with different braking force at 1/4 span
(— 0.6 m_cg; - - - 0.4 m_cg; · · · · 0.2 m_cg)

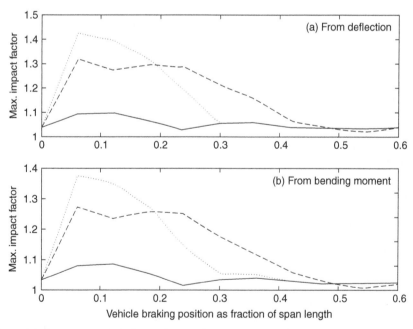

Figure 4.3 Maximum impact factor from different vehicle braking position (— 0.6 s; - - - 0.3 s;
· · · · 0.0 s)

Table 4.1 Impact factors with different road surface conditions (Single span bridge deck)

Road Class		A	B	C	D	E
1.7 m/s – no braking	Deflection	1.09	1.06	1.14	1.24	1.79
	Moment	1.08	1.05	1.12	1.22	1.71
17 m/s – no braking	Deflection	1.04	1.05	1.13	1.27	1.55
	Moment	1.04	1.04	1.10	1.23	1.49
17 m/s – braking at 1/4 span	Deflection	1.08	1.17	1.32	1.62	2.24
	Moment	1.07	1.15	1.29	1.59	2.20

(b)   Figure 4.3 shows that the duration of braking rise time t_b has very significant effect on the maximum impact factor. Braking within the first quarter span produces large impact factor compared with braking at other locations of the span. The impact factors produced from a hard braking $t_b = 0.3$ s is approximately 1.30 while that from a sharp braking with $t_b = 0.0$ s is approximately 1.42.

(c)   Braking on the road approach is not studied. Braking on the approach produces non-zero initial displacement and velocity of the vehicle at the entry point of the bridge, but the equivalent impulsive force from braking does not act on the bridge span. The dynamic effect depends on the characteristics of the vehicle suspension system, but it would be less severe than braking inside the bridge span in general.

(d)   Table 4.1 shows the maximum impact factor for different road surface conditions when the vehicle is moving at 1.7 and 17 m/s or braking with a braking force of $0.6\ m_c g$ at a rise time of 0.6 s at one-quarter span. There is only a slight difference in the impact factor for the cases travelling at constant velocity on Classes A to D roads. Also braking causes a distinctly higher impact factor when compared with the no braking cases on Classes C to E roads. Road Class E exhibits the worst dynamic responses with or without braking.

(e)   The case of travelling at low velocity on Class E road has a significantly higher impact factor than that with a higher velocity. When the road roughness peaks are defined over a long time interval (lower speed), the vehicle would experience a lower frequency excitation closer to its own natural frequency, thus generating larger excitation force and larger dynamic responses in the bridge deck.

Example 2: A Three-Span Continuous Non-Uniform Bridge Deck

A modern non-uniform three-span box-section bridge deck (Zheng et al., 1998) shown in Figure 4.4 is modeled as a 36 m-48 m-36 m three-span non-uniform continuous beam. The vehicle to bridge mass ratio is 0.0235. The time step required in the calculation of responses is taken as 0.0147 second taking into account the first ten modes of the bridge superstructure in the analysis. The mode shape function for the bridge deck modeled as a continuous beam is referred to in Section 2.2.2.2. The same three-axle vehicle that was used for last example is used in this study.

The dynamic effects are studied with variations in the influencing parameters for comparison. The maximum impact factors are plotted in Figures 4.5 to 4.8 and in

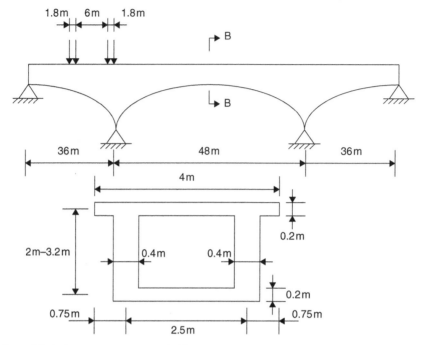

Figure 4.4 A three-span continuous bridge

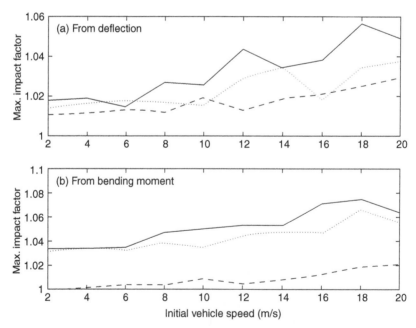

Figure 4.5 Mid-span impact factor from different vehicle speed (— first span; - - - second span;
···· third span)

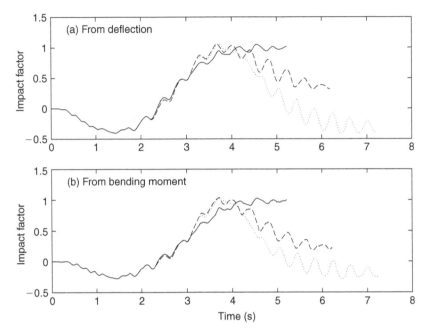

Figure 4.6 Impact factor at middle of second span from different braking position (braking at: — 2/7L; - - - 3/7L; · · · · 4/7L)

Figure 4.10, and the maximum impact factors for different classes of roads are shown in Tables 4.2 and 4.3. The following observations are made:

(a) Figure 4.5 shows the maximum impact factors at the mid-span of each of the three spans for different constant travelling velocities. No road surface roughness is included in the analysis. The impact factors are, in general, small even at high velocities. The impact factor is smallest in the second of the three spans. This may be due to its higher flexibility compared with the other two spans.

(b) Figure 4.6 shows the variation of the impact factors resulting from braking at positions 2/7, 3/7 and 4/7 of the total bridge span, L, with 2/7L almost on the second support, and 3/7L and 4/7L are inside the second span. The vehicle initial velocity is 17 m/s, and the Class B road is considered. The impact factor at the second span is largest among the three spans for these braking locations. The impact factors are, in general, small with the highest value of 1.04 for braking at 3/7 span.

(c) Figure 4.7 shows the variation of impact factor with vehicle braking at 3/7L on a Class B road moving with an initial velocity of 17 m/s. The braking rise time t_b is found to have significant effect on the impact factor with a value of 1.17 for a hard braking at $t_b = 0.3$ s. In the case of a sharp braking at $t_b = 0.0$ s, the impact factor is 1.49. It is noted that the oscillating component in the responses in Figures 4.6 to 4.8 is due to the large pitching action, arising from braking, of the vehicle.

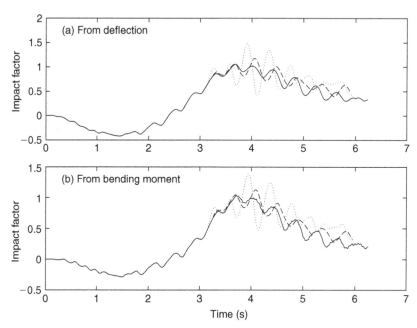

Figure 4.7 Impact factor at middle of second span from different braking rise time and braking at 3/7L (— 0.6 s; - - - 0.3 s; · · · · 0.0 s)

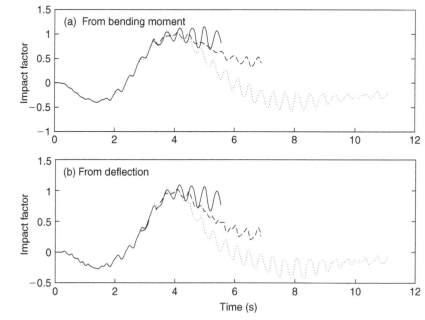

Figure 4.8 Impact factor at middle of second span from different amplitude of braking force (— 0.6 $m_c g$; - - - 0.4 $m_c g$; · · · · 0.2 $m_c g$)

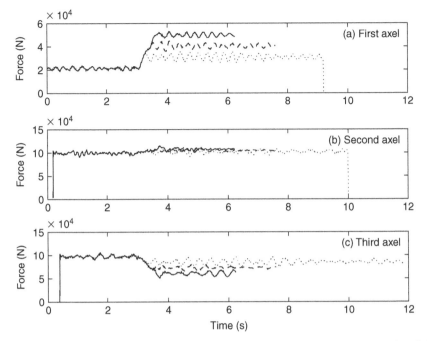

Figure 4.9 Interaction axle forces from different amplitude of braking force (— $0.6\,m_c g$; --- $0.4\,m_c g$; ···· $0.2\,m_c g$)

(d) Figure 4.8 shows the variation of impact factor at the middle of second span from braking at 1/3L (within the second span) on a Class B road with an initial velocity of 17 m/s. The braking force $F_{d\,max}$ has little effect on the maximum impact factor with the largest value of 1.14 for $F_{d\,max} = 0.6\,m_c g$. The corresponding interaction axle forces are shown in Figure 4.9. The curves indicate approximately proportional increase in the interaction forces in the first and the third axles, while that in the second axle exhibits very small change with $F_{d\,max}$. This phenomenon is due to the pitching action of the vehicle.

(e) Figure 4.10 gives the maximum impact factors at the second span from different vehicle velocity and braking positions. Both factors have little effect on the impact factor when braking occurs at 1/7L. This is because when braking occurs at this location in the first span the vehicle would never reach the middle of the second span. When braking occurs at 3/7L, all the impact factors are larger than unity, and it gradually decreases with higher velocity. The maximum is approximately 1.15 at 10 m/s velocity, the lower end of the velocity range. When braking occurs at 2/7L, which is almost on top of the second support, it creates compression in the suspension system of the vehicle. And when it comes into the second span, the vehicle will bounce on top of the bridge deck causing higher impact factors than usual. However their values are not higher than those obtained when braking at the more critical location of 3/7L as seen in Figure 4.10.

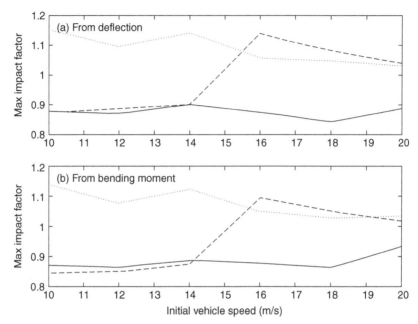

Figure 4.10 Maximum impact factor at middle of the second span from different vehicle velocity and braking position (braking at: — 1/7L; - - - 2/7L; · · · · 3/7L)

(f)   Table 4.2 shows the impact factor for different classes of roads. Only the impact factor in the third span increases very significantly with velocity, and there is very small change in the other spans. Both bending moments and deflections give approximately the same impact factor. Since spans 1 and 3 are identical, this difference may be due to the different initial conditions of the vehicle at entry into the two spans (zero initial conditions for span 1 and non-zero conditions for span 3). This observation can also be found in Figure 4.5 where the maximum impact factors are similar for both spans 1 and 3. However, the impact factors may be larger in the analysis due to the inclusion of road surface roughness.

(g)   Table 4.3 shows the impact factor for the case of $F_{d\max} = 0.6\, m_c g$, $t_b = 0.6$ s and 17 m/s initial velocity with braking at different locations. The maximum impact factor increases slightly with the road surface roughness in Road Classes A to E, and the impact factor from deflection and moment are approximately the same for all Road Classes. Small impact factors are found when braking starts at 1/7L and the reason is as explained for Figure 4.10. Those factors from braking at 3/7L are largest, and those from braking beyond 3/7L are the same. This is similar to the observation for a single span beam where braking within the first quarter span produces the most dynamic responses.

From the above observations, it may be concluded that a more correct definition of impact factor for a continuous beam is needed. The impact factor due to braking

Table 4.2 Impact factor from deflection and moment with different road surface conditions (three-span bridge deck)

Road Class		Span	A	B	C	D	E
17 m/s no braking	Deflection Impact Factor	1	1.01	1.02	1.08	1.19	1.40
		2	1.02	1.05	1.11	1.24	1.50
		3	1.07	1.15	1.32	1.65	2.29
	Moment Impact Factor	1	1.05	1.06	1.12	1.24	1.45
		2	1.02	1.05	1.11	1.23	1.49
		3	1.11	1.20	1.37	1.71	2.33
1.7 m/s no braking	Deflection Impact Factor	1	1.00	1.03	1.10	1.24	1.34
		2	1.02	1.08	1.15	1.30	1.59
		3	1.01	1.05	1.10	1.14	1.46
	Moment Impact Factor	1	1.04	1.06	1.11	1.25	1.35
		2	1.02	1.07	1.12	1.25	1.49
		3	1.04	1.08	1.12	1.17	1.46

Table 4.3 Impact factor from deflection and moment in the second span with different braking position of vehicle on different classes of road

Road Class		A	B	C	D	E
Brake at 1/7L	Deflection	0.43	0.45	0.51	0.69	1.05
	Moment	0.48	0.49	0.52	0.67	0.96
Brake at 2/7L	Deflection	1.03	1.04	1.10	1.24	1.65
	Moment	1.02	1.03	1.09	1.02	1.55
Brake at 3/7L	Deflection	1.03	1.06	1.12	1.26	1.50
	Moment	1.03	1.06	1.11	1.23	1.47
Brake at 4/7L	Deflection	1.02	1.05	1.11	1.24	1.50
	Moment	1.02	1.05	1.11	1.23	1.49
Brake at 5/7L	Deflection	1.02	1.05	1.11	1.24	1.50
	Moment	1.02	1.05	1.11	1.23	1.49
Brake at 6/7L	Deflection	1.02	1.05	1.11	1.24	1.50
	Moment	1.02	1.05	1.11	1.23	1.49

effect should be based on a comparison of the maximum dynamic and static responses at the same span in which braking occurs.

4.3 Dynamic Loads on Continuous Multi-Lane Bridge Decks from Moving Vehicles

This section investigates the dynamic loading on a multi-lane continuous bridge deck due to moving vehicles on top. The bridge is modeled as a multi-span continuous orthotropic rectangular plate with intermediate line rigid supports. The analytical vehicle is simulated as a two-axle three-dimensional vehicle model with seven degrees-of-freedom according to the H20-44 vehicle design loading in AASHTO (2007). The dynamic behavior of the bridge deck under several moving vehicles is analyzed using orthotropic plate theory and modal superposition technique. The effects of multi-lane loading from multiple vehicles on the dynamic impact factor of the bridge deck are discussed. The impact factor is found to be varying in an opposite trend as the dynamic responses for the different loading cases under study.

4.3.1 Bridge Model

The equations of motion for the bridge deck have been given in Section 3.3 as:

$$M_b \ddot{Q} + C_b \dot{Q} + K_b Q = W_b F_b^{\text{int}} \tag{4.19}$$

where M_b, C_b, K_b are the mass, damping and stiffness matrices of the bridge; F_b^{int} is the vector of interaction force under the wheels of the moving vehicles; $\dot{Q}, \ddot{Q}$ are the first and second derivatives of Q; and Q is the vector of modal amplitudes.

4.3.2 Vehicle Model

The mathematical model for the H20-44 truck is shown in Figure 4.11. The model is similar to that employed by Marchesiello et al. (1999). The vehicular body is assigned three degrees-of-freedom, corresponding to the vertical displacement (y), rotation about the transverse axis (pitch or θ_p), and rotation about the longitudinal axis (roll or θ_r). Each wheel/axle set is provided with two degrees-of-freedom in the vertical and roll direction ($y_{a1}, y_{a2}, \theta_{a1}, \theta_{a2}$).

Therefore, the total number of independent degrees-of-freedom is seven. The equations of motion of the vehicle are derived using Lagrange's formulation as follows:

$$M_v \ddot{Z} + C_v \dot{Z} + K_v Z = F_v^{\text{int}} \tag{4.20}$$

where F_v^{int} is the interaction force vector applied on the vehicle; M_v, C_v, K_v are, respectively, the mass, damping and stiffness matrices of the vehicle system and Z is the vector of the degrees-of-freedom of the vehicle. The variables in Equation (4.20) are given as

$$Z = \left\{ y_c, \theta_p, \theta_r, y_{a1}, \theta_{a1}, y_{a2}, \theta_{a2} \right\}^T;$$

$$F_v^{\text{int}} = \left\{ 0, 0, 0, -F_{t1} - F_{t2}, \frac{S_{d1}}{2}(F_{t1} - F_{t2}), -F_{t3} - F_{t4}, \frac{S_{d2}}{2}(F_{t3} - F_{t4}) \right\}^T;$$

$$M_v = diag\{m_c, I_c, I_t, m_{a1}, I_{a1}, m_{a2}, I_{a2}\};$$

$$K_v = \{K_{vij}, K_{vij} = K_{vji}, \quad i = 1, 2, \ldots, 7; \ j = 1, 2, \ldots, 7\};$$

$$K_{v11} = \sum_{i=1}^{4} K_{syi}; \quad K_{v12} = (K_{sy1} + K_{sy2})a_1 S_x - (K_{sy3} + K_{sy4})a_2 S_x;$$

$$K_{v13} = \frac{S_{y1}}{2}(-K_{sy1} + K_{sy2}) + \frac{S_{y2}}{2}(-K_{sy3} + K_{sy4}); \quad K_{v14} = -(K_{sy1} + K_{sy2});$$

$$K_{v15} = \frac{S_{y1}}{2}(K_{sy1} - K_{sy2}); \quad K_{v16} = -K_{sy3} - K_{sy4}; \quad K_{v17} = \frac{S_{y2}}{2}(K_{sy3} - K_{sy4});$$

$$K_{v22} = (K_{sy1} + K_{sy2})a_1^2 S_x^2 + (K_{sy3} + K_{sy4})a_2^2 S_x^2;$$

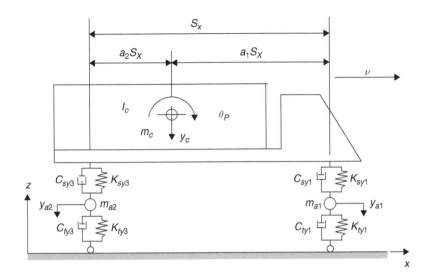

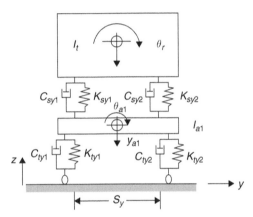

Figure 4.11 Idealization of two-axle vehicle

$$K_{v23} = \frac{1}{2}(-K_{sy1} + K_{sy2})a_1 S_x S_{y1} + \frac{1}{2}(K_{sy3} - K_{sy4})a_2 S_x S_{y2};$$

$$K_{v24} = -(K_{sy1} + K_{sy2})a_1 S_x; \quad K_{v25} = \frac{1}{2}(K_{sy1} - K_{sy2})a_1 S_x S_{y1};$$

$$K_{v26} = (K_{sy3} + K_{sy4})a_2 S_x; \quad K_{v27} = -\frac{1}{2}(K_{sy3} - K_{sy4})a_2 S_x S_{y2};$$

$$K_{v33} = \frac{S_{y1}^2}{4}(K_{sy1} + K_{sy2}) + \frac{S_{y2}^2}{4}(K_{sy3} + K_{sy4});$$

$$K_{v34} = \frac{S_{y1}}{2}(K_{sy1} - K_{sy2}); \quad K_{v35} = -\frac{S_{y1}^2}{4}(K_{sy1} + K_{sy2});$$

$$K_{v36} = \frac{S_{y2}}{2}(K_{sy3} - K_{sy4}); \quad K_{v37} = -\frac{S_{y2}^2}{4}(K_{sy3} + K_{sy4});$$

$$K_{v44} = (K_{sy1} + K_{sy2}); \quad K_{v45} = \frac{S_{y1}}{2}(-K_{sy1} + K_{sy2}); \quad K_{v46} = K_{v47} = 0;$$

$$K_{v55} = \frac{S_{y1}^2}{4}(K_{sy1} + K_{sy2}); \quad K_{v56} = K_{v57} = 0;$$

$$K_{v66} = K_{sy3} + K_{sy4}; \quad K_{v67} = -\frac{S_{y2}}{2}(K_{sy3} - K_{sy4}); \quad K_{v77} = \frac{S_{y2}^2}{4}(K_{sy3} + K_{sy4});$$

where S_{y1}, S_{y2} are the spacing of suspensions in the front and rear axles respectively; S_x is the axle spacing.

$$F_b^{int} = F_g + F_b; \quad F_b = \{F_{t1}, F_{t2}, F_{t3}, F_{t4}\}^T;$$

$$F_g = \{(m_c a_1 + m_{a1})g/2, (m_c a_1 + m_{a1})g/2, (m_c a_2 + m_{a2})g/2, (m_c a_2 + m_{a2})g/2\}^T;$$

where F_g is the force vector caused by the effect of gravitation.

4.3.3 Vehicle–Bridge Interaction

The vehicle–bridge interaction forces for a single vehicle can be written as follows:

$$F_{t1} = K_{ty1}\left(y_{a1} - \frac{1}{2}S_{d1}\theta_{a1} - w_1 - d_1\right) + C_{ty1}\left(\dot{y}_{a1} - \frac{1}{2}S_{d1}\dot{\theta}_{a1} - \dot{w}_1 - \dot{d}_1\right);$$

$$F_{t2} = K_{ty2}\left(y_{a1} + \frac{1}{2}S_{d1}\theta_{a1} - w_2 - d_2\right) + C_{ty2}\left(\dot{y}_{a1} + \frac{1}{2}S_{d1}\dot{\theta}_{a1} - \dot{w}_2 - \dot{d}_2\right);$$

$$F_{t3} = K_{ty3}\left(y_{a2} - \frac{1}{2}S_{d2}\theta_{a2} - w_3 - d_3\right) + C_{ty3}\left(\dot{y}_{a2} - \frac{1}{2}S_{d2}\dot{\theta}_{a2} - \dot{w}_3 - \dot{d}_3\right); \quad (4.21)$$

$$F_{t4} = K_{ty4}\left(y_{a2} + \frac{1}{2}S_{d2}\theta_{a2} - w_4 - d_4\right) + C_{ty4}\left(\dot{y}_{a2} + \frac{1}{2}S_{d2}\dot{\theta}_{a2} - \dot{w}_4 - \dot{d}_4\right);$$

where $\{K_{tyi}, i = 1, 2, 3, 4\}$ are the stiffness of the tyres; $\{C_{tyi}, i = 1, 2, 3, 4\}$ are the damping coefficients of the tyres. S_{d1}, S_{d2} are the wheel spacing of the front and rear axles respectively.

$$w_i = w(\hat{x}_i(t), \hat{y}_i(t), t); \quad d_i = d(\hat{x}_i(t), \hat{y}_i(t)); \quad i = 1, 2, 3, 4 \quad (4.22)$$

where $d(x, y)$ is the surface roughness of the bridge deck; $(\hat{x}_i(t), \hat{y}_i(t))$ is the location of the ith tyre at time t. The vehicle moves along one lane, with $\hat{y}_1(t) = y_0 + S_{d1}/2$, $\hat{y}_2(t) = y_0 - S_{d1}/2$, $\hat{y}_3(t) = y_0 + S_{d2}/2$, $\hat{y}_4(t) = y_0 - S_{d2}/2$, and y_0 is the transverse coordinate of the centre-line of the lane. For the purpose of this study $d(x, y)$ is taken to be one dimensional only with $\hat{y}_i(t)$ the same for both wheels in an axle.

The dynamic responses of the bridge deck under moving vehicles can be calculated from Equations (4.19) to (4.21) using an iterative method (such as the Newmark method) or the algorithm by Henchi et al. (1998).

4.4 Impact Factors

Not much research has been done with multiple vehicles on top of a multi-lane bridge deck. Humar and Kashif (1995) simplified a slab-type bridge deck as a single span orthotropic plate, and the effects of off-center vehicles and two vehicles on the bridge were discussed with a one-quarter vehicle model. The effects of having multiple vehicles on a single span bridge deck with two lanes have been presented by Yener and Chompooming (1994). Mabsout et al. (1999) have also studied the effect of multi-lanes on the wheel load distribution in a steel girder bridge.

A continuous three span multi-girder bridge, as shown in Figure 4.12, is used for this study. There are four equal lanes over the total width of the bridge deck. The parameters of the bridge deck are listed as follows:

Span lengths are 24 m, 30 m, 24 m for the first, second and third spans respectively; distance between two adjacent main girders is 2.743 m; distance between two adjacent diaphragms is 6 m; deck slab thickness is 0.2 m, $b = 13.715$ m, $\rho = 3000$ kg/m^3, $E_x = 4.1682 \times 10^{10}$ N/m^2, $E_y = 2.9733 \times 10^{10}$ N/m^2, $v_{xy} = 0.3$.
For the steel I-beam: web thickness $= 0.01111$ m, web height $= 1.490$ m, flange width $= 0.405$ m, flange thickness $= 0.018$ m.
For the diaphragms: cross-sectional area $= 0.001548$ m^2, $I_y = 0.707 \times 10^{-6}$ m^4, $I_z = 2 \times 10^{-6}$ m^4, $J = 1.2 \times 10^{-7}$ m^4.

The rigidities of the equivalent orthotropic plate can be calculated according to Bakht and Jaeger (1985) with $D_x = 2.415 \times 10^9$ Nm, $D_y = 2.1807 \times 10^7$ Nm, $D_{xy} = 1.1424 \times 10^8$ Nm. The first 13 natural frequencies of the continuous bridge are 4.13, 4.70, 6.31, 6.86, 7.76, 8.20, 15.81, 16.39, 20.84, 22.29, 22.90, 24.31 and 24.86 Hz. The damping coefficients of the bridge deck are taken as 0.02 for all the

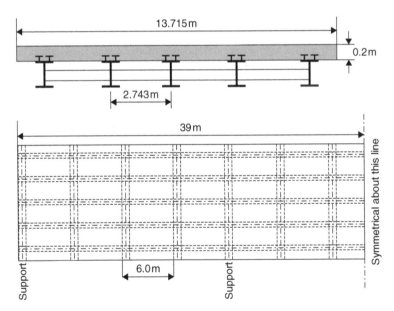

Figure 4.12 Views of the bridge deck

vibration modes. Road Classes A to D according to ISO-8608 (1995) and the case without roughness are used in the simulations. The vehicle is the same as that used in Section 4.3.2 with the following parameters:

$$S_x = 4.73 \, \text{m}, \quad a_1 = 0.67, \quad a_2 = 0.33, \quad S_{d1} = S_{d2} = 2.05 \, \text{m},$$
$$S_{y1} = S_{y2} = 1.41 \, \text{m},$$
$$m_c = 17000 \, \text{kg}, \quad m_{a1} = 600 \, \text{kg}, \quad m_{a2} = 1000 \, \text{kg}, \quad I_c = 9 \times 10^4 \, \text{kgm}^2,$$
$$I_t = 1.3 \times 10^4 \, \text{kgm}^2,$$
$$I_{a1} = 550 \, \text{kgm}^2, \quad I_{a2} = 600 \, \text{kgm}^2,$$
$$K_{sy1} = K_{sy2} = 1.16 \times 10^5 \, \text{N/m}, \quad K_{sy3} = K_{sy4} = 3.73 \times 10^5 \, \text{N/m},$$
$$K_{ty1} = K_{ty2} = 7.85 \times 10^5 \, \text{N/m}, \quad K_{ty3} = K_{ty4} = 1.57 \times 10^6 \, \text{N/m},$$
$$C_{sy1} = C_{sy2} = 2.5 \times 10^4 \, \text{Ns/m}, \quad C_{sy3} = C_{sy4} = 3.5 \times 10^4 \, \text{Ns/m},$$
$$C_{ty1} = C_{ty2} = 100 \, \text{Ns/m}, \quad C_{ty3} = C_{ty4} = 200 \, \text{Ns/m}.$$

I_{a1} and I_{a2} are the torsional moments of inertia of the two axles respectively; S_{y1} and S_{y2} are the spacing of the suspensions of the front and rear axle respectively; the subscript syi refers to the ith suspension of the vehicle. Initial Q_0 and Y_0 in the calculation are set to zero.

4.4.1 Dynamic Loading from a Single Vehicle

The bending moment and shear force in the plate are calculated as:

$$\begin{cases} M_x = -\left(D_x \dfrac{\partial^2 w}{\partial x^2} + v_{xy} D_y \dfrac{\partial^2 w}{\partial y^2} \right) \\ V_x = -\left[D_x \dfrac{\partial^3 w}{\partial x^3} + (v_{xy} D_y + 4 D_{xy}) \dfrac{\partial^3 w}{\partial x \partial y^2} \right] \end{cases} \tag{4.23}$$

The impact factor and wheel-load distribution factor are defined after Huang et al. (1992) as follows:

$$I_p = \left(\frac{R_d}{R_S} - 1 \right) \times 100\%$$
$$\eta = \frac{M_i}{M_t} \tag{4.24}$$

where R_d and R_S are the absolute maximum response from the dynamic and static studies, respectively. Here R_S is obtained with the vehicle moving at a very low speed of 0.1 m/s with 0.1 s for the time step in the calculation and no road surface roughness is included. M_i is the maximum bending moment of one beam at the section; $M_t = M/n$; where M is the sum of the bending moment of all beams at one section; n is the number of wheel loads in the transverse direction.

Table 4.4 shows the static wheel load distribution (Static LDF), dynamic wheel load distribution (Dynamic LDF) and the impact factor (IMP) from bending moment, displacement and strain under different loading cases. The Loading Cases from a single vehicle are shown in Figure 4.13(a). Figure 4.14 shows the time histories of the

Table 4.4 Load distribution factor (LDF) and impact factor (IMP) (single vehicle)

	Span	Load Case	Beam-1	Beam-2	Beam-3	Beam-4	Beam-5
Static LDF	First	1	0.863	0.575	0.320	0.078	0.165
		2	0.586	0.486	0.395	0.309	0.224
		3	0.384	0.393	0.400	0.408	0.415
	Second	1	0.874	0.567	0.319	0.087	0.153
		2	0.564	0.509	0.405	0.305	0.217
		3	0.375	0.407	0.405	0.404	0.409
	Third	1	0.863	0.576	0.319	0.078	0.164
		2	0.587	0.487	0.394	0.309	0.224
		3	0.384	0.393	0.400	0.408	0.415
Dynamic LDF	First	1	0.850	0.568	0.319	0.091	0.172
		2	0.583	0.484	0.395	0.311	0.227
		3	0.384	0.393	0.400	0.408	0.415
	Second	1	0.864	0.561	0.320	0.098	0.157
		2	0.557	0.505	0.405	0.309	0.225
		3	0.376	0.407	0.405	0.404	0.409
	Third	1	0.856	0.572	0.317	0.085	0.170
		2	0.586	0.486	0.394	0.309	0.224
		3	0.384	0.393	0.400	0.408	0.415
IMP(%) (Bending Moment)	First	1	2.54	2.87	3.76	21.90	8.73
		2	2.90	3.16	3.54	4.11	5.15
		3	3.59	3.52	3.50	3.45	3.41
	Second	1	2.91	3.13	4.47	16.60	6.61
		2	2.48	2.92	3.75	5.14	7.51
		3	4.00	3.70	3.66	3.63	3.56
	Third	1	5.07	5.13	5.33	15.87	9.63
		2	5.25	5.25	5.27	5.32	5.40
		3	5.29	5.29	5.25	5.27	5.27
IMP(%) (Displacement)	First	1	2.53	2.81	3.59	21.48	7.26
		2	2.81	3.03	3.36	3.89	4.85
		3	3.42	3.36	3.32	3.29	3.26
	Second	1	3.56	4.04	4.83	14.41	6.41
		2	3.45	3.34	3.97	5.11	7.12
		3	4.19	3.90	3.89	3.90	3.86
	Third	1	4.50	4.51	4.55	17.21	10.89
		2	4.56	4.53	4.50	4.47	5.53
		3	4.49	4.48	4.49	4.49	4.50
IMP(%) (Strain)	First	1	2.54	2.87	3.76	21.85	8.74
		2	2.90	3.16	3.53	4.11	5.15
		3	3.59	3.53	3.49	3.45	3.42
	Second	1	2.91	3.13	4.46	16.61	6.58
		2	2.48	2.92	3.75	5.14	7.52
		3	4.00	3.71	3.66	3.63	3.56
	Third	1	5.08	5.15	5.30	15.98	9.68
		2	5.25	5.25	5.27	5.30	5.40
		3	5.27	5.26	5.26	5.27	5.26

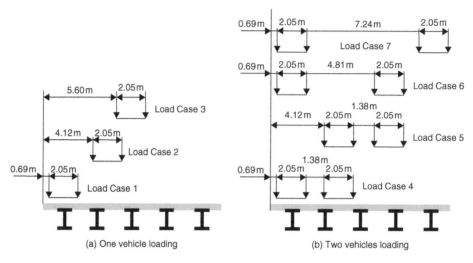

Load Case 7: 0.69 m | 2.05 m | 7.24 m | 2.05 m

Load Case 6: 0.69 m | 2.05 m | 4.81 m | 2.05 m

Load Case 5: 4.12 m | 2.05 m | 1.38 m | 2.05 m

Load Case 4: 0.69 m | 2.05 m | 1.38 m | 2.05 m

Load Case 3: 5.60 m | 2.05 m

Load Case 2: 4.12 m | 2.05 m

Load Case 1: 0.69 m | 2.05 m

(a) One vehicle loading   (b) Two vehicles loading

Figure 4.13 Vehicle Loading

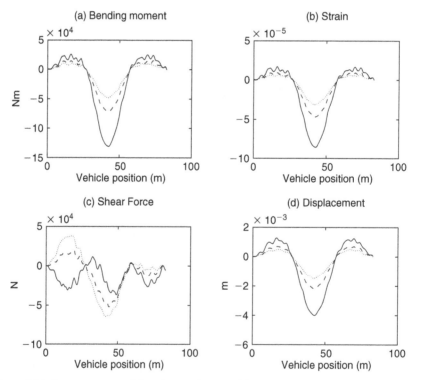

(a) Bending moment

(b) Strain

(c) Shear Force

(d) Displacement

Figure 4.14 Responses at middle of second span of Beam-1 under different loading (— Load
Case 1; --- Load Case 2; ···· Load Case 3)

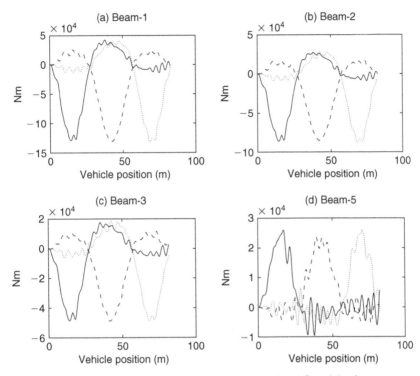

Figure 4.15 Bending moment at middle of each beam under Load Case 1 (——first span; - - - second span; · · · · third span)

bending moments, strains, shear forces and displacements at middle of the second span of beam-1 for Load Cases 1 to 3. Figure 4.15 shows the bending moments at middle of each span of Beams 1, 2, 3 and 5 under Load Case 1. The speed of the vehicle is 30 m/s, and a time step of 0.001 s is used in the calculation. The road surface roughness is of Class B. Figures 4.16 and 4.17 show the effects of moving speed and road surface roughness respectively on the impact factors calculated at different points of the bridge deck under different loading cases. The legends 1~3, 4~6, 7~9, 10~12 and 13~15 define the midpoints of the first, second and third spans on beams 1~5, respectively. The following observations are obtained from these figures and Table 4.4.

1. The dynamic impact factor calculated at different locations on the bridge deck varies in an opposite manner to that for the wheel load distribution factor. The former has a large value whereas the latter has a small value at the same point, and vice versa.
2. The impact factors obtained from different measurable variables, such as bending moment, strain and displacement, are similar, and hence strains can be used in further studies as they can be easily measured.

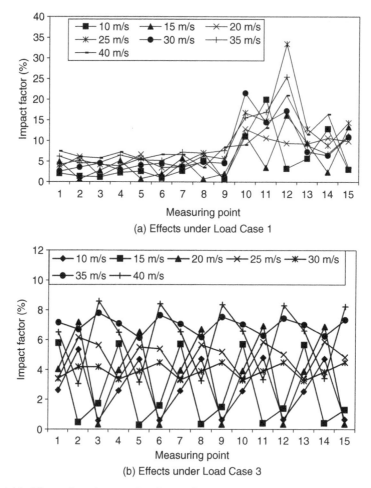

Figure 4.16 Effects of moving speed on impact factor

3. The bending moments in beams 1 and 2 are larger when the vehicle is close to them while those in beams 4 and 5 are smaller. This is because the motion of the vehicle in the outer lane excites the torsional modes which are significant to the responses.

4. The impact factors in the beams near to the path of the moving vehicle are smaller than those in the beams further away. But the dynamic responses behave oppositely, as seen in Figure 4.14.

5. The impact factor is insensitive to the moving speed of the vehicle.

6. In Load Case 1, the impact factors in beams 4 and 5 are larger than those in the other beams. The impact factors in beam 4 are largest in all the cases studied. This may be due to the torsional modes excited in Load Case 1.

7. When the road surface roughness is increasing the impact factor also increases, especially for the case under eccentric Load Case 1. The maximum impact factor is 225 percent, as seen in Figure 4.17.

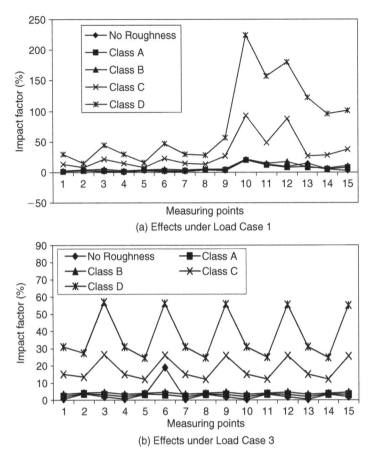

Figure 4.17 Effects of road-surface roughness on impact factor

4.4.2 Dynamic Loading from Multiple Vehicles

Two vehicles moving in different lanes are studied in the simulations. The loading cases are shown in Figure 4.13(b). Load case 4 consists of two vehicles moving in the same direction. load cases 5 to 7 consist of two vehicles moving in opposite directions. Figure 4.18 shows the bending moments, strains, shear forces and displacements at mid-span of span 2 of beam-1 under different loading cases. The speed of moving vehicles is 30 m/s and the two vehicles enter the bridge at the same time. The road surface roughness is of Class B. Time step in the computation is 0.001 s. Figure 4.19 show the bending moments at middle of span 2 of beams 2, 3, 4 and 5 under loading cases 4, 5 and 7 which are more severe. Table 4.5 shows the computed static and dynamic load distribution factor and impact factor. The following observations are made from these figures and Table 4.5.

1. Tables 4.4 and 4.5 show that the impact factor generated from two vehicles is smaller than that from a single vehicle.

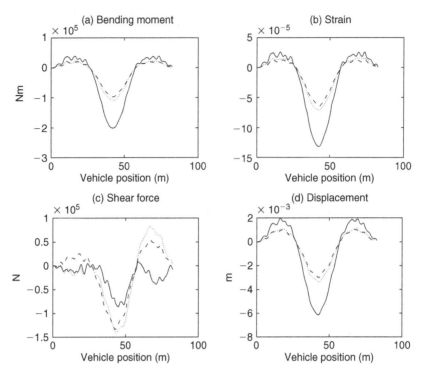

Figure 4.18 Responses at middle of second span of Beam-1 under different loading (— Load Case 4; - - - Load Case 5; · · · · Load Case 7)

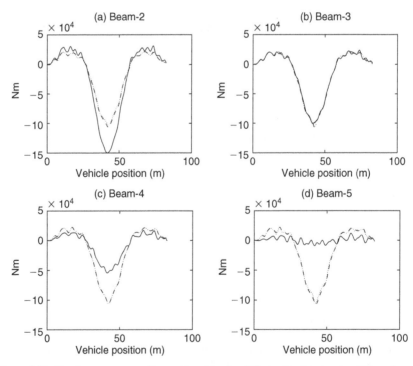

Figure 4.19 Bending moment at Beams 2 to 5 under different loading (— Load Case 4; - - - Load Case 5; · · · · Load Case 7)

Table 4.5 Load distribution factor (LDF) and impact factor (IMP) (two vehicles)

	Span	Load Case	Beam-1	Beam-2	Beam-3	Beam-4	Beam-5
Static LDF	First	4	1.604	1.163	0.771	0.399	0.062
		5	0.896	0.756	0.644	0.783	0.921
		6	1.267	0.858	0.523	0.625	0.726
		7	1.072	0.730	0.460	0.734	1.005
	Second	4	1.593	1.176	0.781	0.407	0.043
		5	0.757	0.815	0.809	0.806	0.814
		6	1.222	0.973	0.781	0.604	0.420
		7	0.852	0.790	0.785	0.791	0.782
	Third	4	1.607	1.167	0.769	0.399	0.059
		5	0.895	0.751	0.648	0.784	0.922
		6	1.268	0.857	0.522	0.626	0.727
		7	1.072	0.725	0.462	0.735	1.007
Dynamic LDF	First	4	1.565	1.138	0.7600	0.400	0.138
		5	0.900	0.765	0.642	0.778	0.916
		6	1.275	0.868	0.519	0.618	0.720
		7	1.071	0.737	0.4600	0.731	1.001
	Second	4	1.558	1.152	0.775	0.420	0.095
		5	0.754	0.810	0.808	0.808	0.820
		6	1.201	0.964	0.780	0.614	0.441
		7	0.838	0.783	0.784	0.797	0.798
	Third	4	1.584	1.151	0.760	0.395	0.110
		5	0.915	0.772	0.658	0.761	0.894
		6	1.296	0.881	0.522	0.604	0.697
		7	1.094	0.747	0.456	0.719	0.985
IMP(%) (Displacement)	First	4	2.66	2.93	3.50	5.27	125.19
		5	0.78	1.26	0.29	0.21	0.16
		6	1.42	1.61	0.31	−0.09	−0.30
		7	0.33	0.91	0.25	−0.12	−0.30
	Second	4	3.02	3.17	4.00	7.17	120.82
		5	6.48	5.99	6.42	6.90	7.20
		6	4.94	5.93	6.88	8.39	11.60
		7	5.14	6.27	6.84	7.55	8.61
	Third	4	4.67	4.68	4.70	5.91	92.60
		5	6.22	6.93	5.34	0.88	0.61
		6	5.64	6.39	2.99	−0.73	0.34
		7	4.81	5.86	1.14	0.12	−0.15

2. The impact factor also behaves in opposite trend when compared with the wheel load distribution factor.

3. The magnitude of bending moments in each beam is closely related to the transverse location of the resultant of the vehicular loads on the bridge deck. It is large when the resultant is close to the beam and small when the force is further away from the beam.

4. Load case 4 gives the largest impact factors in beam-5 which are 125.19 percent, 120.82 percent and 92.60 percent in the middle of spans 1, 2 and 3, respectively.

But the corresponding load distribution factor and the responses in Figure 4.19 are small. This is similar to the observation in Point (6) made for a single vehicle in Section 4.4.1 where the contribution of torsional modes of the bridge deck is suspected. This indicates a common point that is often mixed up in design. The magnitude of the dynamic impact factor is of no significance if it is not related to the magnitude of member stress or capacity. A high impact factor, in general, corresponds to a very low stress level. Therefore, only impact factors that relate to a design configuration are of significance to engineering.

5.   The impact factors differ significantly in the three spans under different loading. This would indicate a need to have different impact factors for the three spans in the design unless a conservative design is desired.

4.5 Summary

The dynamic behavior of a multi-span continuous bridge deck under moving vehicles has been shown in this chapter. The bridge is modeled either as a multi-span continuous beam when the bridge is narrow compared with its length, or as an orthotropic rectangular plate if otherwise. The influence of different parameters such as the road-surface roughness of the bridge deck, the surface condition of the approach, multiple vehicles and their transverse positions, and braking or acceleration on the bridge deck are studied with computational simulations. The vehicle–bridge interaction is discussed in terms of the impact factor in the studies.

Part II

Moving Load Identification Problems

Chapter 5

Moving Force Identification in Frequency–Time Domain

5.1 Introduction

Loads are traditionally measured statically with strain-based or other types of sensors. The technique of indirect load determination is of special interest when the applied loads cannot be measured directly, while the responses can be measured easily (Busby and Trujillo, 1987; Doyle, 1987b; Yen and Wu, 1995; Choi and Chang, 1996; Cao et al., 1998; Kammer, 1998; Moller, 1999; Rao et al., 1999; Liu et al., 2000). It is an ill-posed problem because the response is typically a continuous vector function in the spatial coordinates, and it is only defined at a few points of the structure. Therefore, solutions to the problem are frequently found to be unstable in the sense that small changes in the responses would result in large changes in the calculated load magnitudes. Existing methods, described in Chapter 1, are mainly for the identification of stationary (non-moving) loads. These methods can be classified into time domain method, frequency domain method, state space method, orthogonal function expansion method and neural network method.

The Frequency and Time Domains Approach (FTDM) (Law et al., 1999) performs Fourier transformation on the equations of motions, which are expressed in modal coordinates. The Fourier transforms of the responses and the forces are related in the frequency domain, and the time histories of the forces are found directly by the least-squares method. Results obtained from this method are noise sensitive and they exhibit fluctuations at the beginning and end of the time histories. These moments correspond to the switching of free vibration state of the structure to the forced vibration state, and vice versa, and the solutions are ill-conditioned. Later Law et al. (2001) introduces a regularization method in the ill-conditioned problem to provide bounds to the identified forces. The FTDM is introduced in this chapter and the results are compared with those by the Time Domain Method presented in Section 5.4.3.

5.2 Moving Force Identification in Frequency–Time Domain

5.2.1 Equation of Motion

The following assumptions are made on the system model which, in this case, is a simply supported beam carrying a moving point load:

1. The beam is simply supported.
2. The changes in the system characteristics, i.e. the stiffness, damping and mass matrices before and after the force occurrence are negligible.

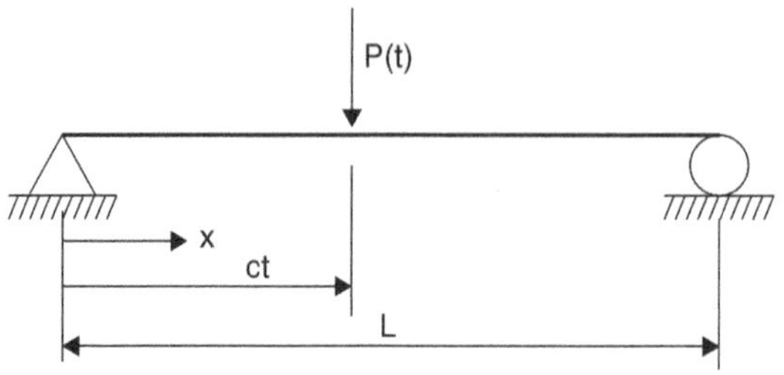

Figure 5.1 Simply supported beam subject to a moving force P(t)

3. The force is a perfect point force represented as a step function in a small time interval as shown in Equation (5.1).
4. Structural damping is included in the analysis.
5. The structure may not be at rest before an external load is applied.

A time-varying force is moving on a simply supported Euler-Bernoulli beam as shown in Figure 5.1. The beam is assumed to be of constant cross-section with constant mass per unit length and, having linear, viscous proportional damping and with small deflections. The effects of shear deformation and rotary inertia are neglected. The force moves from left to right at a constant speed v. The equation of motion can be written as:

$$\rho \frac{\partial^2 w(x,t)}{\partial t^2} + C \frac{\partial w(x,t)}{\partial t} + EI \frac{\partial^4 w(x,t)}{\partial x^4} = \delta(x - vt)P(t) \tag{5.1}$$

where $w(x,t)$ is the beam deflection at point x and time t; ρ is the mass per unit length; C is the viscous damping parameter; E is the Young's modulus of material; I is the second moment of inertia of the beam cross-section; L is the length of the beam; $P(t)$ is the time-varying interaction point force; $\delta(t)$ is the Dirac delta function.

Based on modal superposition, the dynamic deflection $w(x,t)$ can be described as:

$$w(x,t) = \sum_{n=1}^{\infty} \Phi_n(x) q_n(t) \tag{5.2}$$

where n is the mode number; $\Phi_n(x)$ is the mode shape function of the nth mode and $q_n(t)$ is the nth modal amplitudes. Substituting Equation (5.2) into (5.1), and multiplying by $\Phi_j(x)$, and integrating with respect to x between 0 and L, and applying the orthogonality conditions, we obtain:

$$\frac{d^2 q_n(t)}{dt^2} + 2\xi_n \omega_n \frac{d q_n(t)}{dt} + \omega_n^2 q_n(t) = \frac{1}{M_n} p_n(t) \tag{5.3}$$

where ω_n is the modal frequency of the nth mode; ξ_n is the damping ratio of the nth mode; M_n is the modal mass of the nth mode, and $p_n(t)$ is the modal force. These modal parameters are:

$$\omega_n = \frac{n^2\pi^2}{L^2}\sqrt{\frac{EI}{\rho}}$$

$$\Phi_n(x) = \sin\left(\frac{n\pi x}{L}\right) \tag{5.4}$$

$$M_n = \frac{\rho L}{2}$$

$$p_n(t) = P(t)\sin\left(\frac{n\pi v t}{L}\right)$$

For practical structures, the modal parameters can be obtained from the finite element model and/or from the modal testing. Performing the Fast Fourier transformation (FFT) on Equation (5.3), we get:

$$Q_n(\omega) = H_n(\omega)\frac{1}{M_n}P_n(\omega) \tag{5.5}$$

where

$$P_n(\omega) = \frac{1}{2\pi}\int_{-\infty}^{\infty} p_n(t)e^{-i\omega t}\,dt$$

$$Q_n(\omega) = \frac{1}{2\pi}\int_{-\infty}^{\infty} q_n(t)e^{-i\omega t}\,dt \tag{5.6}$$

$$H_n(\omega) = \frac{1}{\omega_n^2 - \omega^2 + 2\xi_n\omega_n\omega} \tag{5.7}$$

$H_n(\omega)$ is the frequency response function of the nth mode. Performing again the Fast Fourier transformation on Equation (5.2), and substituting Equations (5.5) and (5.7) into the resultant equation, the Fourier transform of the dynamic deflection $w(x,t)$ is obtained as:

$$W(x,\omega) = \sum_{n=1}^{\infty}\frac{1}{M_n}\Phi_n(x)H_n(\omega)P_n(\omega) \tag{5.8}$$

5.2.2 Identification from Accelerations

The Fourier transform of the acceleration of the beam at point x is obtained from Equation (5.8) as:

$$\ddot{W}(x,\omega) = -\omega^2\sum_{n=1}^{\infty}\frac{1}{M_n}\Phi_n(x)H_n(\omega)P_n(\omega) \tag{5.9}$$

Substituting the last relationship in Equation (5.4) into (5.9), and rewriting in discrete terms:

$$\ddot{W}(m) = -\sum_{k=0}^{N-1}\sum_{n=1}^{\infty}\frac{\Delta f^3 m^2}{M_n}\Phi_n(x)H_n(m)\Psi_n(m-k)P(k), \quad (m = 0,1,\ldots,N-1)$$

$$\tag{5.10}$$

where Ψ_n is the Fourier transform of the nth mode shape; Δf is the frequency resolution and N is the number of data sample in the FFT; k and m denote the kth and the mth term in the FFT; and P is the Fourier transform of the moving force $P(t)$. Let:

$$\overline{H}_{xn}(m) = -\frac{\Delta f^3 m^2}{M_n} \Phi_n(x) H_n(m) \tag{5.11}$$

Then Equation (5.10) can be rewritten in terms of the real $P_R(k)$ and imaginary $P_I(k)$ parts of the Fourier transforms $P(k)$:

$$\ddot{W}(m) = \sum_{k=0}^{N/2-1} \sum_{n=1}^{\infty} \overline{H}_{xn}(m)\Psi_n(m-k)[P_R(k) + iP_I(k)]$$

$$+ \sum_{k=N/2}^{N-1} \sum_{n=1}^{\infty} \overline{H}_{xn}(m)\Psi_n(m-k)[P_R(N-k) - iP_I(N-k)] \tag{5.12}$$

Considering the periodic property of the Discrete Fourier Transform (DFT), Equation (5.12) can be written as:

$$\ddot{W}(m) = \sum_{n=1}^{\infty} \overline{H}_{xn}(m)\Psi_n(m)[P_R(0) + iP_I(0)]$$

$$+ \sum_{k=1}^{N/2-1} \sum_{n=1}^{\infty} \overline{H}_{xn}(m)[\Psi_n(m-k) + \Psi_n(m+k-N)]P_R(k)$$

$$+ i \sum_{k=1}^{N/2-1} \sum_{n=1}^{\infty} \overline{H}_{xn}(m)[\Psi_n(m-k) - \Psi_n(m+k-N)]P_I(k)$$

$$+ \sum_{n=1}^{\infty} \overline{H}_{xn}(m)\Psi_n(m-N/2)[P_R(N/2) - iP_I(N/2)] \quad (m = 0, 1, \ldots, N-1) \tag{5.13}$$

Rewriting Equation (5.13) in matrix form:

$$\ddot{W}_{(N+2)\times 1} = A_{(N+2)\times(N+2)} P_{(N+2)\times 1} \tag{5.14}$$

where $\ddot{W}$ and P are Fourier transforms of the acceleration vector $\ddot{w}$ and the force vector p respectively. Writing P as its real and imaginary parts P_R and P_I:

$$\ddot{W} = (A_{RR} + iA_{RI})P_R + i(A_{IR} + iA_{II})P_I \tag{5.15}$$

Again, separating $\ddot{W}$ into real and imaginary parts $\ddot{W}_R$ and $\ddot{W}_I$, we have:

$$\left\{ \begin{array}{c} \ddot{W}_R \\ \ddot{W}_I \end{array} \right\}_{(N+2)\times 1} = \left[\begin{array}{cc} A_{RR} & -A_{II} \\ A_{RI} & A_{IR} \end{array} \right]_{(N+2)\times(N+2)} \left\{ \begin{array}{c} P_R \\ P_I \end{array} \right\}_{(N+2)\times 1} \tag{5.16}$$

Since the Fourier transforms of the imaginary parts of vectors $\ddot{w}$ and p equal to zero in the following terms, $P_I(0)=0, P_I(N/2)=0, \ddot{W}_I(0)=0, \ddot{W}_I(N/2)=0,$ Equation (5.16) can be rewritten into a set of N number of simultaneously equations as:

$$\ddot{W}_{RI} = A_D P_{RI} \tag{5.17}$$

Components P_R and P_I can be found from Equation (5.16) by solving the Nth order linear equation. The time history of the moving force $P(t)$ can then be obtained by performing the inverse Fourier transformation on P_{RI}. The solution is obtained in the frequency domain. However the computation cost for solving Equation (5.17) is high as it involves finding the inverse of a full matrix and, therefore, the following procedure in time domain is developed to overcome this difficulty.

5.2.3 Solution in Time Domain

If the DFTs are expressed in matrix form, the Fourier transform P of the force vector p can be written as follows if the terms in p are real (Bendat and Piersol, 1993).

$$P = \frac{1}{N}Up \tag{5.18}$$

where

$$U = e^{-i2k\pi/N} \tag{5.19}$$

and

$$k = \begin{bmatrix} 0 & 0 & 0 & \cdots & 0 & 0 \\ 0 & 1 & 2 & \cdots & N-2 & N-1 \\ 0 & 2 & 4 & \cdots & N-4 & N-2 \\ \vdots & \vdots & \vdots & & \vdots & \vdots \\ 0 & N-2 & N-4 & \cdots & 4 & 2 \\ 0 & N-1 & N-2 & \cdots & 2 & 1 \end{bmatrix}_{N \times N}$$

The matrix U is an unitary matrix, which means:

$$U^{-1} = (U^*)^T \tag{5.20}$$

where U^* is a conjugate of U. Substituting Equation (5.18) into (5.14),

$$\ddot{W} = \frac{1}{N}A\begin{bmatrix} U_B \\ N \times N_B \end{bmatrix} \begin{Bmatrix} p_B \\ 0 \end{Bmatrix} \tag{5.21}$$

or

$$\ddot{W}_{N \times 1} = \frac{1}{N} \underset{N \times N}{A} \underset{N \times N_B}{U_B} \underset{N_B \times 1}{p_B} \tag{5.22}$$

linking the Fourier transform of acceleration $\ddot{W}$ with that of the force vector p_B of the moving forces in the time domain. U_B is the sub-matrix of U. $N_B = L/(v\Delta t)$ is the number of data point on the beam.

Using Equation (5.22) for identification has the advantage of weighting the response data in the frequency domain. The disadvantage is that the noise of the responses during the time interval $N_B \Delta t$ to $N \Delta t$ will affect the accuracy of the identified results. Equation (5.22) can be rewritten using Equation (5.18) to relate the accelerations and force vectors in the time domain as:

$$\underset{N \times 1}{\ddot{W}} = \underset{N \times N}{(U^*)^T} \underset{N \times N}{A} \underset{N \times N_B}{U_B} \underset{N_B \times 1}{p_B} \tag{5.23}$$

If $N = N_B$, p_B can be found by solving the Nth order linear equation in Equation (5.22) or (5.23). If $N > N_B$ or more than one acceleration are measured, the least-squares method can be used to find the time history of the moving force $P(t)$.

If only N_c ($N_c \leq N$) response data points of the beam are used, the equations based on these data points in Equations (5.22) and (5.23) are extracted, and described as:

$$\underset{N_C \times 1}{\ddot{W}} = \frac{1}{N} \underset{N_C \times N_C}{A} \underset{N_C \times N_B}{U_B} \underset{N_B \times 1}{p_B}$$

$$\underset{N_C \times 1}{\ddot{W}_c} = \underset{N_C \times N}{(U_B^*)^T} \underset{N \times N}{A} \underset{N \times N_B}{U_B} \underset{N_B \times 1}{p_B} \tag{5.24}$$

In practice, more than one acceleration measurement can be used to identify a single moving force for a higher accuracy.

5.2.4 Identification from Bending Moments and Accelerations

Similarly, we can find the relationships between the bending moments m, its Fourier transform M, and the moving force p as:

$$\underset{N \times 1}{M} = \frac{1}{N} \underset{N \times N}{B} \underset{N \times N_B}{U} \underset{N_B \times 1}{p_B} \tag{5.25}$$

$$\underset{N \times 1}{m} = \underset{N \times N}{(U^*)^T} \underset{N \times N_B}{B} \underset{N \times N_B}{U_B} \underset{N_B \times 1}{p_B} \tag{5.26}$$

$$\underset{N_C \times 1}{m_B} = \underset{N_C \times N}{(U^*)^T} \underset{N \times N}{B} \underset{N \times N_B}{U_B} \underset{N_B \times 1}{p_B} \tag{5.27}$$

where matrix B is similar to the matrix A in Equations (5.22) and (5.23) for the accelerations. The force vector p_B can be obtained from Equations (5.25) to (5.27) depending on the length of measured responses with respect to N_B. Equations (5.25) to (5.27) can also be combined with Equations (5.22) to (5.24) to form an over-determined set of equations when both bending moments and accelerations are used in the identification. The equations have to be scaled as shown below to have dimensionless unit before they are used.

$$(U^*)^T \begin{bmatrix} B/\|m\| \\ A/\|\ddot{w}\| \end{bmatrix} U_B p_B = \begin{Bmatrix} m/\|m\| \\ \ddot{w}/\|\ddot{w}\| \end{Bmatrix} \tag{5.28}$$

where $\|\bullet\|$ is the norm of the vector.

The above procedure is derived for single force identification. Equation (5.26) can be modified for two forces identification using the linear superposition principle as:

$$
m = (U^*)^T \begin{bmatrix} B_a & 0 \\ B_b & B_a \\ B_c & B_b \end{bmatrix} U_B \begin{Bmatrix} p_1 \\ p_2 \end{Bmatrix}
\tag{5.29}
$$

where $B_a\ [N_s \times (N_B - 1)]$, $B_b[(N - 1 - 2N_s) \times (N_B - 1)]$, and $B_c[N_s \times (N_B - 1)]$ are sub-matrices of matrix B. The first row of sub-matrices in the B matrix describes the state having the first force on beam after its entry. The second and third rows of sub-matrices describe the states having two forces on beam and one force on beam after the exit of the first force respectively. The whole matrix has a dimension of $(N - 1) \times (N_B - 1)$. $N_s = l_s/(v\Delta t)$ and is the number of data sample when only the first force or the second force is on the beam. l_s is the distance between the two axles. The two forces can be identified using more than one measured bending moment measurements. We can also modified Equation (5.23) in a similar way for two forces identification using more than one measured acceleration measurements.

5.2.5 Regularization of the Solution

We take the identification using measured bending moment m as an example. Since the identified force p is not a continuous function of the measured responses at the beginning and end of the time history, solution to Equation (5.26) is ill-conditioned (Morozov, 1984). A regularization method developed by Tikhonov (Tikhonov and Arsenin, 1977) is used to provide bounds to the solution. The Tikhonov regularization method is based on the radical idea that minimizes the deviations of $U^{*T}BU_Bp_B$ from the measured response vector m in Equation (5.26) for a stable solution by means of an auxiliary non-negative parameter. This is equivalent to imposing certain constraints in the form of added penalty terms with adjustable weighting (regularization) parameters to the solution. The Tikhonov function is written as follows:

$$
J(p_B, \lambda) = \left\| Rp_B - m \right\|^2 + \lambda \left\| p_B \right\|^2
\tag{5.30}
$$

where λ is the non-negative regularization parameter, and $R = U^{*T}BU_B$. The solution of Equation (5.26) is obtained in the Tikhonov regularization with the damped least-squares method as (Santantamarina and Fratta, 1998):

$$
p_{B\,N_f \times 1} = (R^T{}_{N \times N_f} R_{N \times N_f} + \lambda I)^{-1} R^T{}_{N \times N_f} m_{N \times 1}
\tag{5.31}
$$

where I is the identity matrix, and singular value decomposition is used in the pseudo-inverse calculation.

The main difficulty of applying the Tikhnov regularization lies in the method to find the optimal regularization parameter λ. Two methods to find the optimal regularization parameter are discussed in this chapter. The use of each method depends upon the availability of the true forces. If the true forces were known, the true force $p_B^{(True)}$ is compared with the identified values $p_B^{(identify)}$, and an error curve, the S-curve (Busby

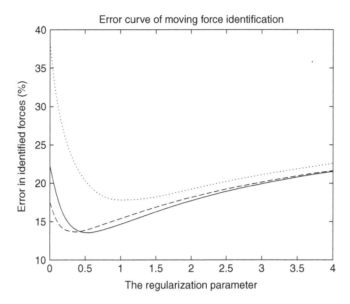

Figure 5.2 Typical S-curve (--- 1% noise; — 5% noise; · · · · 10% noise)

and Trujillo, 1997) can be plotted for different λ as shown in Figure 5.2. The error of identification in the force time history is:

$$error = \frac{\left\| p_B^{(identify)} - p_B^{(True)} \right\|}{\left\| p_B^{(True)} \right\|} \times 100\% \tag{5.32}$$

and $\|\bullet\|$ is the norm of a matrix. It is noted from Figure 5.2 that the optimal value of λ corresponds to the smallest error.

In the more practical case when $p_B^{(True)}$ is unknown as in experiment or in field test, the L-curve proposed by Hansen (1992) is used to determine the optimal λ value. The L-curve is a plot of the semi-norm of the solution against the residual norm. The norm of residuals E of the forces is calculated as:

$$E = \left\| Rp_B^{(identify)} - m \right\| \tag{5.33}$$

and for the zeroth-order regularization proposed by Busby and Trujillo (1997), the semi-norm of the estimated forces is:

$$E1 = \left\| p_{Bj+1}^{(identify)} - p_{Bj}^{(identify)} \right\| \tag{5.34}$$

where $p_{Bj}^{(identify)}$, $p_{Bj+1}^{(identify)}$ are the identified forces with λ_j and $\lambda_j + \Delta\lambda$. Typical L-curves are plotted in Figure 5.3 for different noise level in the measured data, and they all exhibit a corner in each L-curve. The value of λ corresponds to the point immediately to the right of the corner and is the optimal value.

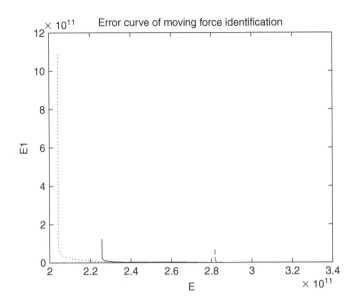

Figure 5.3 Typical L-curves (--- 1% noise; — 5% noise; · · · · 10% noise)

5.3 Numerical Examples

The following parameters of the forces and structure are used in the study:

(a) Type I Force – Single moving force from a quarter vehicle model
The quarter vehicle model has two lump masses $m_t = 4{,}000\,\text{kg}$ and $m_s = 36{,}000\,\text{kg}$ with springs $k_t = 7.2 \times 10^7\,\text{kg/m}$ and $k_s = 1.8 \times 10^7\,\text{kg/m}$ and dampers $c_t = c_s = 14.4 \times 10^4\,\text{kg-sec/m}$ as shown in Figure 5.4. The two natural frequencies of the quarter model are 3.18 Hz and 23.93 Hz which are very close to the natural frequencies of the bridge beam as given below.

(b) Type II Force – Two forces moving at a fixed distance of 4 m

$$P_1(t) = 20000[1 + 0.1\sin(10\pi t) + 0.05\sin(40\pi t)]\ \text{N}$$

$$P_2(t) = 20000[1 - 0.1\sin(10\pi t) + 0.05\sin(50\pi t)]\ \text{N}$$

The parameters of the beam are as follows:

$$EI = 1.274916 \times 10^{11}\,\text{Nm}^2, \quad \rho A = 12000\,\text{kg/m}, \quad l = 40\,\text{m}$$

$$\xi_1 = 0.02, \quad \xi_2 = 0.02, \quad \xi_3 = 0.04, \quad f_1 = 3.2\,\text{Hz}, \quad f_2 = 12.8\,\text{Hz}, \quad f_3 = 28.8\,\text{Hz}.$$

where f_i and ξ_i are the natural frequency and damping ratio of the bridge beam respectively. A moving speed of 40 m/s is studied for both force types. The analysis frequency bandwidth is from 0 Hz to 40 Hz to include the first three modes of the beam in the calculation. Sampling frequency is 100 samples per second with N_B equals 100 points. The record length N is 512 points, and N_c equals 110 points are used in the identification. The moving forces are identified using Equations (5.23), (5.26) and (5.28) without regularization. Please note that in the case of two moving forces an equal but opposite component exists in each force, simulating the effect of pitching motion of a vehicle.

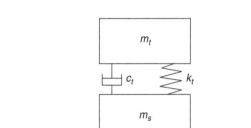

Figure 5.4 Quarter vehicle model

5.3.1   *Single Force Identification*

Bending moment and/or acceleration responses at 1/2 span and/or 1/4 span are used. The effectiveness of nine combinations of the responses in the force identification are studied as follows:

1/2m	1/2a	1/2m & 1/2a
1/4m	1/4a	1/4m & 1/4a
1/2m & 1/4m	1/2a & 1/4a	1/2m & 1/4a

where 1/2 and 1/4 represent the location of span, and m and a represent the bending moment and acceleration responses respectively. Random noise is added to the calculated responses to simulate the polluted measurements as:

$$\ddot{w} = \ddot{w}_{calculated} + Ep \times \|\ddot{w}_{calculated}\| \times N_{oi}$$

where Ep is a specified error level; N_{oi} is a standard normal distribution vector with zero mean and unit standard deviation.

   The simulated bending moment and acceleration at 1/4 span are shown in Figures 5.5 and 5.6 for 10 percent noise level. The Power Spectral Density Functions (PSDs) in the figures indicate that errors due to simulated random noise exist in the higher frequency range. This error level is approximately equivalent to 90 dB dynamic range which is of the same order as those in the measurement system (65 dB). Therefore this error level could represent typical values in practical situations.

   Errors in the simulated forces computed from Equation (5.32) are shown in Table 5.1 for 1 percent, 5 percent and 10 percent noise level. The errors are small for all the cases with acceleration or acceleration and bending moment responses. Those using only bending moments give larger errors in general. This is because the bending moment responses in the high frequency range is very small (referring to Figure 5.5) causing large error. Results using responses from a single measurement point only are less accurate than those using responses from multi-points as some of the modal responses may have not been used in the identification. Some of the identified results are shown in Figures 5.7 and 5.8. The PSDs of the identified forces in these figures are close to

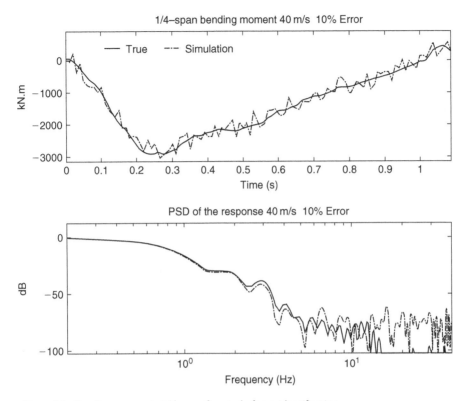

Figure 5.5 Bending moment at ¼ span for single force identification

the true one although the time histories are not. There are large discrepancies between the true and the identified forces from using only the 1/2 span acceleration at the time instance of 0.3 second. Since the measured responses consist of that from the first and third modes only at 1/2 span, this large variation is due to the low sensitivity of the responses to the moving force when it passes over the nodes of the third vibration mode of the beam.

5.3.2 *Two Forces Identification*

Bending moment and/or acceleration responses at 1/4, 1/2 and 3/4 spans in 12 combinations described in Table 5.2 are used to identify the two forces. The error study on the identified individual forces is extended to the combined total force in an attempt to assess the accuracy of weighing-in-motion of the weight of a moving vehicle on top of the beam. The following results are obtained in a similar way as those for the quarter vehicle model.

Errors in the identification result of individual forces and the total force are shown in Table 5.2 for 1 percent, 5 percent and 10 percent noise levels. Sensor combination Case (d) gives the least errors among all the 12 cases. Cases with combined bending moments and accelerations give better results than those having only bending moments

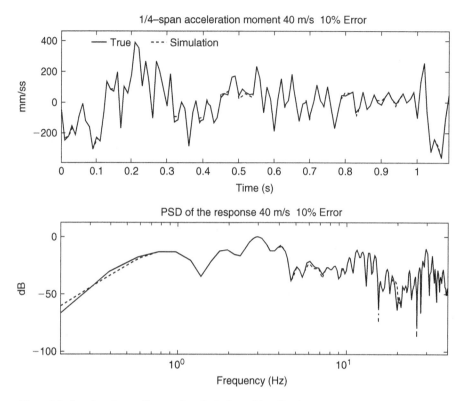

Figure 5.6 Acceleration at ¼ span for single force identification

Table 5.1 Errors in identified force from quarter vehicle model without regularization

Sensor Location	% error in Response		
	1%	5%	10%
1/2m	18.5	87.2	*
1/2a	0.5	4.6	9.2
1/2a&1/2m	0.8	1.1	2.3
1/4m	11.8	75.1	*
1/4a	0.2	1.6	3.2
1/4a&1/4m	0.2	1.3	2.6
1/4m&1/2m	5.8	47.1	94.3
1/4a&1/2a	0.1	1.4	2.8
1/4a&1/4m&1/2a&1/2m	0.2	1.1	2.2

*denotes error exceeds 100%; *m* denotes bending moment; *a* denotes acceleration.

or accelerations. The combinations with at least two acceleration responses give the smaller errors. Table 5.2 also shows that bending moment response at 1/2 span is more useful than that collected at 1/4 span, while it is just the opposite for acceleration responses. The accuracy in the individual force identification is lower than that in the

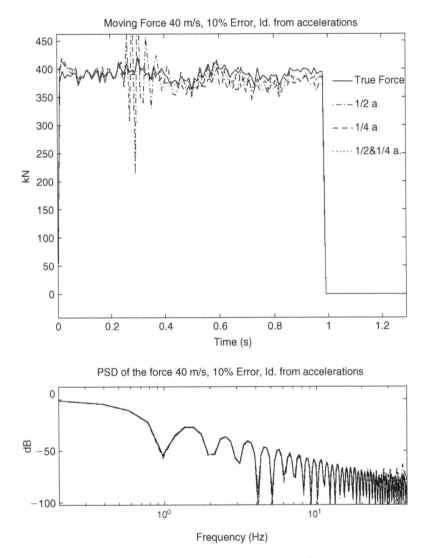

Figure 5.7 Identified force from acceleration in single force identification

combined force identification. One reason is that there is a force component with the same amplitude and opposite phase in the two individual forces.

Regularization is also applied to improve the identified results using the damped least-squares method from Equation (5.31), and the optimal regularization parameter λ is calculated by the S-curve method. Samples of the time histories for combinations (a) (1/2 m, 1/4 m) and (c) (1/2a,1/4a) with and without regularization are shown in Figures 5.9 and 5.10. Regularization has improved the results from strains greatly but not in those from accelerations. This is explained in the following discussions.

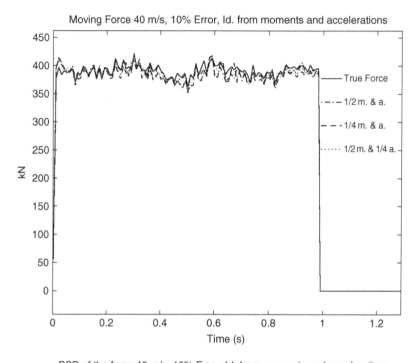

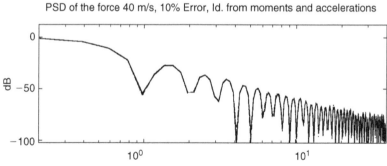

Figure 5.8 Identified force from acceleration and bending moment for single force identification

The solution to Equation (5.31) is given by (Fierro et al., 1997)

$$f_\lambda = \sum_{i=1}^{r} \frac{\sigma_i^2}{\sigma_i^2 + \lambda^2} \frac{u_i^T m}{\sigma_i} v_i \tag{5.35}$$

where u_i and v_i are components of matrices U and V in the singular value decomposition of matrix R; σ_i is the singular value of matrix R. m is the vector of measured bending moments. The components of the solution corresponding to the small singular values of R are suppressed through this solution. When the singular values are in general small in the case from strains, all solutions are affected greatly in this equation

Table 5.2 Errors in identified forces without regularization

Sensor Location Cases	First Force			Second Force			Total Force		
	1%	5%	10%	1%	5%	10%	1%	5%	10%
(a) 1/2m&1/4m	97.6	*	*	*	*	*	74.7	*	*
(b) 1/2m&1/4m&3/4m	30.9	*	*	58.1	*	*	32.6	*	*
(c) 1/2a&1/4a	43.3	*	*	13.9	69.5	*	22.6	*	*
(d) 1/2a&1/4a&3/4a	3.3	16.4	33.0	3.4	17.4	34.8	1.8	9.0	18.0
(e) 1/2m&1/2a	63.4	*	*	49.1	*	*	39.9	*	*
(f) 1/2m&1/4m&1/2a	51.0	*	*	25.2	*	*	28.4	*	*
(g) 1/2m&1/4m&1/2a&1/4a	25.7	*	*	9.5	47.7	95.5	13.6	68.4	*
(h) 1/4m&1/4a	*	*	*	80.6	*	*	*	*	*
(i) 1/4m&1/4a&1/2a	35.2	*	*	10.5	52.6	*	18.3	91.6	*
(j) 1/2m&1/4a	53.1	*	*	55.0	*	*	38.1	*	*
(k) 1/2m&1/4a&1/4m	58.6	*	*	41.0	*	*	35.7	*	*
(l) 1/4a&1/2a&1/2m	34.9	*	*	10.4	52.3	*	18.2	91.3	*

Note: * denotes error exceeds 100 percent.

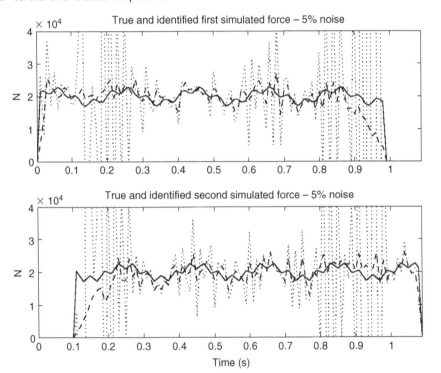

Figure 5.9 Identified forces in two-forces identification from 1/2 m and 1/4 m (— static forces; ···· without regularization; --- with regularization)

with the presence of parameter λ. There is a large smoothing effect to the whole time history as shown in Figure 5.9. The singular values are much larger in the case from accelerations. The solutions are less affected by parameter λ, and the smoothing effect is small, as seen in Figure 5.10.

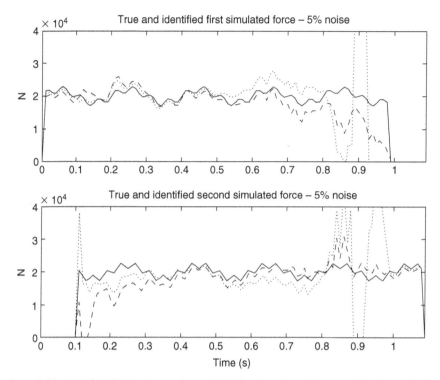

Figure 5.10 Identified forces in two-forces identification from 1/2a and 1/4a (— static forces; · · · · without regularization; --- with regularization)

There is still some noise effect in the results from strains after regularization while the curves from accelerations are relatively smooth. This is because the noise is represented in the term m in Equation (5.35). The variation from noise in the results is dependent on the magnitude of singular value σ_i in the denominator. Matrix R from strains has small singular values in general, and the variation in the identified forces is therefore large.

The errors in the identified forces with and without regularization are compared in Table 5.3 for the case with 5 percent random noise. Significant reductions are observed in all cases particularly those having large errors obtained without regularization. All combinations of sensors give errors of the same order for both the first force and the second force.

5.4   Laboratory Experiments with Two Moving Loads

5.4.1   *Experimental Setup*

The experimental setup is shown diagrammatically in Figure 5.11. The main beam, 3376 mm long with 100 mm × 25 mm uniform cross-section is simply supported. The leading beam is for the acceleration of the model car and the tailing beam is for the slowing down of the car. Some damping material is placed underneath the main beam to improve its damping properties.

Table 5.3 Errors in Identified Forces with 5 percent noise with and without regularization

Sensor Location Cases	First Force Regularization		Second Force Regularization	
	No	Yes	No	Yes
(a) 1/2m&1/4m	455.2(+)	25.0	559.4(+)	23.2
(b) 1/2m&1/4m&3/4m	154.5(+)	19.9	291.4(+)	21.3
(c) 1/2a&1/4a	216.7(222)	22.4	69.5(78.6)	23.9
(d) 1/2a&1/4a&3/4a	16.4(10.7)	8.7	17.4(10.7)	13.3
(e) 1/2m&1/2a	317.3(+)	24.5	245.8(+)	26.5
(f) 1/2m&1/4m&1/2a	255.0(+)	22.7	126.5(474)	20.9
(g) 1/2m&1/4m&1/2a&1/4a	125.6(201)	21.5	47.7(49.7)	17.3
(h) 1/4m&1/4a	991.7(+)	22.6	403.0(+)	20.8
(i) 1/4m&1/4a&1/2a	176.4(206)	21.4	52.6(156)	18.0
(j) 1/2m&1/4a	265.9(+)	24.7	275.4(+)	26.8
(k) 1/2m&1/4a&1/4m	293.4(+)	22.3	205.2(+)	20.7
(l) 1/4a&1/2a&1/2m	174.4(193)	23.5	52.3(46.8)	22.5

Note: (*) is the result from Time Domain Method (Law et al., 1997)
(+) denotes error larger than 1000 percent

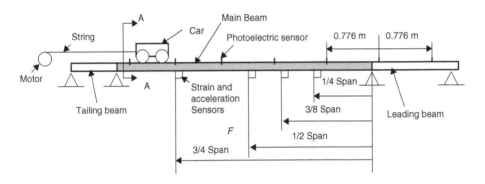

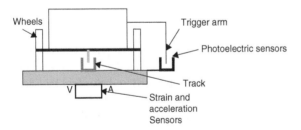

Figure 5.11 Diagrammatic drawing of the experimental setup

An U-shaped aluminum section is glued on the upper surface of the beams as a sort of direction guide for the car. The model car is pulled along the guide by a string wound up to the wheel of an electric motor where rotating speed can be adjusted. Seven photoelectric sensors are mounted on the beams to measure and monitor the moving speed of the car. The second sensor is located at the point where the front

Table 5.4 Modal parameters of the model car

Mode	Frequencies (Hz)	Damping Ratios (%)	Mode Shape
1	27.5	10.5	Bounce
2	42.9	11.7	Pitch
3	69.4	10.8	Roll

Table 5.5 Modal parameters of the main beam

Mode	Frequencies (Hz)	Damping (%)	Mode Shapes	Modal Masses (kg)	Corrected EI (kN-m^2)
1	6.612	2.71	$\sin(\pi x/3.776)$	40.13	63.4
2	18.51	.653	$\sin(2\pi x/3.776)$	38.58	31.1
3	39.45	.199	$\sin(3\pi x/3.776)$	38.65	28.6

wheels of the car just get on the main beam, and the last sensor is located at the point where the rear set of wheels just get off the main beam. They are used to measure the speed of the model car. The others are located on the beams at a spacing of 0.776 m to check on the uniformity of the speed.

Three strain gauges and four accelerators are mounted at the bottom of the main beam to measure the responses. One gauge and one accelerometer are mounted at each cross-section at the 1/4, 1/2 and 3/4 span. The fourth accelerometer is mounted at the 3/8 span. An eight-channel dynamic testing and analysis system (DTAS) is used for data collection and analysis in the experiment. The first channel is used to monitor the signal of the photoelectric sensors. The second, third and fourth channels are used to measure the signal of the strain gauges. The remaining channels are used to measure the acceleration responses. The sampling frequency is 256 Hz. The model car has two axles at a space of 0.203 m and it runs on four rubber wheels. The mass of the whole car is 7.1 kg. The average speed of the vehicle is 3.102 m/s.

5.4.2   Experimental Procedure

1. Performing a modal test on the model car and the modal parameters obtained are shown in Table 5.4.
2. Performing a modal test on the main beam, and the modal parameters obtained are shown in Table 5.5. The modal shapes were obtained through curve fitting of the measured shapes. The modal masses were obtained and checked by additive masses. The results show that the modal frequencies are different with the calculated results for a simply supported beam and, therefore, the flexural stiffness EI of the beam was corrected from Equation (5.4).
3. Adjusting the output of channels two, three and four for the strain gauges to zero when the main beam was unloaded. This takes care of the fact that the signals from the strain gauges are very small and are usually with zero-shift phenomenon. Removing the zero-shift portion in the output increases the dynamic range of the measurement.
4. Calibrating the strain gauges by adding masses at the middle of the main beam. The average sensitivities were found to be 2.243, 2.532 and 2.259 mV/N-m

respectively for the 1/4, 1/2 and 3/4 span gauges. During the calibration, the signal of the strain gauges were found not very stable and repeatability was not completely satisfactory. It is noted that this will lead to calibration errors in the identified results.

5. Placing the car at the left end of the leading beam, and setting the DTAS in pre-trigger state at channel 1. The power for the motor is turned on and the car moved on top of the beams. Eight channels of signal were acquired.

6. Calculating the speed and checking the uniformity of the speed.

7. If the speed was stable, repeating Steps (2) to (4) to check whether the properties of the structure and measurement system were changed or not. If no significant change was found, the recorded data was accepted.

8. Removing the zero-shift in the measured signals, and calibrating the signals with measured channel sensitivities. The point in the signals when the front wheel of the car just got on the main beam is identified.

9. Interpolating between two measured points in the time histories of the responses if necessary to ensure a sampling frequency at least five times the highest analysis frequency of interest.

10. Identifying the interaction forces as a single moving force using the bending moment and acceleration responses at the 1/4 and 1/2 span and, indirectly, checking the identified results by comparing the measured bending moment and acceleration at the 3/4 span with the responses calculated from the identified force.

11. Identifying the interaction forces as two moving forces using the bending moment and acceleration responses at the 1/4, 1/2 and 3/4 span and, indirectly, checking the identified results by comparing the measured acceleration at the 3/8 span with the responses calculated from the identified forces. Due to limitation of available measurement channels, only the acceleration at the 3/8 span was used for checking.

5.4.3 *Experimental Results*

The first three modes are used in the identification. The moving forces are identified from 1/4a, 1/2a and 1/2m, with and without regularization, and the optimal regularization parameter λ is obtained using the L-curve method. The time histories of the identified forces, with and without regularization, are shown in Figure 5.12 and the resultant of the two forces is shown in Figure 5.13. The results from regularization vary around the static value and a clear pitching motion of the vehicle can be observed from the time histories. The identified resultant force varies around the total static force of the vehicle with some high frequencies due to the measurement noise. This type of error can be removed by pre-processing of the measured responses before the identification. Figure 5.13 shows that the resultant force can be used to closely estimate the total weight of the passing vehicle.

The TDM (Law et al., 1997) described in Chapter 6 is also used to obtain another set of forces with and without regularization and the results are shown in Figures 5.14 and 5.15. The FTDM method gives slightly better results than those obtained from the TDM. This is again explained by the same reason, presented for the two forces identification in simulation in Section 5.3.2.

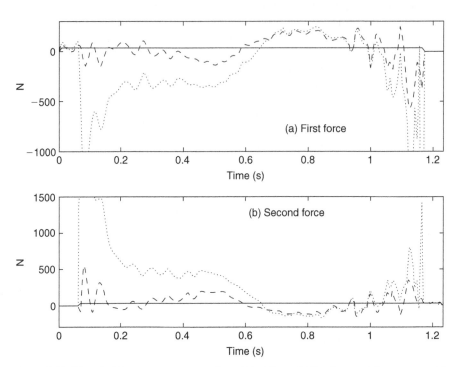

Figure 5.12 Identified forces in experiment from 1/4a, 1/2a and 1/2 m (— static forces; · · · without regularization; --- with regularization)

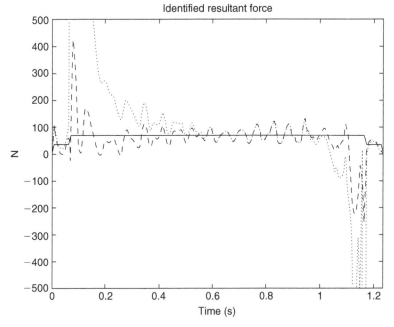

Figure 5.13 Identified resultant force in experiment (— static forces; · · · · without regularization; --- with regularization)

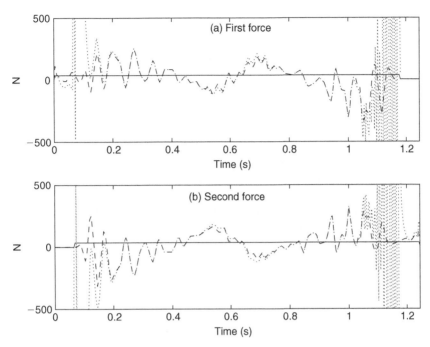

Figure 5.14 Identified forces in experiment from 1/4a, 1/2a and 1/2 m by the Time Domain Method (— static forces; · · · · without regularization; --- with regularization)

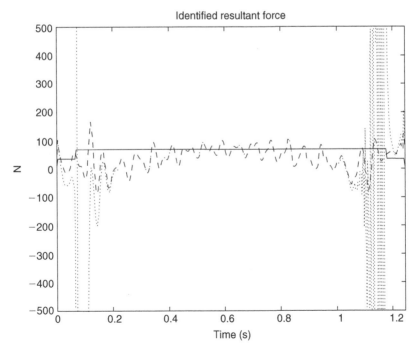

Figure 5.15 Identified resultant force in experiment from Time Domain Method (— static forces; · · · · without regularization; --- with regularization)

A comparison between Figures 5.12 and 5.14 shows that the FTDM gives less accurate results than the TDM in the first half of the time histories. Regularization improves the FTDM curves but not the TDM curves. Since regularization only provides bounds to the ill-conditioned solution without any smoothing effect on the measurement noise, this observation means the TDM is better than the FTDM in solving the ill-posed problem.

5.5 Summary

The Frequency and Time Domain Method in the moving force identification has been discussed in this chapter and the regularization method is applied to solve the problem. Results obtained are greatly improved over those without regularization with acceptable errors from using different combinations of measured responses. Time Domain Method in Chapter 6 is found to be better than the Frequency and Time Domain Method in solving the ill-posed problem. Both simulation and laboratory test results indicate that the total weight of a vehicle can be estimated indirectly using moving force identification methods with some accuracy.

Moving Force Identification in Time Domain

6.1 Introduction

The dynamic behavior of a continuous beam and an orthotropic plate under moving loads has been analyzed in Chapter 2. This chapter is on the development of the moving force identification method in time domain on these structures without knowledge of the load characteristics. Law et al. (1997) modeled the moving loads on a simply supported Euler beam as step functions in a small time interval. Based on system identification theory and modal superposition, the moving forces are identified by using the least-squares method in time domain. Chan, Yu and Law (2000) compared this method with other moving load identification methods using laboratory results and they found that this method has the best performances except for the computation time. However, the identified results are noise sensitive and exhibit fluctuations at the beginning and the end of the time histories. These moments correspond to the switching of free vibration state of the structure to the forced vibration state, and vice versa, and the solutions are ill-conditioned. Law et al. (2001) introduced the regularization technique in the ill-posed problem to provide bounds to the identified forces. The results obtained are greatly improved over those without regularization with acceptable errors from using different combinations of measured responses.

Another time domain method (Zhu and Law, 2002b) was developed to give exact solution to the forces with improved formulation over existing methods for a more efficient computation. Least-Squares solution technique is applied to the Time Domain Method, while Tikhonov regularization procedure is applied to the Exact Solution Technique Method (EST) to provide bounds to the solution in time domain. The latter method also features an improved formulation with the large matrices in TDM breaking up into smaller matrices in the solution, leading to enhanced computation efficiency. The continuous beam model and the orthotropic plate model are used in the identification. Numerical examples are used to demonstrate the feasibility and accuracy of the method, and factors affecting the errors in the identification are discussed. Laboratory results are also reported to verify the effectiveness and validity of the proposed method. Finally a comparison on the computation accuracy of the FTDM, TDM and EST method is presented in Section 6.3.3.

6.2 Moving Force Identification – The Time Domain Method (TDM)

6.2.1 Theory

6.2.1.1 Equation of Motion and Modal Superposition

The time-varying force moving on a simply supported beam, as shown in Figure 5.1, is used to demonstrate the method. The parameters of the beam are the same as those described in Section 5.2. The force moves from left to right at a constant speed c. The equation of motion of the beam can be written as:

$$\rho \frac{\partial^2 v(x,t)}{\partial t^2} + C \frac{\partial v(x,t)}{\partial t} + EI \frac{\partial^4 v(x,t)}{\partial x^4} = \delta(x - ct)f(t) \tag{6.1}$$

where $v(x,t)$ is the dynamic beam deflection at point x and time t. Based on modal superposition, the dynamic deflection $v(x,t)$ can be described as follows:

$$v(x,t) = \sum_{n=1}^{\infty} \Phi_n(x)q_n(t) \tag{6.2}$$

where n is the mode number; $\Phi_n(x)$ is the mode shape function of the n-th mode and $q_n(t)$ is the n-th modal amplitudes. Substituting Equation (6.2) into Equation (6.1), and multiplying by $\Phi_j(x)$, integrating with respect to x between 0 and l, and applying the orthogonality conditions, we obtain:

$$\frac{d^2 q_n(t)}{dt^2} + 2\xi_n \omega_n \frac{dq_n(t)}{dt} + \omega_n^2 q_n(t) = \frac{1}{M_n} p_n(t) \tag{6.3}$$

where ω_n is the modal frequency of the n-th mode; ξ_n is the damping ratio of the n-th mode; M_n is the modal mass of the n-th mode, and $p_n(t)$ is the modal force. Based on assumptions for the Euler-Bernoulli beam, the modal parameters of the beam can be calculated as follows:

$$\omega_n = \frac{n^2 \pi^2}{l^2} \sqrt{\frac{EI}{\rho}} \tag{6.4}$$

$$\Phi_n(x) = \sin\left(\frac{n\pi x}{l}\right) \tag{6.5}$$

$$M_n = \frac{\rho l}{2} \tag{6.6}$$

$$p_n(t) = f(t) \sin\left(\frac{n\pi ct}{l}\right) \tag{6.7}$$

For practical structures, the modal parameters can be obtained from finite element model and/or modal testing.

Equation (6.3) can be solved in the time domain by the convolution integral, and yields:

$$q_n(t) = \frac{1}{M_n} \int_0^t h_n(t-\tau)p(\tau)d\tau \qquad (6.8)$$

where

$$h_n(t) = \frac{1}{\omega_n'} e^{-\xi_n \omega_n t} \sin(\omega_n' t), \quad t \geq 0 \qquad (6.9)$$

$$\omega_n' = \omega_n \sqrt{1 - \xi_n^2} \qquad (6.10)$$

Substituting Equations (6.5) and (6.8) into Equation (6.2), the dynamic deflection of the beam at point x and time t can be found as:

$$v(x,t) = \sum_{n=1}^{\infty} \frac{2}{\rho l \omega_n'} \sin \frac{n\pi x}{l} \int_0^t e^{-\xi_n \omega_n (t-\tau)} \sin \omega_n'(t-\tau) \sin \frac{n\pi c\tau}{l} f(\tau)d\tau \qquad (6.11)$$

6.2.1.2 Force Identification from Bending Moments

The bending moment of the beam at point x and time t is:

$$m(x,t) = -EI \frac{\partial^2 v(x,t)}{\partial x^2} \qquad (6.12)$$

Substituting Equation (6.11) into Equation (6.12) gives:

$$m(x,t) = \sum_{n=1}^{\infty} \frac{2EI\pi^2 n^2}{\rho l^3 \omega_n'} \sin \frac{n\pi x}{l} \int_0^t e^{-\xi_n \omega_n (t-\tau)} \sin \omega_n'(t-\tau) \sin \frac{n\pi c\tau}{l} f(\tau)d\tau \qquad (6.13)$$

Assuming the force $f(t)$ is a step function in a small time interval, Equation (6.13) can be rewritten in discrete terms as:

$$m(i) = \frac{2EI\pi^2}{\rho l^3} \sum_{n=1}^{\infty} \frac{n^2}{\omega_n'} \sin \frac{n\pi x}{l} \sum_{j=0}^{i} e^{-\xi_n \omega_n \Delta t(i-j)} \sin \omega_n' \Delta t(i-j) \sin \frac{n\pi c \Delta t j}{l} f(j) \Delta t$$

$$(i = 0, 1, 2, \ldots, N) \qquad (6.14)$$

where Δt is the sample interval and $N+1$ is the number of sample points. Let

$$C_{xn} = \frac{2EI\pi^2}{\rho l^3} \frac{n^2}{\omega_n'} \sin \frac{n\pi x}{l} \Delta t, \qquad (6.15)$$

$$E_n^k = e^{-\xi_n \omega_n \Delta t k},$$

$$S_1(k) = \sin(\omega_n' \Delta t k),$$

$$S_2(k) = \sin\left(\frac{n \pi c \Delta t}{l} k\right)$$

(6.16)

Arranging Equation (6.14) into matrix form

$$
\begin{Bmatrix} m(0) \\ m(1) \\ m(2) \\ \vdots \\ m(N) \end{Bmatrix} = \sum_{n=1}^{\infty} C_{xn}
\begin{bmatrix}
0 & 0 & 0 & \cdots & 0 \\
0 & 0 & 0 & \cdots & 0 \\
0 & E_n^1 S_1(1) S_2(1) & 0 & \cdots & 0 \\
\vdots & \vdots & \vdots & \cdots & \vdots \\
0 & E_n^{N-1} S_1(N-1) S_2(1) & E_n^{N-2} S_1(N-2) S_2(2) & \cdots & E_n^{N-N_B} S_1(N-N_B) S_2(N_B)
\end{bmatrix}
$$

$$
\times \begin{Bmatrix} f(0) \\ f(1) \\ f(2) \\ \vdots \\ f(N_B) \end{Bmatrix}
$$

(6.17)

where

$$N_B = \frac{l}{c \Delta t}$$

Assuming

$$f(0) = 0, \quad f(N_B) = 0$$

(6.18)

at the entry and exit of vehicle, we have from Equation (6.17)

$$m(0) = 0, \quad m(1) = 0$$

(6.19)

Equation (6.17) can then be condensed as:

$$
\begin{Bmatrix} m(2) \\ m(3) \\ \vdots \\ m(N) \end{Bmatrix} = \sum_{n=1}^{\infty} C_{xn}
\begin{bmatrix}
E_n^1 S_1(1) S_2(1) & 0 & \cdots & 0 \\
E_n^2 S_1(2) S_2(1) & E_n^1 S_1(1) S_2(2) & \cdots & 0 \\
\vdots & \vdots & \vdots & \vdots \\
E_n^{N-1} S_1(N-1) S_2(1) & E_n^{N-2} S_1(N-2) S_2(2) & \cdots & b_{ee}
\end{bmatrix}
$$

$$
\times \begin{Bmatrix} f(1) \\ f(2) \\ \vdots \\ f(N_B - 1) \end{Bmatrix}
$$

(6.20)

where

$$b_{ee} = E_n^{N-N_B+1} S_1(N - N_B + 1) S_2(N_B - 1)$$

Equation (6.20) is simply rewritten as:

$$\underset{(N-1)\times 1}{\mathbf{m}} = \underset{(N-1)\times(N_B-1)}{\mathbf{B}} \underset{(N_B-1)\times 1}{\mathbf{f}} \tag{6.21}$$

If $N = N_B$, matrix $\mathbf{B}$ is a lower triangular matrix. We can directly find the force vector $\mathbf{f}$ by solving Equation (6.21). If $N > N_B$, and/or N_l bending moments ($N_l > 1$) are measured, least-squares method can be used to find the force vector $\mathbf{f}$ from

$$\begin{Bmatrix} m_1 \\ m_2 \\ \vdots \\ m_{N_1} \end{Bmatrix} = \begin{bmatrix} B_1 \\ B_2 \\ \vdots \\ B_{N_1} \end{bmatrix} \mathbf{f} \tag{6.22}$$

The above procedure is derived for single force identification. Equation (6.21) can be modified for two-forces identification using the linear superposition principle as:

$$\mathbf{m} = \begin{bmatrix} B_a & 0 \\ B_b & B_a \\ B_c & B_b \end{bmatrix} \begin{Bmatrix} f_1 \\ f_2 \end{Bmatrix} \tag{6.23}$$

where $\mathbf{B_a}$ [$N_s \times (N_B - 1)$], $\mathbf{B_b}$ [$(N - 1 - 2N_s) \times (N_B - 1)$], and $\mathbf{B_c}$ [$N_s \times (N_B - 1)$] are sub-matrices of matrix $\mathbf{B}$. The first row of sub-matrices in the first matrix describes the state having the first force on beam after its entry. The second and third row of sub-matrices describe the states having two-forces on beam and one force on beam after the exit of the first force. The whole matrix has a dimension of $(N - 1) \times (N_B - 1)$. $N_s = l_s/(c\Delta t)$, and l_s is the distance between two forces. The two forces can be identified using more than one measured bending moment measurements.

6.2.1.3 Identification from Accelerations

The acceleration at the point x and time t is:

$$\ddot{v}(x, t) = \sum_{n=1}^{\infty} \frac{1}{M_n} \Phi_n(x) \left[p_n(t) + \int_0^t \ddot{h}_n(t - \tau) p_n(\tau) d\tau \right] \tag{6.24}$$

where

$$\ddot{h}_n(t) = \frac{1}{\omega_n'} e^{-\xi_n \omega_n t} \{ [(\xi_n \omega_n)^2 - \omega_n'^2] \sin \omega_n' t + [-2\xi_n \omega_n \omega_n'] \cos \omega_n' t \} \tag{6.25}$$

Equation (6.24) can be rewritten in discrete terms:

$$\ddot{v}(i) = \frac{2}{\rho l} \sum_{n=1}^{\infty} \sin \frac{n\pi x}{l} \left[\sin \frac{n\pi c \Delta t i}{l} f(i) + \sum_{j=0}^{i} \ddot{h}(i-j) \sin \frac{n\pi c \Delta t j}{l} f(j) \Delta t \right] \tag{6.26}$$

The response of mode n is:

$$\ddot{v}(i)_n = \frac{2}{\rho l} \sin \frac{n\pi x}{l} \left[\sin \frac{n\pi c \Delta t i}{l} f(i) + \frac{1}{\omega'_n} \sum_{j=0}^{i} \ddot{h}(i-j) \sin \frac{n\pi c \Delta t j}{l} f(j) \Delta t \right] \tag{6.27}$$

Let

$$D_{xn} = \frac{2}{\rho l} \sin \frac{n\pi x}{l}, \tag{6.28}$$

$$H_n(k) = \frac{\Delta t}{\omega'_n} \ddot{h}(k),$$

$$S_2(k) = \sin \left(\frac{n\pi c \Delta t}{l} k \right) \tag{6.29}$$

Arranging Equation (6.27) into matrix form:

$$
\begin{Bmatrix} \ddot{v}(0) \\ \ddot{v}(1) \\ \ddot{v}(2) \\ \vdots \\ \ddot{v}(N) \end{Bmatrix}_n = D_{xn}
\begin{bmatrix}
0 & 0 & 0 & \cdots & 0 \\
0 & S_2(1)(1+H_n(0)) & 0 & \cdots & 0 \\
0 & H_n(1)S_2(1) & S_2(2)(1+H_n(0)) & \cdots & 0 \\
\vdots & \vdots & \vdots & \cdots & \vdots \\
0 & H_n(N-1)S_2(1) & H_n(N-2)S_2(2) & \cdots & H_n(N-N_B)S_2(N_B)
\end{bmatrix}
$$

$$
\times \begin{Bmatrix} f(0) \\ f(1) \\ f(2) \\ \vdots \\ f(N_B) \end{Bmatrix} \tag{6.30}
$$

Assuming

$$f(0) = 0, \quad f(N_B) = 0 \tag{6.31}$$

we have

$$\ddot{v}(0) = 0 \tag{6.32}$$

Equation (6.30) can be condensed as:

$$
\begin{Bmatrix} \ddot{v}(1) \\ \ddot{v}(2) \\ \vdots \\ \ddot{v}(N) \end{Bmatrix}_n = D_{xn} \begin{bmatrix} (1 + H_n(0))S_2(1) & 0 & \cdots & 0 \\ H_n(1)S_2(1) & (1 + H_n(0))S_2(2) & \cdots & 0 \\ \vdots & \vdots & \vdots & \vdots \\ H_n(N-1)S_2(1) & H_n(N-2)S_2(2) & \cdots & H_n(N-N_B+1)S_2(N_B-1) \end{bmatrix}
$$

$$
\times \begin{Bmatrix} f(1) \\ f(2) \\ \vdots \\ f(N_B - 1) \end{Bmatrix} \tag{6.33}
$$

Equation (6.33) is simply rewritten as:

$$
\underset{N\times 1}{\ddot{\mathbf{v}}_n} = \underset{N\times(N_B-1)}{\mathbf{A}_n} \underset{(N_B-1)}{\mathbf{f}} \tag{6.34}
$$

If $N = N_B - 1$, matrix $\mathbf{A}_n$ is a lower triangular matrix. From Equation (6.34), we can directly find the force vector $\mathbf{f}$:

$$
\mathbf{f} = \left(\sum_{n=1}^{N_m} \mathbf{A}_n \right)^{-1} \ddot{\mathbf{v}} \tag{6.35}
$$

If $N > N_B - 1$ and/or more than one bending moments are measured ($N_L > 1$), the least-square method can be used to find the force vector $\mathbf{f}$:

$$
\mathbf{f} = \begin{bmatrix} \left(\sum_{n=1}^{N_m} D_{xn}\mathbf{A}_n \right)_1 \\ \left(\sum_{n=1}^{N_m} D_{xn}\mathbf{A}_n \right)_2 \\ \vdots \\ \left(\sum_{n=1}^{N_m} D_{xn}\mathbf{A}_n \right)_{N_L} \end{bmatrix}^{+} \begin{Bmatrix} \ddot{\mathbf{v}}_1 \\ \ddot{\mathbf{v}}_2 \\ \vdots \\ \ddot{\mathbf{v}}_{N_L} \end{Bmatrix} \tag{6.36}
$$

6.2.1.4 Identification from Bending Moments and Accelerations

If the bending moments and accelerations responses are measured at the same time, both of them can be used together to identify the moving force. The vector $\mathbf{m}$ in Equation (6.21), and $\ddot{\mathbf{v}}$ in Equation (6.34), should be scaled to have dimensionless unit. The two equations are then combined together to have:

$$
\begin{bmatrix} \mathbf{B}/\|\mathbf{m}\| \\ \mathbf{A}/\|\ddot{\mathbf{v}}\| \end{bmatrix} \mathbf{f} = \begin{Bmatrix} \mathbf{m}/\|\mathbf{m}\| \\ \ddot{\mathbf{v}}/\|\ddot{\mathbf{v}}\| \end{Bmatrix} \tag{6.37}
$$

where $\|\bullet\|$ is the norm of the vector.

Table 6.1 Errors on two-forces identification (in percent)

Response Combinations Location & Response	1% Error in Response		5% Error in Response	
	First force	Second force	First force	Second force
1/2m & 1/4m	421	620	*	*
1/2m & 1/4m & 3/4m	428	397	*	*
1/2a & 1/4a	26.9	12.4	222	78.6
1/2a & 1/4a & 3/4a	3.26	6.34	10.7	10.7
1/2m & 1/2a	266	580	*	*
1/2m & 1/4m & 1/2a	199	360	*	474
1/2m & 1/4m & 1/2a & 1/4a	26.9	13.7	201	49.7
1/4m & 1/4a	813	402	*	*
1/4m & 1/4a & 1/2a	28.2	13.9	206	156
1/2m & 1/4a	162	353	*	*
1/2m & 1/4a & 1/4m	171	224	*	*
1/4a & 1/2a & 1/2m	27.1	13.5	193	46.8

Note: * indicate the error is larger than 1000.

6.2.2 Simulation Studies

The beam described in Section 5.3, with two moving forces on top, is used for the studies. The parameters of the beam are the same as those stated in Section 5.3. The two moving forces described in Section 5.3 are at a fixed spacing of 4 m.

$$f_1(t) = 20000[1 + 0.1\sin(10\pi t) + 0.05\sin(40\pi t)] \text{ N}$$

$$f_2(t) = 20000[1 - 0.1\sin(10\pi t) + 0.05\sin(50\pi t)] \text{ N}$$

The moving speed is 40 m/s. The analysis frequency bandwidth is from 0 Hz to 40 Hz and therefore the first three modes of the beam are included in the calculation. Sampling frequency f_s is 100 Hz and N_B equals to 100. The record length N is 512, and 110 points are used in the identification.

White noise is added to the calculated responses to simulate the polluted measurements. The errors of identification are defined as Eq. (5.32) and they are calculated by the following equation:

$$Error = \frac{\|f_{identified} - f_{true}\|}{\|f_{true}\|} \times 100\%$$

Bending moment and/or acceleration responses at 1/4, 1/2 and 3/4 spans in 12 combinations described in Table 6.1 are used to identify the two forces. Similar to single force identification, the following results are obtained:

- If $E_p = 0$, i.e., when no noises are added into the measured responses, accurate results are obtained. This means the proposed method and algorithms for two forces identification are correct.
- For $E_p = 1$ percent and $E_p = 5$ percent, errors on the identification results are shown in Table 6.1. Samples of the time history and PSDs of the identified forces

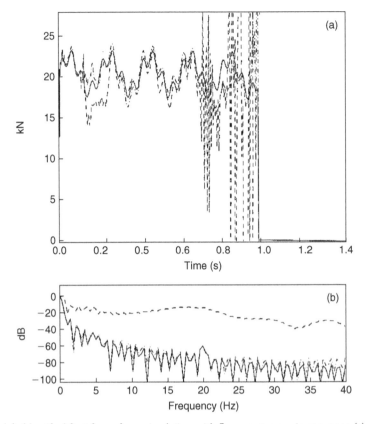

Figure 6.1 Identified first force from simulation with 5 percent error in response (a) the moving force; (b) the PSD of the force (— True force; - - - 1/4a and 1/2a; -.-.- 1/4a, 1/2a and 3/4a)

are shown in Figures 6.1 and 6.2. The results obtained from using 1/4a and 1/2a varies greatly close to the beam ends and at the nodal points of the third mode of the beam, while that obtained from 1/4a, 1/2a and 3/4a gives satisfactory results except at both ends of the beam. The results show that they are very noise sensitive. The identified forces are only close to the true forces when they are in the middle length of the beam.

- Other results, not included here, show that acceleration measurement gives much better results than bending moment measurements in the identification.
- The results also show that accuracy in two forces identification is lower than single force identification. One reason is that there is a force component with same amplitude and opposite phase in the two identified forces. This results in large errors in the time domain. Moreover, Figures 6.1 and 6.2 show that there are large errors in the time duration from 0.1 second to 0.3 second and from 0.8 second to 1.0 second. This is due to the low sensitivity of the responses to the forces at the beginning and end of the beam. This leads to large errors in the PSDs and time histories of the forces.

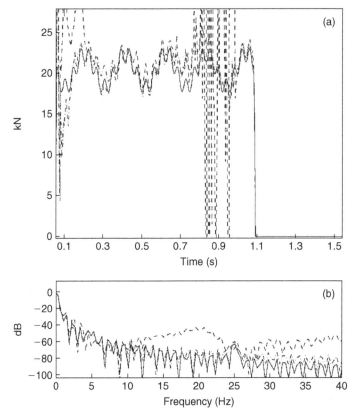

Figure 6.2 Identified second force from simulation with 5 percent error in response (a) the moving force; (b) the PSD of the force (Legends are the same as those for Figure 6.1)

6.2.3 *Experimental Studies*

The experimental setup has been shown diagrammatically in Figure 5.11. The measured data obtained, as described in Section 5.4, are used for the following studies with two moving forces identifications.

Twelve combinations of the measured responses are used to identify the two forces. The acceleration of the main beam at the 3/8 span is calculated using the identified forces and compared with the measured response. Correlation coefficients are calculated between the calculated and measured responses and they are shown in Table 6.2. Some of the identified results are shown in Figures 6.3 to 6.5.

Results from Table 6.2 show that acceleration response alone or combined bending moment and acceleration responses are more suitable for the two-forces identification. The reconstructed response at 3/8 span matched the measured response closely in Figure 6.5 for the cases of 1/4a and 1/2a and 1/4a, 1/2a and 3/4a. Figures 6.2 and 6.3 show large discrepancies in the identified results at around 0.065 second and 1.025 second. These two moments correspond to the entry of the second axle and exit of the

Table 6.2 Correlation coefficients between measured
and calculated responses

Response Combinations Location & Response	Correlation Coefficient
1/2m & 1/4m	0.640
1/2m & 1/4m & 3/4m	0.708
1/2a & 1/4a	0.933
1/2a & 1/4a & 3/4a	0.994
1/2m & 1/2a	0.989
1/2m & 1/4m & 1/2a	0.856
1/2m & 1/4m & 1/2a & 1/4a	0.883
1/4m & 1/4a	0.865
1/4m & 1/4a & 1/2a	0.870
1/2m & 1/4a	0.874
1/2m & 1/4a & 1/4m	0.856
1/4a & 1/2a & 1/2m	0.863

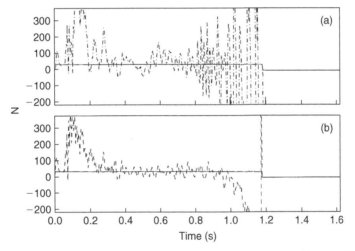

Figure 6.3 Identified first force from experiment (a) – static force; -.-.- 1/4a and 1/2a. (b) – static
force; -.-.- 1/4a, 1/2a and 3/4a

first axle to the bridge, and the forcing system switches from a single force excitation to a two forces excitation and vice versa. The large discrepancy after 1 second corresponds to the incorrect identification by the proposed method in identifying non-existing forces on the bridge deck from the free oscillation of the structure.

There is also a large local discrepancy between the identified results and the true force using 1/2 and 1/4 accelerations around 0.87 second in Figures 6.3 to 6.5. This occurs as the vehicle is located on the nodal point of the third mode shape of the beam, and the noise from the third modal frequency bandwidth would dramatically affect the identified result. Similar phenomenon in the simulation results is found in Figures 6.1

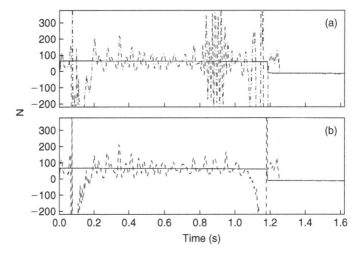

Figure 6.4 Identified second force from experiment (Legends same as Figure 6.3)

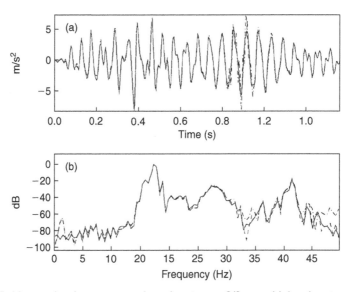

Figure 6.5 Measured and reconstructed accelerations at 3/8 span. (a) Accelerations; (b) the PSD of the response (— measured; -.-.- 1/4a and 1/2a; — 1/4a, 1/2a and 3/4a)

and 6.2. Identified results using 1/4, 1/2 and 3/4 accelerations are close to the true static force within the time range of 0.3 to 0.9 second. This strongly suggests using responses from at least three locations for the identification to reduce these local discrepancies disregard the good correlation shown in Table 6.2 for cases using responses from two locations.

By comparing Figures 6.3 and 6.4, we find that there is a component with the same amplitude and opposite phase in the two identified forces. This is the component due

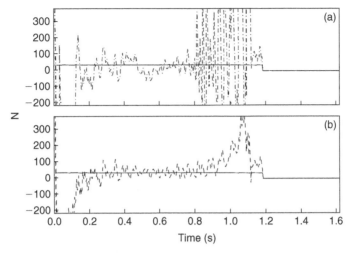

Figure 6.6 Identified resultant force from experiment (a) — static force; -.-.- 1/4a and 1/2a; (b) – static force; -.-.- 1/4a, 1/2a and 3/4a

to the pitching motion of the model car. The two identified forces are added to obtain a resultant force as shown in Figure 6.6. This resultant force matches very closely with the true force. This resulting force is a good estimate of the total dead weight force of the vehicle.

6.2.4 Discussions

- The vehicle and suspension characteristics (mass and stiffness) are not required as only the interaction forces are identified.
- Results show fairly accurate estimates of the interaction forces and this method can be used for WIM (weigh-in-motion) of passing vehicles with no constraints on the type of vehicle and its suspension characteristics.
- This method is beneficial to identify only vehicles with a limited number of axles since each additional moving force adds another set of equations for the solution. For the case of multiple vehicle or vehicle with multi-axle, much more computational cost will be involved.

6.3 Moving Force Identification – Exact Solution Technique (EST)

6.3.1 Beam Model

A continuous beam subject to a system of moving forces P_l $(l = 1, 2, \ldots, N_p)$, as shown in Figure 2.1, is reproduced in Figure 6.7. The beam is assumed to be a simply supported Euler-Bernoulli beam. The equation of motion in Equation (2.1) is reproduced below as (Zhu and Law, 2002b):

$$\rho A \frac{\partial^2 w(x,t)}{\partial t^2} + c \frac{\partial w(x,t)}{\partial t} + EI \frac{\partial^4 w(x,t)}{\partial x^4} = \sum_{l=1}^{N_p} P_l(t)\delta(x - \hat{x}_l(t)) \qquad (6.38)$$

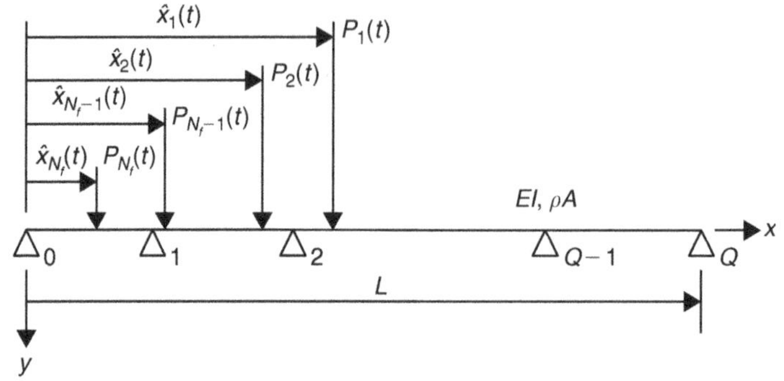

Figure 6.7 A continuous beam subject to moving loads

where L is the total length of the beam; A is the cross-sectional area; E is the Young's modulus; I is the moment of inertia of the beam cross-section; ρ, c and $w(x,t)$ are the mass per unit volume, the damping and the displacement function of the beam respectively; $\hat{x}_l(t)$ is the location of moving force $P_l(t)$ at time t; $\delta(t)$ is the Dirac delta function and N_p is the number of forces.

Based on modal superposition, Equation (6.38) can be re-written as:

$$\frac{d^2 q_i(t)}{dt^2} + 2\xi_i\omega_i\frac{dq_i(t)}{dt} + \omega_i^2 q_i(t) = \frac{1}{M_i}\sum_{l=1}^{N_p} P_l(t)\phi_i(\hat{x}_l(t)) \tag{6.39}$$

where $\phi_i(x)$ is the mode shape function of the ith mode and it can be calculated by the method shown in Chapter 2; $q_i(t)$ is the ith modal amplitude. ω_i, ξ_i, M_i are the modal frequency, the damping ratio and the modal mass of the ith mode. The displacement of the beam at a point x and time t can be found from Equation (6.39) as follows assuming the beam is initially at rest.

$$w(x,t) = \sum_{i=1}^{\infty}\frac{\phi_i(x)}{M_i}\int_0^t h_i(t-\tau)\sum_{l=1}^{N_p} P_l(\tau)\phi_i(\hat{x}_l(\tau))d\tau \tag{6.40}$$

where

$$h_i(t) = \frac{1}{\omega_i'}e^{-\xi_i\omega_i t}\sin\omega_i't; \quad \omega_i' = \omega_i\sqrt{1-\xi_i^2} \tag{6.41}$$

and ω_i' is the damped modal frequency of the ith mode.

6.3.1.1   Identification from Strains

The strain in the beam at a point x and time t can be written as:

$$\varepsilon(x,t) = -H\frac{\partial^2 w(x,t)}{\partial x^2} \tag{6.42}$$

where H is the distance between the bottom surface and the neutral flexural surface of the beam.

Substitute Equation (6.40) into Equation (6.42), and re-write in discrete form:

$$\varepsilon(x_s, m) = -\sum_{i=1}^{N} \frac{H\phi_i''(x_s)\Delta t}{M_i} \sum_{j=0}^{m} h_i(m-j) \sum_{l=1}^{N_p} P_l(j)\phi_i(\hat{x}_l(j))$$

$$(m = 0, 1, 2, \ldots, N_t; \quad s = 1, 2, \ldots, N_s) \tag{6.43}$$

where Δt is the time interval; N is the number of vibration modes; N_t is the number of data points; x_s is the location of the measuring point; N_s is the number of measuring points; and $\phi_i''(x_s)$ is the second derivative of $\phi_i(x_s)$. The discrete form of $h_i(t)$ in Equation (6.41) can be written as:

$$h_i(j) = \frac{1}{\omega_i'} e^{-\xi_i \omega_i j \Delta t} \sin \omega_i' j \Delta t \tag{6.44}$$

Equation (6.43) can be re-written in matrix form:

$$\varepsilon = BP \tag{6.45}$$

where ε is $(N_t * N_s) \times 1$ matrix; B is $(N_t * N_s) \times (N_t * N_p)$ matrix; P is $(N_t * N_p) \times 1$ matrix,

$$\varepsilon = \{\varepsilon(x_1, 1), \varepsilon(x_2, 1), \ldots \varepsilon(x_{N_s}, 1), \varepsilon(x_1, 2), \ldots \varepsilon(x_{N_s}, N_t)\}^T;$$
$$P = \{p_1(0), p_2(0), \ldots, p_{N_p}(0), p_1(1), \ldots, p_{N_p}(N_t - 1)\}^T. \tag{6.46}$$

When the measured data is more than the number of unknown force data points, Equation (6.45) can be solved using the least-squares method. However the solution involves the computation of the inverse of matrix B which would be very inefficient when the measured data is large (Law et al., 1997; 2001). Matrix B can be split into smaller sub-matrices to improve the computation efficiency as follows.

$$B = \begin{bmatrix} B_{10} & 0 & \cdots & 0 \\ B_{20} & B_{21} & \cdots & 0 \\ \vdots & \vdots & \vdots & \vdots \\ B_{N_t 0} & B_{N_t 1} & \cdots & B_{N_t N_t - 1} \end{bmatrix}_{(N_s \cdot N_t) \times (N_p \cdot N_t)} ; B_{mj} = \begin{bmatrix} b_{11} & b_{12} & \cdots & b_{1N_p} \\ b_{21} & b_{22} & \cdots & b_{2N_p} \\ \cdots & \cdots & \vdots & \cdots \\ b_{N_s 1} & b_{N_s 2} & \cdots & b_{N_s N_p} \end{bmatrix}_{N_s \times N_p}$$

$$b_{sl} = -H\Delta t \sum_{i=1}^{N} \frac{\phi_i''(x_s)}{M_i} h_i(m-j)\phi_i(\hat{x}_l(j))$$

$$(m = 1, 2, 3, \ldots, N_t; \ j = 0, 1, 2, \ldots, N_t - 1; \ s = 1, 2, \ldots, N_s; \ l = 1, 2, \ldots, N_p)$$

$$\tag{6.47}$$

6.3.1.2   *Identification from Accelerations*

The acceleration at a point x and time t can be obtained from Equation (6.40) as:

$$\ddot{w}(x,t) = \sum_{i=1}^{\infty} \frac{\phi_i(x)}{M_i} \left[\sum_{l=1}^{N_p} P_l(t)\phi_i(\hat{x}_l(t)) + \int_0^t g_i(t-\tau)\sum_{l=1}^{N_p} P_l(\tau)\phi_i(\hat{x}_l(\tau))d\tau \right] \qquad (6.48)$$

where

$$g_i(t) = \frac{1}{\omega_i'} e^{-\zeta_i \omega_i t}\{[(\zeta_i \omega_i)^2 - \omega_i'^2]\sin \omega_i' t - 2\zeta_i \omega_i \omega_i' \cos \omega_i' t\} \qquad (6.49)$$

The acceleration at a measuring point x_s is written in discrete form to include the N modes.

$$\ddot{w}(x_s, m) = \sum_{i=1}^{N} \frac{\phi_i(x_s)}{M_i} \left[\sum_{l=1}^{N_p} P_l(m)\phi_i(\hat{x}_l(m)) + \sum_{j=0}^{m} g_i(m-j)\sum_{l=1}^{N_p} P_l(j)\phi_i(\hat{x}_l(j))\Delta t \right]$$

$$(s = 1, 2, \ldots, N_s; \quad m = 1, 2, \ldots, N_t) \qquad (6.50)$$

where $N_t + 1$ is the total number of sampling points, and

$$g_i(j) = \frac{1}{\omega_i'} e^{-\zeta_i \omega_i j \Delta t}\{[(\zeta_i \omega_i)^2 - \omega_i'^2]\sin \omega_i' j\Delta t - 2\zeta_i \omega_i \omega_i' \cos \omega_i' j\Delta t\}$$

$$(i = 1, 2, \ldots, N; \quad j = 1, 2, \ldots, N_t) \qquad (6.51)$$

Rewrite Equation (6.50) in matrix forms.

$$\ddot{w} = DP \qquad (6.52)$$

where

$$\ddot{w} = \{\ddot{w}(x_1, y_1, 1), \ddot{w}(x_2, y_2, 1), \ldots, \ddot{w}(x_{N_s}, y_{N_s}, 1),$$
$$\ddot{w}(x_1, y_1, 2), \ldots, \ddot{w}(x_{N_s}, y_{N_s}, N)\}^T; \qquad (6.53)$$
$$P = \{p_1(0), p_2(0), \ldots, p_{N_p}(0), p_1(1), \ldots, p_{N_p}(N-1)\}^T;$$

and again matrix D can be split into smaller sub-matrices as follows:

$$D = \begin{bmatrix} D_{10} & 0 & \cdots & 0 \\ D_{20} & D_{21} & \cdots & 0 \\ \vdots & \vdots & \vdots & \vdots \\ D_{N_t0} & D_{N_t1} & \cdots & D_{N_tN_t-1} \end{bmatrix}_{(N_s \cdot N_t)\times(N_p \cdot N_t)} ; \quad D_{mj} = \begin{bmatrix} d_{11} & d_{12} & \cdots & d_{1N_p} \\ d_{21} & d_{22} & \cdots & d_{2N_p} \\ \cdots & \cdots & \vdots & \cdots \\ d_{N_s1} & d_{N_s2} & \cdots & d_{N_sN_p} \end{bmatrix}_{N_s \times N_p}$$

$$d_{sl}^* = \Delta t \sum_{i=1}^{N} \frac{\phi_i(x_s)}{M_i} h_i''(m-j)\phi_i(\hat{x}_l(j))$$

$$(m = 1, 2, 3, \ldots, N_t; \quad j = 0, 1, 2, \ldots, N_t - 1; \quad s = 1, 2, \ldots, N_s; \quad l = 1, 2, \ldots, N_p)$$

When $j < m$,

$$d_{sl} = d_{sl}^*;$$

When $j = m$,

$$d_{sl} = d_{sl}^* + \sum_{i=1}^{N} \frac{\phi_i(x_s)}{M_i} \phi_i(\hat{x}_l(m)) \qquad (6.54)$$

The computation with smaller matrices, as shown above, is much more efficient than other methods (Law et al., 1997; 2001) working with a single large matrix.

6.3.1.3 Statement of the Problem

The natural frequencies and mode shapes obtained from modal testing and modal analysis are subject to measurement errors. Noise contamination in the test data has adverse effect on the accuracy of the identified moving loads (Law et al., 1997). If $N_s \geq N_p$, the moving loads can be identified from Equations (6.45) and (6.52) by least-squares method. However, the solutions would be unstable in the sense that small perturbations in the responses would result in large deviation from their exact solutions, due to the ill-conditioning of matrices B or D, which increases as the dimension of the problem increases. Hence general regularization methods based on singular value decomposition, cross-validation (Golub et al.,1979) and L-Curves (Hansen, 1992; Hansen and O'Leavy, 1993) are studied with an attempt to overcome the ill-conditioned problems. The problem can then be formulated as:

$$\text{minimize } (r - AP)^T R(r - AP)$$
$$\text{subject to } (SP)^T SP = e \qquad (6.55)$$

where R is an error-weighting matrix depending on the measured information. Since some measured data are of small amplitudes they are easily affected by systematic errors. Also some data may be collected at a time when the background noise is high. R is usually taken as the inverse of the covariance of the measured data. S is a smoothing matrix, which is typically either the identity matrix or a discrete approximation to a derivative operator (Santantamarina and Fratta, 1998). e denotes a residual scalar estimation error. r is the measured response, which may be ε or $\ddot{w}$. A is the coefficient matrix B or D.

The Lagrangian expression on the problem then becomes:

$$J(P, \lambda) = (r - AP, R(r - AP)) + \lambda(SP, SP) \qquad (6.56)$$

where λ is the regularization parameter, and the first term on the right-hand-side is the Euclidean scalar product.

6.3.2 Plate Model

The orthotropic plate supporting a group of moving loads in Figure 3.1 is reproduced here as Figure 6.8. The plate is simply supported along $x = 0$ and $x = a$ with the other

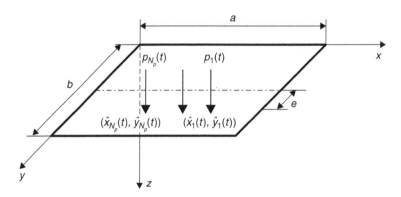

Figure 6.8 Orthotropic plate under moving loads

two edges free. The equations of motion of the orthotropic plate under moving loads can be expressed as follows:

$$D_x \frac{\partial^4 w}{\partial x^4} + 2D_{xy}\frac{\partial^4 w}{\partial x^2 \partial y^2} + D_y \frac{\partial^4 w}{\partial y^4} + C\frac{\partial w}{\partial t} + \rho h \frac{\partial^2 w}{\partial t^2}$$

$$= \sum_{l=1}^{N_p} p_l(t)\delta(x - \hat{x}_l(t))\delta(y - \hat{y}_l(t)) \tag{6.57}$$

where D_x, D_y are the flexural rigidities of the orthotropic plate in the x- and y-directions, respectively. D_{xy} is the torsional rigidity of plate. C is the damping coefficient. h is the thickness of the plate. ρ is the mass density of plate material. $\{p_l(t), l = 1, 2, \ldots, N_p\}$ are the moving loads and they are moving as a group at a fixed spacing. $(\hat{x}_l(t), \hat{y}_l(t))$ is the position of the moving load $p_l(t)$. $\delta(x), \delta(y)$ are the Dirac functions.

The derivation of the lateral displacement of the orthotropic plate is the same as described in Section 3.2.2, and it is given as follows for the sake of clarity of this method. By modal superposition, the displacement of the orthotropic plate can be written as follows:

$$w(x, y, t) = \sum_{m,n} W_{mn}(x, y)q_{mn}(t) \tag{6.58}$$

where $W_{mn}(x, y) = Y_{mn}(y)\sin(m\pi x/a)$ is the mode shape of the orthotropic plate which can be obtained from the formulation in Chapter 2. $q_{mn}(t)$ is the corresponding modal coordinate.

Substituting Equation (6.58) into Equation (6.57) results in:

$$\ddot{q}_{mn}(t) + 2\zeta_{mn}\omega_{mn}\dot{q}_{mn}(t) + \omega_{ij}^2 q_{mn}(t)$$

$$= \frac{2}{\rho h a \int_0^b Y_{mn}^2(y)dy} \sum_{l=1}^{N_p} p_l(t)Y_{mn}(\hat{y}_l(t)) \sin\left(\frac{m\pi}{a}\hat{x}_l(t)\right) \quad (m, n = 1, 2, \ldots) \tag{6.59}$$

where $\zeta_{mn} = C/2\rho h\omega_{mn}$. a, b are the dimensions of the orthotropic plate in x- and y-directions respectively. Equation (6.59) can be solved in the time domain by the convolution integral with the plate initially at rest, yielding:

$$q_{mn}(t) = \frac{1}{M_{mn}} \int_0^t H_{mn}(t - \tau) f_{mn}(\tau) d\tau \qquad (6.60)$$

where

$$M_{mn} = \frac{\rho h a}{2} \int_0^b Y_{mn}^2(y) dy$$

$$H_{mn}(t) = \frac{1}{\omega'_{mn}} e^{-\zeta_{mn}\omega_{mn}t} \sin(\omega'_{mn}t), \quad t \geq 0$$

$$\qquad (6.61)$$

$$f_{mn}(t) = \sum_{l=1}^{N_p} p_l(t) Y_{mn}(\hat{y}_l(t)) \sin\left(\frac{m\pi}{a} \hat{x}_l(t)\right)$$

$$\omega'_{mn} = \omega_{mn}\sqrt{1 - \zeta_{mn}^2}$$

Substituting Equation (6.60) into Equation (6.58), the displacement of the orthotropic plate at point (x, y) and time t can be found as:

$$w(x, y, t) = \sum_{m=1}^{\infty} \sum_{n=1}^{\infty} Y_{mn}(y) \sin\left(\frac{m\pi}{a} x\right) \frac{1}{M_{mn}} \int_0^t H_{mn}(t - \tau) f_{mn}(\tau) d\tau \qquad (6.62)$$

6.3.2.1 Identification from Strains

The strains in the orthotropic plate at point (x, y) and time t are:

$$\varepsilon_x(x, y, t) = z_t \sum_{m=1}^{\infty} \sum_{n=1}^{\infty} \left(\frac{m\pi}{a}\right)^2 Y_{mn}(y) \sin\left(\frac{m\pi}{a} x\right) \frac{1}{M_{mn}} \int_0^t H_{mn}(t - \tau) f_{mn}(\tau) d\tau$$

$$\qquad (6.63)$$

$$\varepsilon_y(x, y, t) = -z_t \sum_{m=1}^{\infty} \sum_{n=1}^{\infty} Y''_{mn}(y) \sin\left(\frac{m\pi}{a} x\right) \frac{1}{M_{mn}} \int_0^t H_{mn}(t - \tau) f_{mn}(\tau) d\tau$$

where $\varepsilon_x(x, y, t), \varepsilon_y(x, y, t)$ are the strains at the bottom surface of the plate along x- and y-directions, respectively, and z_t is the distance from the neutral surface to the bottom tension surface. The strains at measuring point (x_s, y_s) can be written in discrete form including the $MM \times NN$ modes along the x- and y-directions respectively.

$$\varepsilon_x(x_s, y_s, mm) = z_t \sum_{m=1, n=1}^{MM \ NN} \left(\frac{m\pi}{a}\right)^2 Y_{mn}(y_s) \sin\left(\frac{m\pi}{a} x_s\right)$$

$$\times \frac{1}{M_{mn}} \sum_{k=0}^{mm} H_{mn}(mm - k) f_{mn}(k) \Delta t \qquad (6.64)$$

$$\varepsilon_y(x_s, y_s, mm) = -z_t \sum_{m=1, n=1}^{MM \ NN} Y''_{mn}(y_s) \sin\left(\frac{m\pi}{a} x_s\right) \frac{1}{M_{mn}} \sum_{k=0}^{mm} H_{mn}(mm - k) f_{mn}(k) \Delta t$$

$$(s = 1, 2, \ldots, N_s; \quad mm = 1, 2, \ldots, N)$$

where Δt is the time step; $(N+1)$ is the number of sampling points; N_s is the number of measuring points and

$$H_{mn}(k) = \frac{1}{\omega'_{mn}} e^{-\zeta_{mn}\omega_{mn}k\Delta t} \sin(\omega'_{mn}k\Delta t)$$

$$f_{mn}(k) = \sum_{l=1}^{N_p} p_l(k\Delta t) Y_{mn}(\hat{y}_l(k\Delta t)) \sin\left(\frac{m\pi}{a}\hat{x}_l(k\Delta t)\right)$$

$$(m = 1, 2, \ldots, MM; \quad n = 1, 2, \ldots, NN) \tag{6.65}$$

Equation (6.64) is rewritten in matrix form (only the x-direction strains are presented since those for the y-direction strains are similar).

$$\boldsymbol{\varepsilon}_x = \boldsymbol{BP} \tag{6.66}$$

where $\boldsymbol{\varepsilon}_x$ is a $(N * N_s) \times 1$ matrix; $\boldsymbol{B}$ is a $(N * N_s) \times (N * N_p)$ matrix and $\boldsymbol{P}$ is a $(N * N_p) \times 1$ matrix.

$$\boldsymbol{\varepsilon}_x = \{\varepsilon_x(x_1, y_1, 1), \varepsilon_x(x_2, y_2, 1), \ldots \varepsilon_x(x_{N_s}, y_{N_s}, 1),$$

$$\varepsilon_x(x_1, y_1, 2), \ldots \varepsilon_x(x_{N_s}, y_{N_s}, N)\}^T;$$

$$\boldsymbol{P} = \{p_1(0), p_2(0), \ldots, p_{N_p}(0), p_1(1), \ldots, p_{N_p}(N-1)\}^T; \tag{6.67}$$

$$\boldsymbol{B} = \begin{bmatrix} \boldsymbol{B}_{10} & 0 & \cdots & 0 \\ \boldsymbol{B}_{20} & \boldsymbol{B}_{21} & \cdots & 0 \\ \vdots & \vdots & \vdots & \vdots \\ \boldsymbol{B}_{N0} & \boldsymbol{B}_{N1} & \cdots & \boldsymbol{B}_{NN-1} \end{bmatrix}_{(N_s \cdot N) \times (N_p \cdot N)} ; \quad \boldsymbol{B}_{mmk} = \begin{bmatrix} b_{11} & b_{12} & \cdots & b_{1N_p} \\ b_{21} & b_{22} & \cdots & b_{2N_p} \\ \cdots & \cdots & \vdots & \cdots \\ b_{N_s1} & b_{N_s2} & \cdots & b_{N_sN_p} \end{bmatrix}_{N_s \times N_p}$$

$$b_{sl} = z_t \Delta t \sum_{m=1}^{MM} \sum_{n=1}^{NN} \frac{1}{M_{mn}\omega'_{mn}} \left(\frac{m\pi}{a}\right)^2 Y_{mn}(y_s) \sin\left(\frac{m\pi}{a}x_s\right) \tag{6.68}$$

$$\times e^{-\zeta_{mn}\omega_{mn}(mm-k)\Delta t} \sin(\omega'_{mn}(mm-k)\Delta t) Y_{mn}(\hat{y}_l(k\Delta t)) \sin\left(\frac{m\pi}{a}\hat{x}_l(k\Delta t)\right)$$

$$(mm = 1, 2, 3, \ldots, N; \quad k = 0, 1, 2, \ldots, N-1;$$

$$s = 1, 2, \ldots, N_s; \quad l = 1, 2, \ldots, N_p)$$

Since the unknown force vector $\boldsymbol{P}$ is not a continuous function of the measured data, a regularization method is used to solve this ill-posed problem. The load identification problem can be formulated in a similar form as the damped least-squares problem of Equation (6.56) as:

$$\min J(\boldsymbol{P}, \lambda) = (\boldsymbol{\varepsilon}_x - \boldsymbol{BP}, R(\boldsymbol{\varepsilon}_x - \boldsymbol{BP})) + \lambda(\boldsymbol{P}, \boldsymbol{P}) \tag{6.69}$$

where λ is the non-negative regularization parameter in the form of a diagonal matrix with each one of the diagonal element corresponds to one of the forces to be identified. R is a weight matrix determined from the measured information.

6.3.2.2 Identification from Accelerations

The acceleration at a point (x, y) and time t obtained from Equation (6.62) is:

$$\ddot{w}(x,y,t) = \sum_{m=1}^{\infty} \sum_{n=1}^{\infty} Y_{mn}(y) \sin\left(\frac{m\pi}{a}x\right) \frac{1}{M_{mn}} \left[f_{mn}(t) + \int_0^t \ddot{H}_{mn}(t-\tau) f_{mn}(\tau) d\tau \right]$$

(6.70)

where

$$\ddot{H}_{mn}(t) = \frac{1}{\omega'_{mn}} e^{-\zeta_{mn}\omega'_{mn}t} \{[(\zeta_{mn}\omega_{mn})^2 - \omega'^2_{mn}] \sin \omega'_{mn} t - 2\zeta_{mn}\omega_{mn}\omega'_{mn} \cos \omega'_{mn} t\}$$

(6.71)

The acceleration at measuring point (x_s, y_s) can be written in discrete form including the $MM \times NN$ modes as:

$$\ddot{w}(x_s, y_s, mm) = \sum_{m=1}^{MM} \sum_{n=1}^{NN} Y_{mn}(y_s) \sin\left(\frac{m\pi}{a}x_s\right) \frac{1}{M_{mn}}$$

$$\times [f_{mn}(mm) + \sum_{k=0}^{mm} \ddot{H}_{mn}(mm-k) f_{mn}(k)\Delta t]$$

$$(s = 1, 2, \ldots, N_s; \quad mm = 1, 2, \ldots, N)$$

(6.72)

$$\ddot{H}_{mn}(k) = \frac{1}{\omega'_{mn}} e^{-\zeta_{mn}\omega_{mn}k\Delta t} \{[(\zeta_{mn}\omega_{mn})^2 - \omega'^2_{mn}] \sin(\omega'_{mn}k\Delta t)$$

$$-2\zeta_{mn}\omega_{mn}\omega'_{mn} \cos (\omega'_{mn}k\Delta t)\}$$

$$f_{mn}(k) = \sum_{l=1}^{N_p} p_l(k\Delta t) Y_{mn}(\hat{y}_l(k\Delta t)) \sin\left(\frac{m\pi}{a}\hat{x}_l(k\Delta t)\right)$$

$$(m = 1, 2, \ldots, MM; \quad n = 1, 2, \ldots, NN; \quad k = 0, 1, 2, \ldots, N-1) \quad (6.73)$$

Equation (6.72) can also be written in matrix forms as follow.

$$\ddot{w} = DP$$

(6.74)

where

$$\ddot{w} = \{\ddot{w}(x_1, y_1, 1), \ddot{w}(x_2, y_2, 1), \ldots, \ddot{w}(x_{N_s}, y_{N_s}, 1),$$

$$\ddot{w}(x_1, y_1, 2), \ldots, \ddot{w}(x_{N_s}, y_{N_s}, N)\}^T;$$

$$P = \{p_1(0), p_2(0), \ldots, p_{N_p}(0), p_1(1), \ldots, p_{N_p}(N-1)\}^T; \tag{6.75}$$

$$D = \begin{bmatrix} D_{10} & 0 & \cdots & 0 \\ D_{20} & D_{21} & \cdots & 0 \\ \vdots & \vdots & \vdots & \vdots \\ D_{N0} & D_{N1} & \cdots & D_{NN-1} \end{bmatrix}_{(N_s \cdot N) \times (N_p \cdot N)}; \quad D_{mmk} = \begin{bmatrix} d_{11} & d_{12} & \cdots & d_{1N_p} \\ d_{21} & d_{22} & \cdots & d_{2N_p} \\ \cdots & \cdots & \vdots & \cdots \\ d_{N_s 1} & d_{N_s 2} & \cdots & d_{N_s N_p} \end{bmatrix}_{N_s \times N_p}$$

$$d_{sl}^* = \Delta t \sum_{m=1}^{MM} \sum_{n=1}^{NN} \frac{1}{M_{mn}\omega_{mn}'} Y_{mn}(y_s) \sin\left(\frac{m\pi}{a} x_s\right) e^{-\zeta_{mn}\omega_{mn}(mm-k)\Delta t}$$

$$\times \{\sin(\omega_{mn}'(mm-k)\Delta t)[(\zeta_{mn}\omega_{mn})^2 - \omega_{mn}'^2]$$

$$-2\zeta_{mn}\omega_{mn}\omega_{mn}' \cos(\omega_{mn}'(mm-k)\Delta t)\} Y_{mn}(\hat{y}_l(k\Delta t)) \sin\left(\frac{m\pi}{a}\hat{x}_l(k\Delta t)\right)$$

$$(mm = 1, 2, 3, \ldots, N; \quad k = 0, 1, 2, \ldots, N-1;$$

$$s = 1, 2, \ldots, N_s; \quad l = 1, 2, \ldots, N_p) \tag{6.76}$$

When $k < mm$,

$$d_{sl} = d_{sl}^*;$$

When $k = mm$,

$$d_{sl} = d_{sl}^* + \sum_{m=1}^{MM} \sum_{n=1}^{NN} Y_{mn}(y_s) \sin\left(\frac{m\pi}{a}\right) Y_{mn}(\hat{y}_l(mm\Delta t)) \sin\left(\frac{m\pi}{a}\hat{x}_l(mm\Delta t)\right)/M_{mn}$$

Again the load identification problem can be formulated as a damped least-squares problem.

$$\min J(P, \lambda) = (\ddot{w} - DP, R(\ddot{w} - DP)) + \lambda(P, P) \tag{6.77}$$

The moving loads are determined from Equations (6.69) and (6.77) using either strains or accelerations or both. The methods to determine the optimal regularization parameter λ are referred to in Section 5.2.5 of Chapter 5.

6.3.2.3 Computation Algorithm

The computational process is implemented as follows:

1. The sampling frequency is determined basing on the maximum exciting frequency generated by the moving loads, the number of mode shapes $MM \times NN$ of the supporting structure and the number of measuring points N_s;
2. The natural frequencies ω_{mn} and the mode shapes $W_{mn}(x, y)$ of the orthotropic rectangular plate are calculated according to the formulation in Chapter 2;

3.   Matrix B_{mmk} and D_{mmk} are calculated from Equations (6.68) and (6.76);
4.   Set initial regularization parameter λ equals to zero.
5.   Calculate $P(0)$ from

$$P(0) = (B_{10}^T B_{10} + \lambda I)^{-1} B_{10}^T \boldsymbol{\varepsilon}_x(1) \quad \text{or}$$
$$P(0) = (D_{10}^T D_{10} + \lambda I)^{-1} D_{10}^T \ddot{w}(1)$$

(6.78)

where

$$\boldsymbol{\varepsilon}_x(j) = \{\varepsilon_x(x_1, y_1, j), \varepsilon_x(x_2, y_2, j), \ldots, \varepsilon_x(x_{N_s}, y_{N_s}, j)\}^T;$$
$$\ddot{w}(j) = \{\ddot{w}(x_1, y_1, j), \ddot{w}(x_2, y_2, j), \ldots, \ddot{w}(x_{N_s}, y_{N_s}, j)\}^T; \quad (j = 1, 2, \ldots, N)$$
$$P(i) = \{p_1(i), p_2(i), \ldots, p_{N_p}(i)\}^T; \quad (i = 0, 1, 2, \ldots, N-1)$$

6.   Calculate $P(k)$ from:

$$P(k) = (B_{(k+1)K}^T B_{(k+1)k} + \lambda I) B_{(k+1)k}^T \left(\boldsymbol{\varepsilon}_x(k+1) - \sum_{i=0}^{k-1} B_{(i+1)i} P(i) \right) \quad \text{or}$$

$$P(k) = (D_{(k+1)K}^T D_{(k+1)k} + \lambda I) D_{(k+1)k}^T \left(\ddot{w}(k+1) - \sum_{i=0}^{k-1} D_{(i+1)i} P(i) \right)$$

(6.79)

$$(k = 1, 2, \ldots, N-1)$$

Calculate one of the following parameters: error in the identification, generalized cross validation (GCV) value, or the curvature of L-curve.
7.   Calculate the parameter in Steps 5 to 7 for an increment of regularization parameter until the error or GCV value is smaller than a given value or the curvature of the L-curve is larger than a given value. The optimal regularization parameter is obtained.
8.   The moving loads can then be calculated from Steps 5 and 6 with the optimal regularization parameter.

6.3.3   Numerical Examples

6.3.3.1   Beam Model

The same beam and moving forces described in Section 5.3 are used for this study. The forces are moving on top of a 40 m long simply supported beam and they are repeated below for clarity of presentation.

$$\begin{cases} p_1(t) = 20000[1 + 0.1 \sin(10\pi t) + 0.05 \sin(40\pi t)] \\ p_2(t) = 20000[1 - 0.01 \sin(10\pi t) + 0.05 \sin(50\pi t)] \end{cases}$$

(6.80)

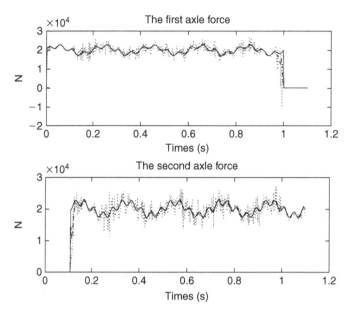

Figure 6.9 Identified forces from accelerations and strains with 5 percent noise (— true forces; - - - from accelerations; · · · · from strains)

The parameters of the beam are the same as those shown in Section 5.3. White noise is added to the calculated responses of the beam to simulate the polluted measurements and 1, 5, 10 and 20 percent noise levels are studied with:

$$
\begin{cases}
\varepsilon = \varepsilon_{calculated} + E_P * N_{oise} * \sigma(\varepsilon_{calculated}) \\
\ddot{w} = \ddot{w}_{calculated} + E_P * N_{oise} * \sigma(\ddot{w}_{calculated})
\end{cases}
\tag{6.81}
$$

where $\ddot{w}$ and ε are the acceleration and strain, respectively. E_p is the noise level. N_{oise} is a standard normal distribution vector with zero mean value and unit standard deviation. $\ddot{w}_{calculated}, \varepsilon_{calculated}$ are the calculated acceleration and strain, and $\sigma(\ddot{w}_{calculated}), \sigma(\varepsilon_{calculated})$ are their standard deviations.

Since the true forces are known, the errors in the identified forces are calculated as:

$$
Error = \frac{\|P_{identified} - P_{True}\|}{\|P_{True}\|} \times 100\%
\tag{6.82}
$$

The first six modes are used in the simulation. The time interval between two adjacent data is 0.002 s. Six measuring points are evenly distributed on the beam at $1/7L$ spacing, in which both acceleration and strain are measured. Figure 6.9 shows the identified results with 5 percent noise level in the responses. The curves obtained from accelerations are noted to be more accurate than those from strain with less fluctuation.

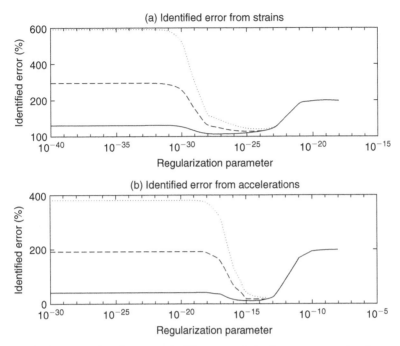

Figure 6.10 Errors in identification with different regularization parameters (— 1 percent noise; --- 5 percent noise; ···· 10 percent noise)

The error in the identified forces is plotted against the regularization parameter in Figure 6.10. The error increases for a given λ with an increase in the noise level. The optimal value of λ increases with increasing noise level. This means that a larger regularization parameter should be used to reduce large noise effect in the solution. Since the true loads are known, the curves in Figure 6.10 can be used to determine the optimal regularization parameter. Figures 6.11 and 6.12 show the GCV function and the L-curve, respectively. A wide range of minimum is available from the GCV function, but the optimal value can be obtained easily from the corner of the L-curve. All the three methods above yield approximately the same regularization parameter. In practice, the GCV and L-curve criteria can be used to obtain the optimal value and they are also found robust against random noise.

Table 6.3 shows the errors in the identified forces obtained from the Time Domain Method (TDM) (Law et al., 1997), Frequency and Time Domain Method (FTDM) (Law et al., 1999) and the EST method presented in this chapter. The example with the two moving forces in Section 5.3.1 is adopted in the comparison with the first three vibration modes of the beam. It is also noted that no regularization has been used in the first two methods. The bending moments, strains and accelerations at $1/4L$, $1/2L$ and $3/4L$ are used in the identification. The errors from using TDM or FTDM are found in general larger and more sensitive to the noise level in the responses than the EST method. The regularization technique is seen to be an effective tool to solve the ill-posed problem in the moving load identification.

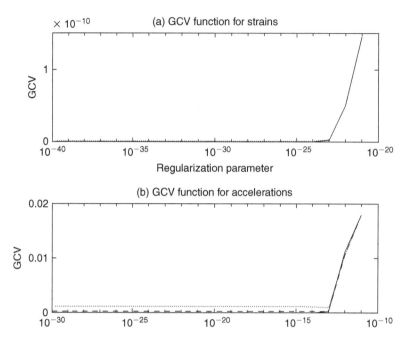

Figure 6.11   GCV function (— 1 percent noise; - - - 5 percent noise; · · · · 10 percent noise)

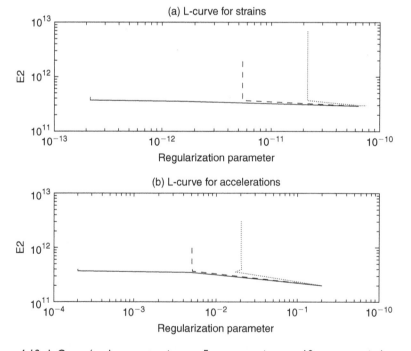

Figure 6.12   L-Curve (— 1 percent noise; - - - 5 percent noise; · · · · 10 percent noise)

Table 6.3 Errors (percent) in identification from using different methods

Noise Level	Loads	TDM		FTDM		EST method	
		M	A	M	A	S	A
1%	P_1 (t)	230.48	1.76	41.84	1.41	14.23	3.94
	P_2 (t)	157.52	4.91	44.01	1.97	14.41	4.87
5%	P_1 (t)	+	18.79	232.61	6.33	21.36	10.05
	P_2 (t)	+	16.54	397.20	11.89	21.24	11.41
10%	P_1 (t)	+	40.19	224.76	32.26	25.06	11.33
	P_2 (t)	+	33.52	630.01	49.34	25.53	13.38
20%	P_1 (t)	+	+	49.21	65.11	30.65	15.49
	P_2 (t)	+	72.68	882.70	71.85	31.02	18.38

Note: M – Bending Moment; A – Acceleration; S – Strain. + denotes error larger than 1000 percent.

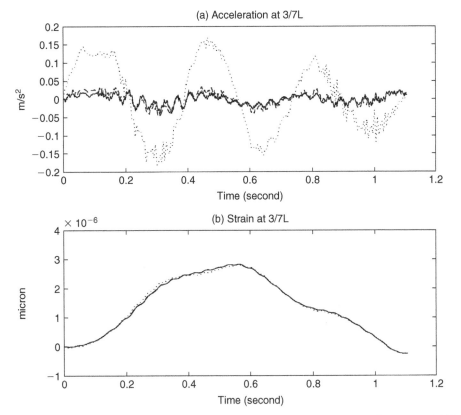

Figure 6.13 Acceleration and strain from different time intervals (— 0.001 s; --- 0.002 s; ···· 0.005 s)

Study 1: Effect of Sampling Frequency

The sampling frequency is an important parameter in the identification. Figure 6.13 shows the accelerations and strains computed from using different time intervals in

Table 6.4 Errors in identification with different number of modes

Data Type	Noise Level (%)	Loads	Number of Modes					
			3	4	5	6	7	8
Strain	1	$P_1(t)$	14.23	9.38	4.02	2.30	1.52	1.49
		$P_2(t)$	14.41	9.88	4.01	2.07	1.72	1.67
	5	$P_1(t)$	21.36	16.60	12.26	7.99	6.84	6.28
		$P_2(t)$	21.24	16.54	12.98	8.17	6.88	6.88
	10	$P_1(t)$	25.06	20.52	16.18	13.58	12.20	11.32
		$P_2(t)$	25.53	20.59	17.03	14.10	12.97	12.42
	20	$P_1(t)$	30.65	25.81	21.40	18.78	17.47	16.83
		$P_2(t)$	31.02	26.35	22.58	19.56	18.06	17.59
Acceleration	1	$P_1(t)$	3.94	9.50	0.87	0.76	0.49	0.58
		$P_2(t)$	4.87	7.90	0.94	0.65	0.46	0.46
	5	$P_1(t)$	10.05	11.20	3.66	2.17	1.62	1.40
		$P_2(t)$	11.41	10.18	3.75	2.05	1.52	1.37
	10	$P_1(t)$	11.33	12.04	5.60	3.14	2.86	2.39
		$P_2(t)$	13.38	10.73	5.68	3.15	2.59	2.55
	20	$P_1(t)$	15.49	13.90	8.78	5.18	4.88	4.31
		$P_2(t)$	18.38	11.74	8.50	5.45	4.82	4.84

Equations (6.45) and (6.52) respectively. The strain time histories remain almost the same as long as the sampling frequency is larger than or equal to twice the maximum frequency of the modes used in the calculation and the frequency components in the responses. But the acceleration time histories are different with different sampling frequency. This is because the dynamic behavior of the beam under moving loads is dominated by transient responses. The impulsive response components in the strains in Equation (6.44) decreases quickly as the frequency increases, but that in the acceleration in Equation (6.49) is different due to the presence of ω_i on the right-hand-side of the equation. The series in Equation (6.48) is divergent and truncation of the series with only a few modes would omit the significant contribution from the frequency components in the higher modes. This means large sampling frequency should be used with accelerations. But in practice, the signal to noise ratio is very small in the high frequency range and the errors in the identified forces are therefore larger than those identified from strains. This will be discussed further in the next study.

Study 2: Effect of different amount of modal information in identification

The effect of using different amount of modal information in the responses on the identification is studied. Table 6.4 shows the errors in the moving loads identified using different number of modes in the responses. The number of vibration modes in the responses is taken equal to the number of modes in the identification in Equations (6.47) and (6.54). The number of measuring points selected match the number of modes and they are evenly located on the beam. The errors in the identified moving loads decrease as the number of modes in the identification increases, and no further

Table 6.5 Errors (percent) in identification with different number of mode shapes in responses

Number of Modes	Data Type	Varying Component		Constant Component		Total Load	
		First load	Second load	First load	Second load	First load	Second load
6	S	0.06	0.12	0.20	0.40	0.20	0.39
5	S	0.72	0.84	0.95	0.85	0.73	1.14
4	S	1.16	0.97	1.04	0.75	1.02	0.73
3	S	1.12	1.49	1.56	1.72	1.55	1.70
2	S	106.02	92.92	2.59	3.41	2.36	3.27
6	A	0.06	0.07	0.20	0.20	0.20	0.20
5	A	5.27	7.30	3.81	4.08	3.44	3.95
4	A	13.47	15.98	13.15	14.58	14.82	15.47
3	A	24.19	25.77	10.60	8.08	10.06	9.06
2	A	18.62	37.44	69.93	77.54	68.87	78.79

Note: S – identified from strains; A – identified from accelerations.

significant reduction in the error can be achieved with the number of sensors greater than seven. This residual amount of error is due to the existence of large variations at both the beginning and the end of the identified load time histories.

Study 3: Modal Truncation in computation

In practice the number of vibration modes in the responses is much larger than the number of modes used in the identification in Equations (6.47) and (6.54), and the effect of modal truncation in the computation is studied. The 'measured' data contains responses from the first six modes and no noise effect is included. Six measuring points are used and they are evenly located on the beam. Table 6.5 shows the identified errors from strains and accelerations with different number of mode shapes in the computation. When six modes are used in the identification, close to zero errors can be found in the identified forces indicating the accuracy of the proposed method as an exact solution. The optimal regularization parameter λ equals 0.0 in these cases indicating the results are the least-squares solution (the small error is due to discretization of the mode shapes and the zero force assumption at the first and last point of the force history).

The modal truncation effect is studied in terms of the identification of a constant force, a force varying about the zero mean and an ordinary force as described in Equation (6.80). The identified results on the constant component and the varying component of the forces in Equation (6.80) and the whole forces are shown in Figure 6.14. Only the first three modes are used in the computation.

When the number of the mode shapes included in the identification is not less than three, acceptable results can be obtained from both strains and accelerations. This is because the third natural frequency (28.8 Hz) is already larger than the highest frequency of the moving loads (25 Hz). The first three modes constitute the main components in the responses. The number of modes included in the identification in

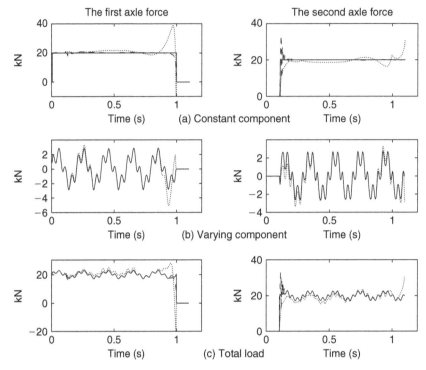

Figure 6.14 Identified forces from three modes (— true forces; - - - from strains; · · · · from accelerations)

Equations (6.47) and (6.54) should be at least equal to the number of contributing modes in the responses in order to capture all the available information.

The errors in the identified loads from accelerations are more sensitive to the number of modes than that from strains. In practice, more modes are excited in the responses with impulses at the beginning and end of the time histories, especially in the case of constant moving loads. Therefore more mode shapes should be used in the identification from accelerations.

Figure 6.14 shows that there are errors in the identified results, even if the first three modes have covered the frequency range of the moving loads. This is due to the fact that the responses from moving loads consist of transient responses represented by the exponential terms in Equations (6.44) and (6.49). This is different from, and more complicated when compared with, general stationary load identification.

Study 4: Effect of Modeling Errors

In practice, the modal parameters are obtained from modal testing. If they are obtained through finite element analysis, they should be updated using the measured data, but some errors should have remained in the modal parameters. Figure 6.15 shows the identified results from strains and accelerations with 1 percent random noise in the

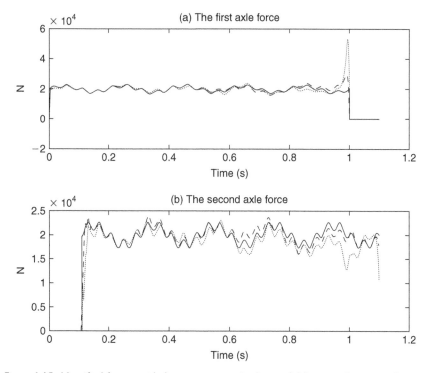

Figure 6.15 Identified forces with 1 percent error in the model (— true forces; — from strains; ···· from accelerations)

Table 6.6 Errors in identification with different errors in model

Data Type	Model Error (%)	Noise Level							
		0%		1%		5%		10%	
Strains	1	4.88	3.66	5.09	4.12	9.12	8.71	13.70	14.15
	5	15.22	14.90	15.24	14.96	15.38	15.88	16.17	17.50
Accelerations	1	6.24	7.82	6.06	7.45	5.22	6.14	4.95	5.14
	5	24.61	36.56	24.48	36.33	23.99	35.41	23.43	34.31

modal frequencies and mode shapes. This is considered sufficient since the eigenvalues and eigenvectors are not sensitive to local errors in the model. The first six modes are used in the identification. The figure shows that acceptable results can still be obtained when the error in the model is small, but identification from accelerations would suffer a larger reduction in the accuracy. Table 6.6 shows the errors in the identification with different magnitudes of the modeling error and different noise levels in the responses. The effect of error in the model is seen to be more significant than the noise level in the responses. A relatively accurate model is therefore a requirement in the moving load identification.

Table 6.7 Natural Frequencies (Hz) of beam and plate with different width

Width of plate	x-mode number	Mode type	M 1	2	3	4	5	6
Beam model (0.1 m width)		B	4.32	17.27	38.86	69.08	107.94	155.43
Plate model	0.1 m	B	4.22	16.89	38.01	67.59	105.66	152.23
	0.2 m	B	4.22	16.90	38.07	67.76	106.03	152.94
		1st T					107.54*	
	0.3 m	B	4.22	16.92	38.14	67.97	106.50	153.80
		1st T				71.77*	144.35*	
	0.4 m	B	4.23	16.94	38.24	68.23	107.01	154.68
		1st T			53.91*		108.89*	
	1.8 m	B	4.27	17.35	39.38	70.36	110.29	159.17
		1st T	12.72*	29.68*	51.57*	85.26*	125.55*	
		2nd T			53.46*	97.36*	128.93*	
		3rd T			70.22*		132.16*	
		4th T					146.95*	

Note: * denotes the natural frequency corresponds to mode shape mainly in y-direction. (B: bending mode; 2nd T: second torsional mode)

6.3.3.2 Two-dimensional Plate Model

The beam model versus the plate model

A beam model for the bridge deck can be derived from Equations (6.61) and (6.62). The displacement at a point along the central line of the bridge deck with the loads moving along $y = e$ can be obtained from Equation (6.62) as follows:

$$w\left(x, \frac{b}{2}, t\right) = \sum_{m=1}^{\infty} \sum_{n=1}^{\infty} C_{bmn} \sin\left(\frac{m\pi}{a}x\right) \frac{1}{M_b} \int_0^t H_{mn}(t - \tau) f_{bm}(\tau) d\tau \tag{6.83}$$

with

$$C_{bmn} = \frac{b Y_{mn}\left(\frac{b}{2}\right) Y_{mn}(e)}{\int_0^b Y_{mn}^2(y) dy}; \quad M_b = \frac{\rho h a b}{2};$$

$$f_{bm} = \sum_{l=1}^{N_p} p_l(t) \sin\left(\frac{m\pi}{a} \hat{x}_l(t)\right) \tag{6.84}$$

where e is the eccentricity of the moving load, and h is the thickness of the beam.

If $n = 1$, i.e. the torsional modes are not considered, Equation (6.83) is the same as that for the displacement of an equivalent beam. Therefore the identification can be simplified using a beam model when $n = 1$. Table 6.7 shows the natural frequencies of an isotropic plate with two simply supported edges and two free edges. The length of the plate is 3.678 m and the thickness is 0.025 m. The width of the plate varies from 0.1 m to 1.8 m. The lowest several modes of the plate mainly consist of longitudinal modes in the x-direction with $n = 1$.

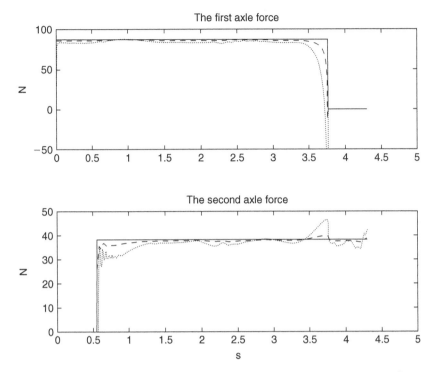

Figure 6.16 Identified forces by beam model and plate model (— Plate model; - - - Beam model (0.1 m wide); ······ Beam model (0.4 m wide))

Two constant forces 87.25 N and 38.25 N are moving across the plate along $y = 0$ at 1.0 m/s at a fixed spacing of 0.557 m. The sampling rate is 100 Hz. Figure 6.16 shows the identified forces from accelerations at 1/4a, 1/2a, 3/4a and 1/2b of a 0.1 m and 0.4 m wide plates using the beam and the plate models, and the lowest three modes with $n = 1$ are used. The resulting curves for the plate model overlap and exactly match those of the true forces without error. The three sets of curves are very close to each other, except near the start and end of the time histories. The plate model can identify exactly the forces using the proposed method, while the result from the beam model deteriorates for a larger width of the plate used in the model. The modal frequencies in Table 6.7 indicate that the beam model would be accurate enough for identifying the moving loads when the set of modes of the plate for identification consists of bending modes only with $n = 1$.

Two moving loads identification with a plate model

The simply supported beam-slab bridge deck model shown in Figure 3.2 is reproduced in Figure 6.17. The parameters of the bridge deck are the same as those shown in Section 3.2.3.2.

The two moving loads to be identified are:

$$\begin{cases} p_1(t) = 150000 * (1 + 0.1 \sin 10\pi t + 0.05 \sin 40\pi t) \text{ N}; \\ p_2(t) = 150000 * (1 - 0.1 \sin 10\pi t + 0.05 \sin 50\pi t) \text{ N}. \end{cases} \tag{6.85}$$

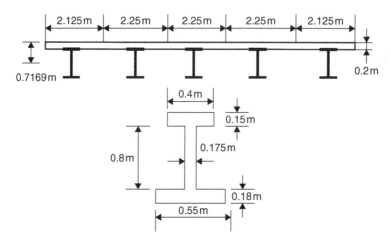

Figure 6.17 Cross-section of the orthotropic plate

Note that there is an out-of-phase component in the forces simulating the pitching motion of a vehicle. White noise is added to the calculated displacements due to the moving loads to simulate the polluted measurement as:

$$\begin{pmatrix} \ddot{w} = \ddot{w}_{calculated}(1 + E_p * N_{oise}) \\ \varepsilon = \varepsilon_{calculated}(1 + E_p * N_{oise}) \end{pmatrix} \tag{6.86}$$

Calculations are made for the loads moving at a fixed spacing of 4 m along the central line and along $y = 3/8b$ respectively. The moving speed is 10 m/s and the sampling rate is 100 Hz. The lowest nine vibration modes are used in the simulation and nine measurement points are located at 1/4a, 1/2a and 3/4a on the second, third and the fourth I-beams. The number of measuring points is taken equal to the number of the vibration modes (Zhu and Law, 1999).

Figure 6.18 shows the identified forces moving along the central line from the accelerations and the strains with 1 percent noise level. Very good results are achieved except at the start and end of the time histories. There is a large deviation between the true load and the curves from strain when the load is near the mid-span of the beam. This is due to the noise effect that increases with the small response when the force traverses the mid-span of the structure where the second longitudinal modal responses are smallest. The observation contrasts with the identified results when there is no noise in the responses and the identified forces exactly match those of the true forces. The curves from acceleration exhibit no such large differences, because the acceleration responses remain relative stable throughout the duration. Those from accelerations almost match the true curves perfectly. Table 6.8 shows the errors in the identified forces with no smoothing on data at different noise levels. Accelerations give much better results than strains at different noise levels, and the identification of eccentric load using this set of sensors gives slightly larger errors than the loads along the central line. This may be due to the smaller responses at the sensor locations caused by the eccentric loads.

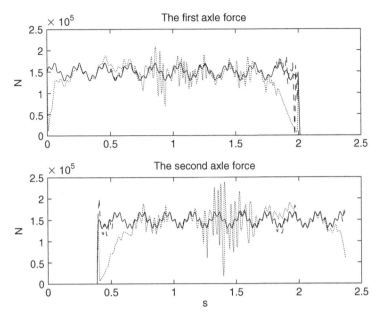

Figure 6.18 Identified forces from accelerations and strains (— True Load; --- Acceleration; ···· Strains)

Table 6.8 Errors in the identified forces with different noise levels

Eccentricity	Responses	N_{oise} (%)	First force (%)	Second force (%)
0	Acceleration	1	2.74	2.53
		5	12.30	11.25
		10	18.97	17.81
	Strain	1	15.92	17.69
		5	27.44	29.42
		10	34.71	37.00
1/8b	Acceleration	1	4.30	4.16
		5	15.29	14.20
		10	22.44	21.24
	Strain	1	16.60	18.48
		5	28.51	30.48
		10	35.97	38.13

Further work has to be done on the best sensor locations for identifying loads moving on different paths.

Figure 6.19 shows the identified loads moving along the central line from the strains only with or without three-points smoothing on the measured data with 1 percent noise level. It is seen that smoothing before the identification can improve the results significantly especially on the variation in the middle of the time histories. Table 6.9

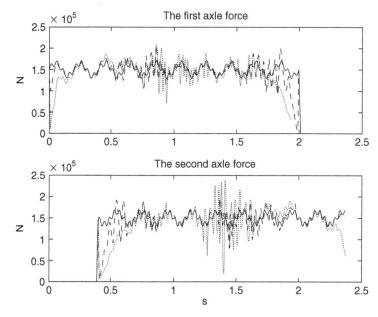

Figure 6.19 Identified forces from strains with or without smoothing (— True Load; - - - with smoothing; · · · · without smoothing)

Table 6.9 Errors in the identified forces with or without smoothing

Responses		N_{oise} (%)	First force (%)	Second force (%)
No Smoothing	Acceleration	1	2.74	2.53
		5	12.30	11.25
		10	18.97	17.81
	Strain	1	15.92	17.69
		5	27.44	29.42
		10	34.71	37.00
Smoothing	Acceleration	1	2.72	3.14
		5	9.24	9.43
		10	13.59	13.59
	Strain	1	9.07	10.06
		5	18.60	20.51
		10	24.76	27.01

also shows reduction in the errors over the whole time period when smoothing is used. The improvement is larger in the strains than in the accelerations.

Study 1: Effect of sensor locations on identification errors

The following two moving loads at 4 meter spacing are used in this study:

$$\begin{cases} p_1(t) = 150000 * (1 + 0.1 \sin 10\pi t + 0.05 \sin 40\pi t) \text{ N}; \\ p_2(t) = 150000 * (1 - 0.1 \sin 10\pi t + 0.05 \sin 50\pi t) \text{ N}. \end{cases} \tag{6.87}$$

Table 6.10 Errors (percent) in the identified forces from different measuring locations

Sensor Set	Noise Level %	From strains		From accelerations	
		First force	Second force	First force	Second force
I	0	0.00	0.49	0.00	0.00
	1	16.58	18.55	4.49	4.30
	5	28.79	30.92	14.92	14.62
	10	36.53	38.73	22.59	22.04
II	0	0.00	0.49	0.00	0.00
	1	16.60	18.48	4.30	4.16
	5	28.51	30.48	15.29	14.20
	10	35.97	38.13	22.44	21.24
III	0	0.00	0.49	0.00	0.00
	1	16.09	17.56	2.74	2.58
	5	27.59	29.54	12.60	17.56
	10	34.80	37.13	19.19	18.02

The group of loads is moving along $y = 3/8b$ at a speed of 10 m/s. The effectiveness of different sensor locations are studied with the following sets of sensors to provide data for the identification.

Set I: Nine sensors on the left three I-beams looking in the direction of travel;
Set II: Nine sensors on the middle three I-beams; and
Set III: Nine sensors on the right three I-beams.

The sensors are located at 1/4a, 1/2a and 3/4a on the I-beams. Table 6.10 also shows the errors in the identification with different noise levels. It is noted that the errors in the identified forces from the accelerations and the strains are very small when there is no noise in the responses. It shows that the proposed method and algorithm are correct, and the accelerations give more accurate results than strains. However the identified results are very noise sensitive. The errors in the identified forces are more or less the same from different sets of sensors, but slightly smaller errors are found from the measurement locations that are further away from the moving loads. This may be due to the higher sensitivity of these sensors to loads along $y = 3/8b$, which may in turn depend on the properties of the plate, such as the torsional rigidity D_{xy}. In fact a set of optimal sensor locations can be sought considering the properties of the plate and the signal to noise ratio of the measured responses.

Study 2: Identification of eccentric group of moving loads

Calculation is performed for the group of moving loads described in Equation (6.87) moving along paths at an eccentricity of 0.0, 1/8b and 3/8b separately at a speed of 10 m/s. Set II sensors described above are used. The errors in the identified forces from responses with different noise levels are shown in Table 6.11. Errors in the identified forces moving at an eccentric path are slightly larger than those for the loads moving along the centerline of the deck. This may be due to the smaller responses at these sensors. Figure 6.20 shows the identified results with 1 percent noise in the measured accelerations. The results match the true forces very closely, except at the beginning and end of the time histories. This is due to the discontinuity of the forces at these two points leading to large fluctuations in the identified results. If a smaller regularization parameter is used for the time segments at the beginning and the end, the errors at these two periods will be reduced (Choi and Chang, 1996).

Table 6.11 Errors (percent) in identified forces from Set II sensor locations

Path Eccentricity	Noise Level (%)	From strains		From accelerations	
		First force	Second force	First force	Second force
0	0	0.00	0.49	0.00	0.00
	1	16.46	18.25	3.79	3.60
	5	28.22	30.13	14.35	13.31
	10	35.55	37.73	21.65	20.16
1/8b	0	0.00	0.49	0.00	0.00
	1	16.60	18.48	4.30	4.16
	5	28.51	30.48	15.29	14.20
	10	35.97	38.13	22.44	21.24
3/8b	0	0.00	0.49	0.00	0.00
	1	16.90	18.96	4.76	5.05
	5	29.15	31.22	15.32	15.78
	10	36.92	39.19	23.46	23.04

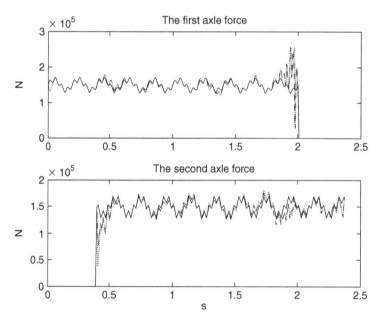

Figure 6.20 Identified results for groups of loads moving at different eccentricity (— True force; --- e = 1/8b; ···· e = 3/8b)

Study 3: Identification of multiple groups of loads

Nowak et al. (1993) have inspected three patterns of multiple vehicles on a bridge deck in his studies of transverse load distribution. They are:

1. In-lane: two vehicles following each other at a distance less than 15 m;
2. Side by side in tandem: two vehicles in adjacent lanes traveling with the front axles adjacent to each other; and
3. Side by side and behind: two vehicles in adjacent lanes but one behind the other with the distance between front axles less than 15 m.

The following loads are assumed to be moving on the bridge deck in a group spaced at 4 meters.

$$\begin{cases} p_1(t) = 75000 * (1 + 0.1 \sin 10\pi t + 0.05 \sin 40\pi t) \text{ N}; \\ p_2(t) = 75000 * (1 - 0.1 \sin 10\pi t + 0.05 \sin 50\pi t) \text{ N}. \end{cases} \quad (6.88)$$

Two groups of these forces are moving along paths at an eccentricity of $e = 1/8b$ and $e = 3/8b$ respectively. Three patterns of vehicles on the bridge deck are studied. They are:

Case 1: Side by side in tandem: Two groups of forces get on the bridge at the same time with the same speed of 10 m/s;

Case 2: Side by side and behind: Two groups of forces get on the bridge at the same time with different speeds of 10 m/s and 12 m/s along paths $e = 3/8b$ and $e = 1/8b$ respectively; and

Case 3: Side by side and behind: Two groups of forces get on the bridge with the group along $e = 3/8b$ one meter behind the group along $e = 1/8b$ and with speeds of 10 m/s and 12 m/s respectively.

Set II sensors and the lowest nine vibration modes are also used for the force identification.

The in-lane case of Nowak et al. (1993) is considered as a special case of Case 3. Table 6.12 shows the errors in the identified forces with different noise level in the

Table 6.12 Errors (percent) in the identified forces of two groups of moving loads

Vehicle pattern	Noise level (%)	From strains				From accelerations			
		First Group		Second Group		First Group		Second Group	
		1st axle	2nd axle	1st axle	2nd axle	1st axle	2nd axle	1st axle	2nd axle
Case 1	0	0.995	0.991	0.995	0.991	0.995	0.500	0.995	0.500
	1	19.0	19.3	18.2	18.3	24.0	19.9	10.55	10.69
	5	30.8	32.2	34.0	35.0	60.8	70.4	57.3	59.5
Case 2	0	0.995	0.991	0.597	0.001	0.994	0.500	0.597	0.588
	1	19.3	20.1	18.5	18.8	27.7	21.2	13.6	14.0
	5	30.6	33.4	33.7	34.0	102.3	77.9	36.8	35.9
Case 3	0	0.995	0.991	0.540	0.001	0.995	0.500	0.540	0.000
	1	19.4	19.9	18.6	18.8	19.9	15.7	19.0	10.4
	5	30.9	33.3	33.4	34.1	96.3	73.9	53.7	41.1

measured responses. The time histories of identified forces for all three cases are similar, the errors increase significantly with a small increase in noise and errors are more or less the same, except those from accelerations with 5 percent noise. This is due to significant coupling of the identified forces in each group, as seen in Figure 6.21 for

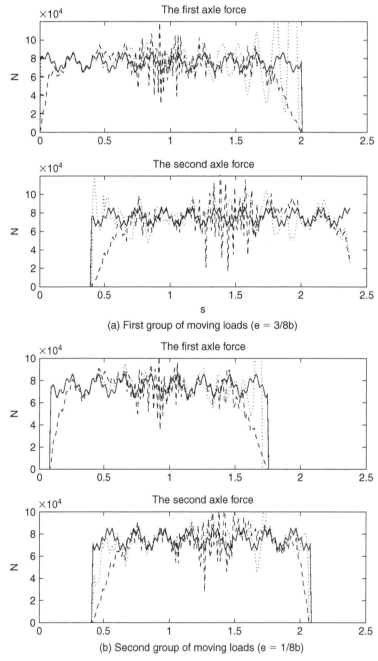

(a) First group of moving loads (e = 3/8b)

(b) Second group of moving loads (e = 1/8b)

Figure 6.21 Identified results of two groups of moving loads (pattern case 3, 1 percent noise) (— True force; - - - from strains; · · · · from accelerations)

Case 3 with 1 percent noise. Significant pitching motion of both groups of forces is found in the curves from accelerations between 1.3 s to 2.0 s in both groups. These errors contribute greatly to the overall errors in Table 6.12. Most of the errors in the curves from strains come from the variations when the forces traverse the mid-span of the bridge deck. This is due to the small responses in the sensors when the force traverses this point where the vibration from the second longitudinal bending modes is smallest.

Another computation is made for case 3 with three-point smoothing on the measured data containing 5 percent errors and the curves are shown in Figure 6.22. The smoothing is effective on the strains but not on the acceleration measurements, since the latter contains large coupling motion in the two forces in a group which cannot be eliminated through this smoothing process. Only the forces identified from strains are shown in Figure 6.22. The smoothing could significantly improve the accuracy of the results.

The ill-conditioned solution from accelerations could be improved either with a better regularization strategy on the time histories (Choi and Chang, 1996), or use an optimal set of sensors with maximum sensitivity of response to the moving loads. A sensor optimization technique has to be developed to meet this need.

Study 4: Identification of a group of four moving loads

The group of loads includes four moving loads representing the four wheel loads of a vehicle with an axle spacing of 4 meters and a wheel spacing of 2 meters.

$$
\begin{cases}
p_1(t) = 75000 * (1 + 0.1 \sin 10\pi t - 0.1 \sin 20\pi t + 0.05 \sin 40\pi t); \text{ N} \\
p_2(t) = 75000 * (1 + 0.1 \sin 10\pi t + 0.1 \sin 20\pi t + 0.05 \sin 40\pi t); \text{ N} \\
p_3(t) = 75000 * (1 - 0.1 \sin 10\pi t - 0.1 \sin 20\pi t + 0.05 \sin 40\pi t); \text{ N} \\
p_4(t) = 75000 * (1 - 0.1 \sin 10\pi t + 0.1 \sin 20\pi t + 0.05 \sin 40\pi t); \text{ N}
\end{cases}
\tag{6.89}
$$

where $p_1(t)$ and $p_2(t)$ are the left and right wheel loads of the front axle looking in the direction of the travelling path, and $p_3(t)$ and $p_4(t)$ are the left and right wheel loads of the second axle. The loads are travelling as a group along the central line of the bridge deck at a speed of 10 m/s. Sensor Set II and the lowest nine vibration modes are used for the identification. Figure 6.23 shows the identified forces with 1 percent noise. Both accelerations and strains can give acceptable results for all the four forces in the same group, and accelerations give better results than strains. Comparison with results from multiple group identification in last study shows that the interaction between adjacent forces is more serious in the multiple group identification than in the present four-loads identification.

6.3.4 Laboratory Studies

6.3.4.1 Beam Model

The cases with the vehicle moving at a uniform speed and with the vehicle braking on top of the supporting beam are studied.

Experimental Setup

The experimental setup is shown diagrammatically in Figure 6.24. It is similar to the setup shown in Figure 5.11 but with a different sensor layout. The main beam, 3678 mm long with a 100 mm × 25 mm uniform cross-section, is simply supported.

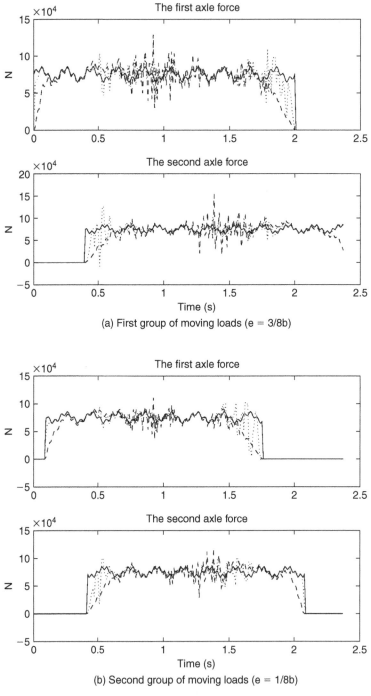

Figure 6.22 Identified results from strains of two groups of moving loads (pattern case 3, 1 percent noise) (— True force; - - - without smoothing; · · · · with smoothing)

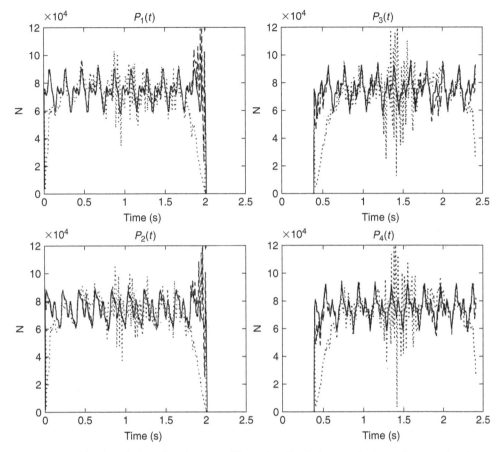

Figure 6.23 Identified results of a group of four wheel loads (— true force; - - - from accelerations; · · · · from strains)

There is a leading beam for accelerating the vehicle and a tailing beam to accept the vehicle when it comes out of the main span. A U-shaped aluminum section is glued to the upper surface of the beams as a direction guide for the car. The model car is pulled along the guide by a string wound around the drive wheel of an electric motor. Thirteen photoelectric sensors are mounted on the beams at an approximately equal spacing to measure and monitor the moving speed of the car. Seven strain gauges are evenly distributed on the beam at $1/8L$ spacing. A TEAC 14-channels magnetic tape recorder and an 8-channel dynamic testing and analysis system are used for data collection and analysis in the experiment. The responses are sampled at 2000 Hz with anti-aliasing filter. The recorded length of each test lasts for six seconds. The model car has two axles at a spacing of 0.557 m and it runs on four steel wheels with rubber band on the outside. The mass of the whole car is 16.6 kg with the front axle load and the rear axle load weighing 9.8 kg and 6.8 kg respectively. Braking force is applied with a set of rubber bands placed transversely in front of the vehicle approximately at the level of its centroid, and the braking force was tuned by adjusting the tension in the rubber

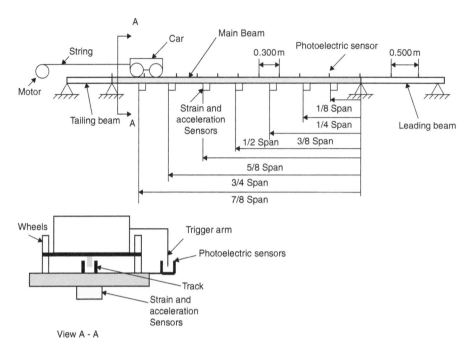

Figure 6.24 Diagrammatic drawing of experimental setup

band. The effect of non-uniform speed on the identified results when identified using a constant speed is investigated. Since the use of acceleration requires a large number of measured modes and measuring points (Zhu and Law, 2001), only the measured strains are used in the following studies.

Experimental Results

Identification from strains with uniform speed
The first experiment is conducted with the vehicle moving approximately at 1.25 m/s, and the first three modes of the beam are used in the identification. The collected strains are re-sampled at 200 Hz to include the first three vibration modes of the beam at 3.67 Hz, 16.83 Hz, and 37.83 Hz respectively. Figure 6.25 shows the identified results with sets of strains at 1/4L, 1/2L and 3/4L and from 1/8L, 1/2L and 7/8L respectively. The reconstructed strains at 5/8L are shown in Figure 6.26. The following observations are made:

1. The mean values of the identified combined loads are close to and varying around the static load. The fluctuations in the combined load are typically arising from the inertia effect of the moving masses.
2. The reconstructed strains at 5/8L are close to the measured strains. The correlation coefficients between them are 0.986 and 0.968 respectively for the two sets of strains. These also indicate that the method is correct.

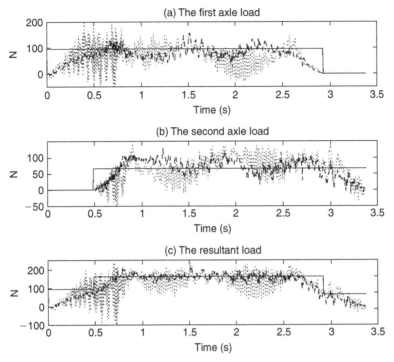

Figure 6.25 Identified results from strains at different measuring points (— static force; - - - from strains at 1/4L, 1/2L, 3/4L; · · · · from strains at 1/8L, 1/2L, 7/8L)

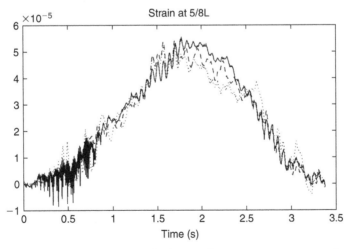

Figure 6.26 Reconstructed strain at 5/8L from different measuring points (— static force; - - - from strains at 1/4L, 1/2L, 3/4L; · · · · from strains at 1/8L, 1/2L, 7/8L)

Table 6.13 Transient speed in experiment

Range of distance (m)	0.0 ~ 0.478	0.478 ~ 0.878	0.878 ~ 1.178	1.178 ~ 1.478	1.478 ~ 1.778
Velocity (m/s)	1.215	1.305	1.271	1.250	1.245
Range of distance (m)	1.778 ~ 2.078	2.078 ~ 2.378	2.378 ~ 2.678	2.678 ~ 3.178	3.178 ~ 3.678
Velocity (m/s)	1.139	1.154	1.149	1.129	1.119

3. The peak of the reconstructed strain is smaller than that of the measured strain. This is because only a small number of modes are used in the identification as discussed in the numerical simulations in Section 6.3.
4. The identified results from strains at 1/4L, 1/2L and 3/4L are similar to those from 1/8L, 1/2L and 7/8L but with less fluctuation. This is because the signal to noise ratios at 1/8L and 7/8L are lower than those at 1/4L and 3/4L. Therefore the selection of measuring point is very important for an accurate identification.

Identification from strains with braking
The parameters are the same as for the above experiment. The moving car starts braking at 0.878 m. The actual speed between adjacent pairs of photoelectric sensors is shown in Table 6.13 with an average speed of 1.19 m/s. The moving car accelerates before braking and it crosses the beam completely with deceleration. Figure 6.27 shows the identified loads from strains at 1/4L, 1/2L and 3/4L for the car moving at the true and the averaged speed. The reconstructed strains at 5/8L are also compared with the measured strain. The correlation coefficients between them are respectively 0.983 and 0.985 for the cases of true and averaged speeds. The following observations are made:

1. Large fluctuations are found in the identified loads in Figure 6.27. This is due to the pitching motion induced by the eccentric braking force. Impulsive interaction forces generated from braking would also cause these fluctuations.
2. The combined load is close to the static load after braking as shown. This shows that braking mainly induces the pitch motion of the car through a re-distribution of the axle loads.
3. The identified individual loads differ slightly when the true or average speed is used, but there is virtually no difference in the combined load. This means that the proposed method can be used to identify accurately the combined load of a moving vehicle from the bridge responses even with braking on the bridge.
4. The second axle load identified from using averaged speed is always smaller than that from using the true speed, while that for the first axle load is reversed (Figure 6.27). This is because the true resultant load is always in front of the location at which the combined force is computed in the case of deceleration. The identified combined force tends to match the true force to have minimum error in the identification, and this means a shift of the centroid forward causing an increase

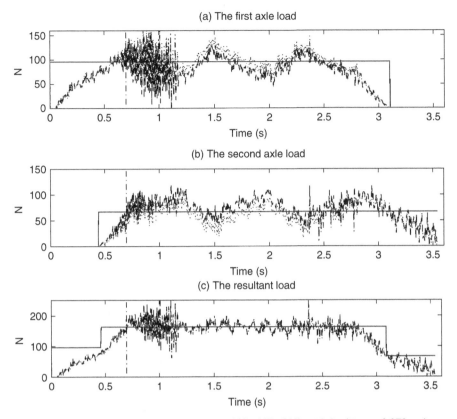

Figure 6.27 Identified results from strains at 1/4L, 1/2L, 3/4L with braking at 0.878 m (— static force; - - - identified from varying speed; · · · · identified from average speed)

in the first axle load and a decrease in the second axle load. There is very small difference in the curves close to the ends of the beam because the difference in the locations between the true and the identified combined forces is very small.

6.3.4.2 Plate Model

Experimental Set-up

An experimental system design includes the model design and measuring system design. The experimental model in this study simulates the vehicle–bridge system which includes the interaction between the vehicle and the bridge. A one-tenth scale model bridge deck is therefore designed to simulate the single span bridge deck described by Fafard and Mallikarjuna (1993). According to the similarity theory, the length and width of the model are selected as 2.4384 m and 1.2192 m respectively. It is a beam-slab type bridge deck composing of five rectangular-section steel ribs and a steel deck. According to AASHTO specifications (H20-44 or H15-44), the ratio of the wheel

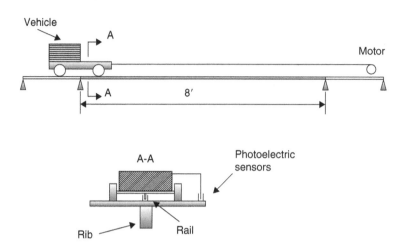

Figure 6.28 Diagrammatic view of the vehicle moving on the bridge deck

spacing and the axle spacing and the ratio between the front axle weight and the back axle weight are selected as 3:7.

The final model vehicle–bridge system fabricated in the laboratory is shown diagrammatically in Figure 6.28. The bridge deck consists of a uniform steel plate (2.4384 m × 1.2192 m × 6.35 m) stiffened with five rectangular ribs (25.4 mm × 12.5 mm) welded underneath the plate and simply supported at the two ends as shown in Figure 6.29. The bridge deck is supported on two steel I-beams, and the I-beams are fixed to the ground through bolts. Moreover, spherical metal balls are placed in between the supporting I-beams and the ribs of the bridge deck to simulate the point supports. At the entrance end of the deck, metal balls are welded and connected to both the I-beams and the ribs along the axis of rotation. However, at the exit end, the metal balls are welded on the ribs only.

Three U-shaped aluminum sections are glued on the upper surface of the deck as direction guides for the car. It is located at the 1/8b, 3/8b and 1/2b where b is the width of the deck. The model car is pulled along the guide by a string wound around the drive wheel of an electric motor. During the test, a leading beam and a tailing beam are provided for the acceleration and deceleration of the model car as shown in Figure 6.28. The leading beam and tailing beam are independently supported on the ground, and there is no other excitation to the bridge deck as the model car moves on the top of the beams.

The model car has four rubber wheels with an axle spacing of 0.457 m and wheel spacing of 0.2 m. The front and the rear axles weigh 5.2 kg and 14.7 kg respectively. The mass ratio between the model car and the bridge deck is 0.12. The ratio between the wheel spacing and the axle spacing is 0.44. At the bottom of the car, there are two studs to guide the car moving along the rails during the tests. Moreover a trigger arm is extended from the car to detect the car location and the moving speed as shown in Figure 6.28.

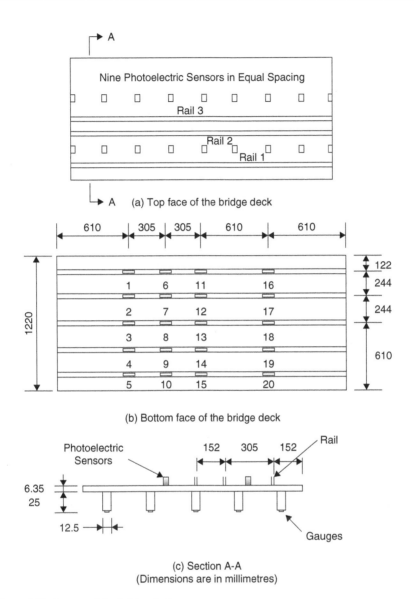

Figure 6.29 Layout of the bridge deck

Nine photoelectric sensors are mounted evenly in a line on the deck to monitor the speed of the car. They are located on the plate, at roughly equal spacing of 30 cm to check on the uniformity of the speed as shown in Figure 6.29(a). Twenty strain gauges are located at the bottom of the beam ribs to measure the responses of the plate. Their locations are shown in Figure 6.29(b). A 16-channel data acquisition system DASP-INV303E and a KYOWA data recorder model RTP800A are used for data collection in the experiments.

Table 6.14 Natural frequencies of the bridge deck and model car (Hz)

m	1				2			3		
n	1	2	3	4	1	2	3	1	2	3
Deck Test	9.42*	12.76	27.01	70.46	34.90*	38.46	49.77	73.20*	76.71	86.12
FEM(P)	9.12	15.10	30.10	64.60	36.10	43.20	57.70	79.50	–	–
FEM(S)	9.13*	12.10	28.10	63.70	36.30*	39.30	52.80	80.40*	–	–
EST method	9.27*	12.89	29.65	65.35	37.14*	41.15	56.41	83.58*	87.65	102.07
Exp. Damping ratio (%)	0.6875	0.2196	0.2544	0.2662	0.2606	0.1719	0.1285	0.3356	0.3339	0.1949
Mode no. of Car	1	2	3	4						
Freq. (Hz)	27.04	44.22	58.74	87.38						

Note: * the longitudinal bending mode. FEM(P) denotes results from finite element method with point supports. FEM(S) denotes results from finite element method with simple supports along edge.

Experimental procedure

The testing procedure consists of four main steps. The first step is the calibration of the measuring system by a static test. Loads from 0.0 to 30 kg are added to the middle points of the three rails. The corresponding strain values at the measuring points are recorded. At the same time, the strains at the measuring points under these loads are calculated using the finite element method by SAP2000 software package. Comparison between the measured values and the calculated strains gives the sensitivity coefficients for each of these measuring points.

Next, the vibration mode shapes of the bridge deck and the model car are obtained from modal test. Table 6.14 shows the identified frequencies and the damping ratios of the bridge deck. By Bakht's simplified method (Bakht and Jaeger, 1985), the rigidities of the equivalent orthotropic plate are calculated as $D_x = 7.3677 \times 10^4$ Nm, $D_y = 4.2696 \times 10^3$ Nm, $D_{xy} = 8.6018 \times 10^3$ Nm. The natural frequencies of the plate are calculated by the method given in Zhu and Law (2000), and they are compared with the measured natural frequencies obtained from the accelerations of the bridge deck as shown in Table 6.14. The measured frequencies are close to those from finite element computation and from the formulation given in Chapter 3.

In the third step, the responses of the bridge deck are measured when the model car moves along different rails. The responses are sampled at 1000 Hz with anti-aliasing filter, and the number of data in each recorded segment is 7680.

Finally, the moving loads on the bridge deck are identified from the measured responses, and they are compared with the static loads to verify the performances of the proposed methods.

Experimental Results

Two Moving Loads Identification from Strains

The first nine vibration modes with $(m = 3, n = 3)$ shown in Table 6.14 and the strains at nine or 15 measuring points are used in the identification. $(m = i, n = j)$ denotes the

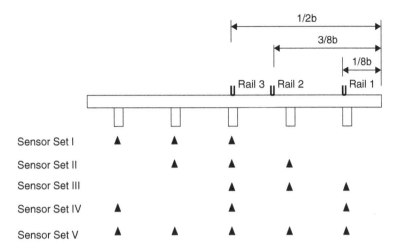

Figure 6.30 Sensor sets for moving load identification

mode number (m, n) with $(m = 1, 2, \ldots, i)$ or $(n = 1, 2, \ldots, j)$. The measured signals are re-sampled to have a time interval of 0.005s to reduce the computation time at the expense of accuracy. As the model car is moving along the centerline (Rail 3), Rail 1 or Rail 2 in turn, the strains at 1/4a, 1/2a and 3/4a of each beam are used to identify the moving axle loads. Five sensor sets shown in Figure 6.30 are used for a comparative study on the sensor location selection. Sensor set I consists of the strains at the three beams on the left. Sensor set II consists of the nine measuring points on the middle three beams, and sensor set III consists of the nine measured strains on the three beams on the right. Sensor set IV consists of the strains at Beams 1, 3 and 5. Sensor set V consists of the strains at all the five beams.

Study 1: Effect of Number of Modes

Figure 6.31 shows the identified results with different modes. The model car is moving along Rail 3 and the speed is 1.1079 m/s. Sensor set V is used in the identification and the sampling frequency is 200 Hz. Table 6.15 shows the correlation coefficients between the reconstructed and measured strains at 3/8a on the five beams. From this figure and Table 6.15, the following observations can be obtained:

1. The identified axle loads are close to the static loads and the correlation coefficients between the reconstructed and measured strains at 3/8a on each beams are all over 0.9. This shows that the method is effective to identify the moving vehicular axle loads from bridge strains and acceptable results can be obtained.
2. Figure 6.31 shows that the identified results with three longitudinal bending modes $(m = 3; n = 1)$ is close to that from nine modes. The legend [4:3:2] in Figure 6.31 indicates that nine mode shapes are used including four modes with $m = 1$, three modes with $m = 2$ and two modes with $m = 3$. This shows that the effect of torsional vibration is small when the model car moves along Rail 3, and there is little difference in the results when the torsional modes are included in

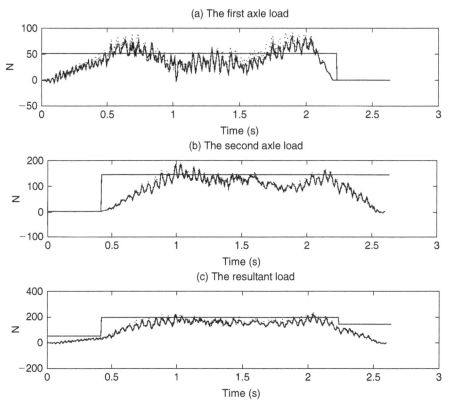

Figure 6.31 Identified axle loads along Rail 3 using different modes (— static forces; - - [3;3;3]; - - - [4;3;2]; · · · · [1;1;1])

Table 6.15 Correlation coefficients at 3/8a on the five beams

Modes	Correlation Coefficients				
	Beam 1	Beam 2	Beam 3	Beam 4	Beam 5
3;3;3	0.9790	0.9821	0.9601	0.9785	0.9498
4;3;2	0.9819	0.9829	0.9606	0.9812	0.9561
1;1;1	0.9668	0.9756	0.9486	0.9775	0.9686

the calculation. The axle loads can therefore be identified approximately using an equivalent beam model when the vehicle is moving along the central line of the bridge deck.

Study 2: Effect of Measuring Locations

The sensor sets and the parameters of the system are the same as above. Figure 6.32 shows the identified loads obtained from the strains at different measuring points as the model car moves along Rail 3. The correlation coefficients between the reconstructed and measured strains at 3/8a of the beams are shown in Tables 6.16.

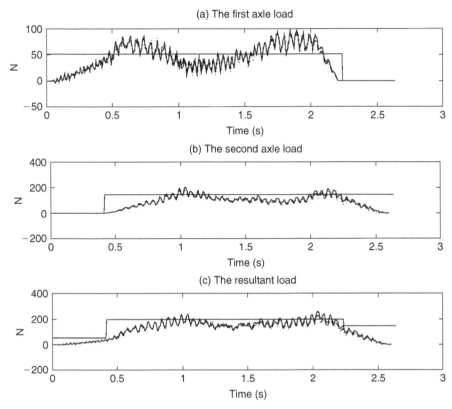

Figure 6.32 Identified axle loads along Rail 3 using different sensor sets (— static forces; - - identified with Set I; - - - identified with Set II; · · · · Identified with Set III)

Table 6.16 Correlation coefficient between reconstructed and measured strains at 3/8a

Sensor Set	Rail Number	Moving Speed (m/s)	Correlation Coefficients				
			Beam 1	Beam 2	Beam 3	Beam 4	Beam 5
I	1	1.09	0.0090	0.0085	0.0792	0.0833	0.0940
	2	1.09	0.9375	0.9753	0.9565	0.9465	0.9755
	3	1.11	0.9763	0.9594	0.9271	0.9582	0.9424
II	1	1.09	0.0722	0.8623	0.9024	0.9093	0.8907
	2	1.09	0.9548	0.9796	0.9597	0.9500	0.9751
	3	1.11	0.9708	0.9609	0.9299	0.9559	0.9314
III	1	1.09	0.1543	0.9001	0.9759	0.9828	0.9572
	2	1.09	0.9578	0.9766	0.9552	0.9461	0.9695
	3	1.11	0.9648	0.9482	0.9147	0.9462	0.9262
IV	1	1.09	0.0709	0.8705	0.9691	0.9772	0.9546
	2	1.09	0.9527	0.9729	0.9494	0.9404	0.9660
	3	1.11	0.9656	0.9482	0.9147	0.9462	0.9262
V	1	1.09	0.0006	0.1036	0.2890	0.02867	0.2810
	2	1.09	0.9114	0.9758	0.9689	0.9609	0.9801
	3	1.11	0.9790	0.9821	0.9601	0.9785	0.9498

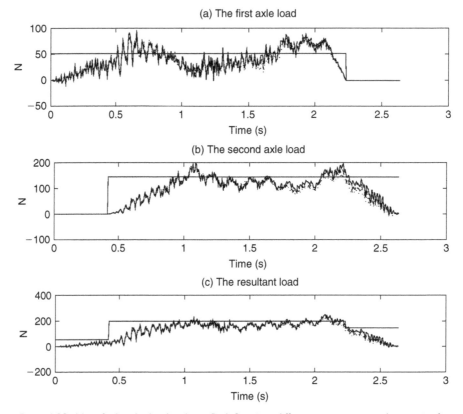

Figure 6.33 Identified axle loads along Rail 2 using different sensor sets (— static forces; -- identified with Set I; --- identified with Set II; ···· Identified with Set III)

Figures 6.32 to 6.34 show the identified results with different sensor sets as the model car moves along Rails 3, 2 and 1 respectively.

The correlation coefficients are all above 0.9 as the model car is moving along the central line (Rail 3) or Rail 2. But the correlation coefficients are very small as sensor set I or V is used with the model car moving along Rail 1. This is because the measuring points are far away from the moving model car and they cannot pick up the dominating bending modes, whereas the torsional modal responses are small, as found in Study 1. Figure 6.34 shows the extreme case with sensor set I as the model car moves along Rail 1. It is concluded that the identified force time histories are satisfactory with different sensor sets when the car moves along Rails 2 or 3.

Study 3: Effect of Eccentricities

The following discussions refer to Table 6.16, Figures 6.34 and 6.35.

Figure 6.34 shows that the identified results from sensor set III is the largest and that from sensor set II is larger than that from sensor set I as the model car moves along Rail 1. This is because sensor set III is close to the moving loads and is more subject

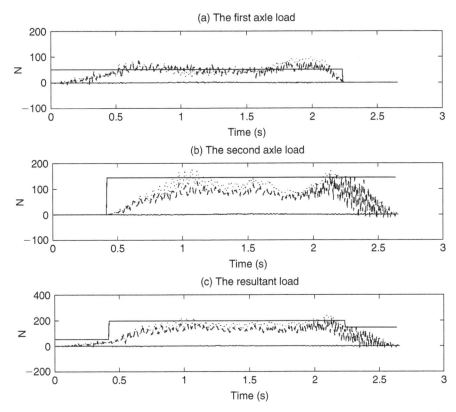

Figure 6.34 Identified axle loads along Rail I using different sensor sets (— static forces; -- identified with Set I; - - - identified with Set II; · · · · Identified with Set III)

to the bending effect of the bridge deck, and the signal to noise ratios of the measured strains are larger than that from sensor sets I or II. So the identification from sensor sets I and II is over-smoothed and the identified results are smaller than those from sensor set III.

When either one of the sensor sets II, III or IV is used in the identification, almost all the correlation coefficients between the reconstructed and measured strains are all over 0.9 when the model car is moving on Rail 1 or Rail 2 at an eccentricity of 1/8b and 3/8b, respectively. The identified eccentric loads are close to that with no eccentricity from sensor set II as shown in Figure 6.35.

Four Moving Loads Identification from Strains

Study 4: Effect of Measuring Locations

The sensor sets and the parameters of the system are the same as for the axle load identification. Figure 6.36 shows the identified wheel loads from different sensor sets when the model car moves along Rail 3. Table 6.17 shows the correlation coefficients between the reconstructed and measured strains at 3/8a. The sampling frequency

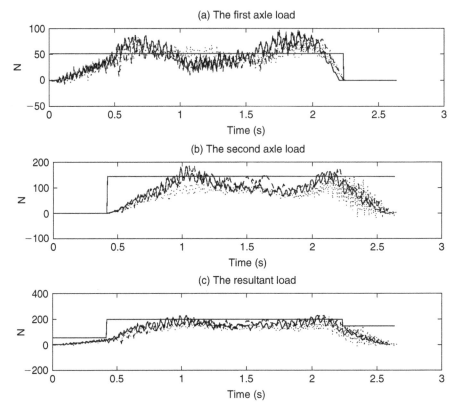

Figure 6.35 Identified axle loads along different rails from Sensor Set II (— static forces; - - Rail 3; - - - Rail 2; · · · · Rail 1)

is 200 Hz. From Figure 6.36 and Table 6.17, the following observations can be obtained:

1. The method is effective to identify the individual dynamic wheel loads from bridge strains and acceptable results can be obtained as indicated by the correlation coefficients in Table 6.17.
2. The identified force time histories from sensor set II are nearly the same as those from sensor set I or III.
3. The left wheel loads (the first and second load) from sensor set I are larger than those from sensor set III and the right wheel loads (the third and fourth load) from sensor set I are smaller than those from sensor set III. No rolling motion of the vehicle is observed in the identified forces. Since there is no spring component in each wheel in the model car and the wheel spacing is very small, the behavior of the four wheel loads system is similar to a single moving mass with some pitching effects. Therefore this difference in the identified forces can only be due to the proximity of the sensors to the loads, i.e. sensor set I is close to the left wheel loads and sensor set III is close to the right wheel loads.

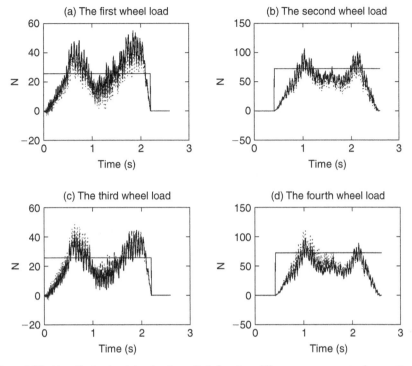

Figure 6.36 Identified wheel loads along Rail 3 using different sensor sets (— static forces; -- identified with Set I; --- identified with Set II; ···· Identified with Set III)

Table 6.17 Correlation coefficients between reconstructed and measured strain at 3/8a for wheel load identification

Sensor Set	Speed (m/s)	Rail	Correlation Coefficients				
			Beam 1	Beam 2	Beam 3	Beam 4	Beam 5
I	1.11	3	0.9769	0.9602	0.9264	0.9584	0.9444
II			0.9759	0.9637	0.9328	0.9613	0.9417
III			0.9700	0.9592	0.9297	0.9566	0.9317
IV			0.9685	0.9513	0.9168	0.9483	0.9290
V			0.9772	0.9791	0.9561	0.9789	0.9558
II	1.09	2	0.9586	0.9798	0.9593	0.9495	0.9753
II	1.09	1	0.1531	0.9124	0.9762	0.9839	0.9632

Study 5: Effect of Eccentricities

The sensor sets and the parameters of the system are the same as above. Figure 6.37 shows the identified results from sensor set II as the model car moves along different rails. The corresponding correlation coefficients between the reconstructed and

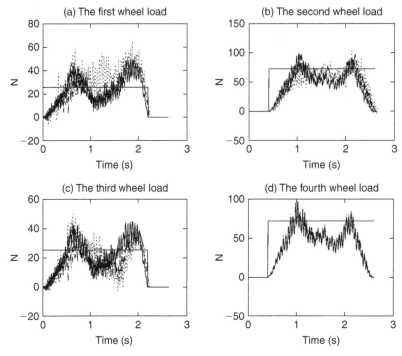

Figure 6.37   Identified wheel loads along different rails from Sensor Set II (— static forces; - - Rail 3; - - - Rail 2; · · · · Rail 1)

measured strain at 3/8a are also shown in Table 6.17. From these results, the following observations can be obtained:

1.   The method is effective to identify the dynamic wheel loads with eccentricity, and acceptable results can be obtained.
2.   The identified results from sensor set II are approximately the same when the model car moves along Rails 2 and 3. The method fails to identify forces along Rail 1.

Study 6: Effect of Number of Modes

The sensor sets and parameters of the system are the same as above. Figure 6.38 shows the identified results with different number of modes when the model car moves along the central line (Rail 3). Sensor set II is used in the identification. The correlation coefficients between the reconstructed and measured strain at 3/8a are listed in Table 6.18. The correlation coefficients are all above 0.9. The identified force time histories in Figure 6.38 are approximately the same when different number of the modes is used in the identification. This shows that the effect of torsional vibration is small when the model car moves along Rail 3, and there is little difference when the torsional

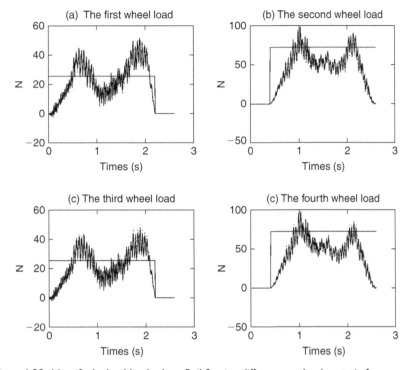

Figure 6.38 Identified wheel loads along Rail 3 using different modes (— static forces; - - [3;3;3]; - - - [4;3;2]; · · · · [1;1;1])

Table 6.18 Correlation coefficients for wheel load identification with different number of modes

Sensor Set	Modes	Correlation Coefficients				
		Beam 1	Beam 2	Beam 3	Beam 4	Beam 5
II	$m = 3; n = 3$	0.9759	0.9637	0.9328	0.9613	0.9417
	$m = 2; n = 2$	0.9586	0.9566	0.9236	0.9602	0.9600
	$m = 1; n = 1$	0.9578	0.9568	0.9237	0.9603	0.9601

modes are included in the calculation. This also supports the findings in studies 1, 2 and 4 that the sensors should be close to the moving loads to pick up the dominating bending modes to have an accurate identified result.

6.4 Summary

Two methods of moving load identification in time domain are presented in this chapter with numerical and experimental studies on continuous beam model and orthotropic plate model for illustration. The TDM method gives the least-squares solution on the forces while the EST method gives highly accurate solution with the application of the regularization technique to provide bounds to the solution. The methods are shown to be effective and accurate to identify moving loads on continuous beam or plate from structural vibration responses, though it fails to identify moving loads at a large eccentricity. Other important factors affecting the accuracy in the identification, such as the sampling frequency, modal truncation in the responses, modeling error, non-uniform speed, sensor locations and the number of modes in the computation, have also been discussed.

Chapter 7

Moving Force Identification in State Space

7.1 Introduction

Dynamic problems can be broadly categorized into the direct problem and the inverse problem. The inverse dynamic problem is one in which measurements are made on some of the state variables of the system to facilitate the solution on the unknown parameters (Trujillo and Busby, 1997). It is, essentially, the solution of an optimization problem. In the problem addressed by this book, a function representing the input is sought such that the discrepancy between the measured and calculated response is minimized. Various solution methods associated with the indirect force measurement have been proposed, e.g. dynamic programming (Busby and Trujillo, 1997; Law and Fang, 2001; Nordstrom, 2006; Law et al., 2007), regularization technique (Zhu and Law, 2002b; Nordberg and Gustafsson, 2006a; Zhu et al., 2006, and etc.).

The moving load identification from measured responses is a typical inverse problem. Chapters 5 and 6 have shown that the moving forces can be identified in the time domain and the frequency and time domains. The regularization technique can enhance the accuracy of the solution in the inverse problem. In this chapter, two state space approaches are presented in Sections 7.2.1 and 7.3.1 with formulations of the equation of motion of the time-varying dynamic system. The time-varying system and moving loads are represented by Markov parameters. The first method described in Section 7.2.4 is a non-iterative recurrence algorithm for the moving force identification in time domain. Section 7.3.2 presents the second method of moving load identification in state space with regularization. The different influencing factors, such as, the combination of strain and velocity measurements, the number of analytical modes included in the identification, the effect of load eccentricity and the choice of regularization parameters are studied in Sections 7.2.5 and 7.3.3. The effect of different regularization parameters on the identified loads is also studied and discussed. Experimental results from laboratory tests will be further used to support the above studies in Sections 7.2.6 and 7.3.4.

7.2 Method I – Solution based on Dynamic Programming

7.2.1 State–Space Model

The finite element representation of a n-DOFs dynamic system is given by:

$$M\ddot{u} + C\dot{u} + Ku = P \tag{7.1}$$

where u is a vector containing all the displacements of the model; $\dot{u}$ is the first derivative of u with respect to time t; M is the system mass matrix; C is the system damping matrix; K is the system stiffness matrix, and P represents the system of exciting forces which is a function of the location and magnitude of the applied forces as follows:

$$P = \phi f \tag{7.2}$$

where P is the nodal force vector and ϕ is the vector on the location of the applied force. For the case of multi-forces acting on the beam element, the global force vector arising from the ith force is represented by:

$$P^i = \phi^i f^i \tag{7.3}$$

where ϕ^i is the vector on the location of the applied force f^i. Using the state space formulation, Equation (7.1) is converted into a set of first order differential equation as follows:

$$\dot{X} = K^* X + BP \tag{7.4}$$

where

$$X = \begin{bmatrix} u \\ \dot{u} \end{bmatrix}_{2n \times 1}, \quad K^* = \begin{bmatrix} 0 & I \\ -M^{-1}K & -M^{-1}C \end{bmatrix}_{2n \times 2n}, \quad B = \begin{bmatrix} 0 \\ -M^{-1} \end{bmatrix}_{2n \times n} \tag{7.5}$$

where X represents a vector of state variables of length $2n$ containing the displacements and velocities of the nodes; n_f is the number of forces, and f is a vector of length n_f representing the unknown applied forces. These differential equations are then rewritten into discrete equations using the standard exponential matrix representation as:

$$X_{j+1} = FX_j + \overline{G}BP_j \tag{7.6}$$

$$F = e^{K^* h} \tag{7.7}$$

and

$$\overline{G} = K^{*-1}(F - I) \tag{7.8}$$

where matrix F is the exponential matrix, and together with matrix $\overline{G}$ they represent the dynamics of the system; $(j+1)$ denotes the value at $(j+1)$th time step of computation; the time step h represents the time difference between the variable states X_j and X_{j+1} in the computation, and $\overline{G}$ is a matrix relating the forces to the system. Substituting Equations (7.5) and (7.8) into Equation (7.6) we have:

$$X_{j+1} = FX_j + G_{j+1} f_j \tag{7.9}$$

where

$$G_{j+1} = \overline{G}_{2n\times 2n} \begin{bmatrix} 0 \\ -M^{-1}\phi_{j+1} \end{bmatrix}_{2n\times nf} \tag{7.10}$$

7.2.2 Formulation of Matrix G for Two Moving Loads Identification

Suppose we have two forces spaced at a constant distance moving across a simply supported beam at a constant speed c. The matrices $\overline{G}$ and G in Equations (7.8) and (7.10) respectively are varying for different location of the forces. Rewrite Equations (7.3) and (7.8) into:

$$P = [\phi^1, \phi^2] \begin{pmatrix} f^1 \\ f^2 \end{pmatrix} \tag{7.11}$$

$$\overline{G} = \begin{bmatrix} \overline{G}_{11} & \overline{G}_{12} \\ \overline{G}_{21} & \overline{G}_{22} \end{bmatrix} \tag{7.12}$$

and

$$\overline{P} = BP = \begin{bmatrix} 0 & 0 \\ -M^{-1}\phi^1 & -M^{-1}\phi^2 \end{bmatrix} \begin{pmatrix} f^1 \\ f^2 \end{pmatrix} \tag{7.13}$$

The discrete representation of the system in Equation (7.6) can then be written as:

$$\begin{aligned}
X_{j+1} &= FX_j + \overline{G}_{j+1}\overline{P}_j \\
&= FX_j + \begin{bmatrix} \overline{G}_{11} & \overline{G}_{12} \\ \overline{G}_{21} & \overline{G}_{22} \end{bmatrix} \begin{bmatrix} 0 & 0 \\ -M^{-1}\phi_{j+1}^1 & -M^{-1}\phi_{j+1}^2 \end{bmatrix} \begin{pmatrix} f_j^1 \\ f_j^2 \end{pmatrix} \\
&= FX_j + \begin{bmatrix} -\overline{G}_{12}M^{-1}\phi_{j+1}^1 & -\overline{G}_{12}M^{-1}\phi_{j+1}^2 \\ -\overline{G}_{22}M^{-1}\phi_{j+1}^1 & -\overline{G}_{22}M^{-1}\phi_{j+1}^2 \end{bmatrix} \begin{pmatrix} f_j^1 \\ f_j^2 \end{pmatrix} \\
&= FX_j + G_{j+1} \begin{pmatrix} f_j^1 \\ f_j^2 \end{pmatrix}
\end{aligned} \tag{7.14}$$

and in matrix form as:

$$X_{j+1} = FX_j + G_{j+1}f_j \tag{7.15}$$

where G_{j+1} is the value at $(j+1)$th time step. Note that Equation (7.15) is the same as Equation (7.9) but for two moving forces.

7.2.3 Problem Statement

The problem here is one in which the system matrices K, C and M are known together with information on some of the displacements and velocities. However the forcing term f is unknown. The goal of this problem is to find the forcing term f that causes the system described in Equation (7.9) to best match the measurement.

In practice it is not possible to measure all the displacements and velocities and, only certain combinations of the variables X_j are measured. The measurement equation is given as

$$d_j = QX_j \tag{7.16}$$

where d_j is a $(m \times 1)$ measurement vector; Q is a $(m \times 2n)$ selection matrix relating the measurements to the state variables and X_j is of dimension $(2n \times 1)$. The actual measurements are represented by a vector Z_j which is of the same dimension as d_j. The number of measured variable m is usually much smaller than the number of state variables (or n DOFs of the system) but larger than or equal to n_f, the length of vector f. In the case of a two-dimensional simply supported beam divided into L elements, $n = 2(L+1) - 2$ including all vertical displacements and rotational displacements at each of its nodes.

When the unknown force f_j is included in Equation (7.9), an exact match of the model with the measured data will usually not work. This is due to the fact that all measurements have some degree of noise. Even the least-squares criteria is not sufficient because a mathematical solution that will minimize the least-squares error E represented by:

$$E = \sum_{j=1}^{N} ((Z_j - QX_j), \, A(Z_j - QX_j)) \tag{7.17}$$

will usually end up with the model exactly matching the data. (x, y) in Equation (7.17) denotes the inner product of two vectors x and y. This situation could be avoided by adding a smoothing term to the least-squares error (Busby and Trujillo, 1993; Santantamarina and Fratta, 1998) to become a non-linear least-squares problem in the Tikhonov (1963) method:

$$E = \sum_{j=1}^{N} ((Z_j - QX_j), \, A(Z_j - QX_j) + (f_j, Bf_j)) \tag{7.18}$$

Parameter B in Equation (7.18) is the regularization parameter. Matrices $A(m \times m)$ and $B(n_f \times n_f)$ are symmetric positive definite weighting matrices that provide the flexibility of weighing the measurements and the forcing terms. Matrix A is usually an identity matrix and matrix B is a diagonal matrix. The second term with the positive parameter B has the effect of smoothing the identified forces. A small value of B causes the solution to match the data closely but produces large oscillatory deviations. A large value of B produces smooth forces that may not match the data well. When B is zero, the solution becomes that for the least-squares problem.

7.2.4 Computation Algorithm

To minimize the least-squares error E in Equation (7.18) over the sequence of the forcing vector f_j, dynamic programming method (Trujillo, 1978) and Bellman's Principle of Optimality (Bellman, 1967) are applied. The minimum value of E at the nth stage for any initial state X is written as:

$$g_n(X) = \min_{f_j} E_n(X, f_j) \tag{7.19}$$

A recursive formula for Equation (7.9) is derived from the Bellman's Principle of Optimality as:

$$g_{n-1}(X) = \min_{f_{n-1}}((Z_{n-1} - QX), A(Z_{n-1} - QX) + (f_{n-1}, Bf_{n-1}) + g_n(FX + G_n f_{n-1})) \tag{7.20}$$

This equation represents the classic dynamic programming structure in that the minimum at any point is determined by selecting the decision f_{n-1} to minimize the immediate cost (the first and second terms) and the remaining cost resulting from the decision (the third term). It is noted that the minimization is performed over a previously determined function g_n. The term f_n and g_n are the optimal forcing term and the optimal cost term respectively. The solution is obtained by starting at the end of the process, $n = N$, and working backward to $n = 1$. At the end point, the minimum is determined from:

$$g_N(X) = \min_{f_N}[(Z_N - QX), A(Z_N - QX) + f_N, Bf_N] \tag{7.21}$$

When $f_N = 0$, Equation (7.21) can be expanded to get the minimum solution as shown in Equation (7.22).

$$g_N(X) = q_N + (X, S_N) + (X, R_N X) \tag{7.22}$$

where

$$q_N = Z_N, AZ_N$$

$$S_N = -2Q_N^T AZ_N$$

$$R_N = Q_N^T AQ_N \tag{7.23}$$

These are the initial conditions for working backward at $n = N$. Substituting Equation (7.23) for the nth and $(n-1)$th steps into Equation (7.20), and expanding the right-hand-side of the equation, we have:

$$q_{n-1} + (X, S_{n-1}) + (X, R_{n-1}X)$$
$$= \min_{f_{n-1}}[(f_{n-1} + V_n X + U_n), H_n(f_{n-1} + V_n X + U_n) + r_{n-1}(X)] \tag{7.24}$$

where

$$H_n = B + G_n^T R_n G_n,$$

$$2H_n V_n = 2G_n^T R_n F, \qquad V_n = H_n^{-1} G_n^T R_n F,$$

$$2H_n U_n = G_n^T S_n, \qquad U_n = (H_n^{-1} G_n^T S_n)/2.$$

$$r_{n-1}(X) = \begin{bmatrix} (q_n + Z_{n-1}^T A Z_{n-1}) + X^T (Q^T A Q + F^T R_n F) X + X^T (F^T S_n - 2Q^T A Z_{n-1}) \\ -X^T V_n^T H_n U_n X - U_n^T H_n U_n - 2X^T V_n^T H_n U_n \end{bmatrix}$$

Minimizing the term on the right-hand-side of Equation (7.24) yields the optimal forcing term as:

$$f_{n-1} = -H_n^{-1} G_n^T \left[R_n F X_{n-1} + \frac{S_n}{2} \right] \tag{7.25}$$

and Equation (7.24) becomes:

$$q_{n-1} + (X, S_{n-1}) + (X, R_{n-1} X) = r_{n-1}(X) \tag{7.26}$$

Equating like powers of X in Equation (7.26) will yield the following relationship:

$$R_{n-1} = Q^T A Q + F^T [I - R_n^T G_n H_n^{-1} G_n^T] R_n F$$
$$S_{n-1} = -2Q^T A Z_{n-1} + F^T [I - R_n^T G_n H_n^{-1} G_n^T] S_n \tag{7.27}$$

These are the recursive formula required to determine the optimal solution of Equation (7.20). The complete sequence of operations is as follows:

Step 1: Matrices Q and Z and the speed of the forces are obtained from measurement. Select a smoothing parameter B;
Step 2: Matrix $\overline{G}$ and hence matrix G are obtained from information on the location of the forces from Equations (7.8) and (7.14);
Step 3: Compute the initial values q_N, R_N and S_N from Equation (7.23); compute H_N from Equation (7.24);
Step 4: Compute S_{n-1} and R_{n-1} from Equation (7.27) for $n = N$ to 1;
Step 5: Initial condition of X is set as zero and compute the responses X_{j+1} from Equation (7.15) for $j = 0$ to N, and compute the forces f_{n-1} from Equation (7.25) for $n = 1$ to N;
Step 6: Steps 1 to 5 are repeated for a different smoothing parameter B. Convergence is reached when the error computed from Equation (7.28) is smaller than a predetermined value.

7.2.5 Numerical Examples

The proposed method is studied for its accuracy and effectiveness in identifying the following simulated forces:

(a)   for single moving force identification

$$f_1(t) = 40000[1 + 0.1 \sin(10\pi t) + 0.05 \sin(40\pi t)] \text{ N}$$

(b)   for two moving forces identification

$$f_1(t) = 20000[1 + 0.1\sin(10\pi t) + 0.05\sin(40\pi t)]\,\text{N}$$
$$f_2(t) = 20000[1 - 0.1\sin(10\pi t) + 0.05\sin(50\pi t)]\,\text{N}$$

The two forces are at a constant spacing of 4 meters apart and they are moving together on a 40 m long, simply supported, beam at a velocity of 40 m/s. The physical parameters of the beam are the same as those shown in Section 5.3.

The finite element model of the beam consists of 10 elements with 11 nodes and 20 rotational and translational DOFs. The state variable matrix X has a dimension of (40×1). A sampling frequency of 200 Hz is used indicating 100 Hz is the upper frequency limit of the study. The time when there are forces on the beam is 1.1 second, and 220 data points are used.

Dynamic analysis was performed on this system to find the velocity and bending moment time histories at specified locations. Five percent root-mean-square, normally distributed, random noise, with zero mean and unit standard deviation, was added to these responses to simulate the polluted measurements.

The error in the force identification is calculated by the following equation where $\|\bullet\|$ is the norm of a matrix:

$$Error = \frac{\|f_{identified} - f_{true}\|}{\|f_{true}\|} \times 100\% \tag{7.28}$$

Since the true force is known, the optimal regularization parameter B is obtained by comparing the true force f_{true} with the identified values $f_{identified}$, and an error curve can be plotted for different value of B. The error of the identified results is calculated from Equation (7.28) with the optimal value of B corresponding to the smallest error.

If both the measured bending moments and velocities are used together to identify the moving forces, the velocity component and the bending moment component in the vector X in Equation (7.9) should be scaled by their respective norm similar to Equation (5.28) to have dimensionless units.

7.2.5.1   Single-Force Identification

Nine combinations of the responses at 1/4 and 1/2 span, as shown in Table 7.1, are used in the identification. Table 7.1 shows that the use of single or multiple responses does not have significant difference in the errors of identification. Results not shown here indicate that the use of additional sensor at 3/4 span does not improve significantly the identified results.

The time histories and the PSDs of the calculated and true forces for sensor combinations of (1/4v,1/2v) and (1/4m,1/4v) are shown in Figure 7.1. The time histories closely match each other except at the start and end of the time duration. It is found that the variation in the identified forces increases with a decrease in matrix B. Inspection of the PSDs of the force reveals that good match between the forces is found around the exciting frequencies but with large noise in the upper frequency range. The random noise in the polluted responses is reflected in the upper frequency range of the identified forces with little adverse effect on the results in the lower frequency range. The use of bending moments seems to have poorer performances compared with the velocities.

Table 7.1 Error in one force identification (in percent) (with 5 percent noise in response)

Sensor location and response type	Moment	Velocity	Both
1/2 span	26.8	24.5	26.9
1/4 span	26.1	21.7	26.0
1/2 span, 1/4 span	26.7	19.8	26.6

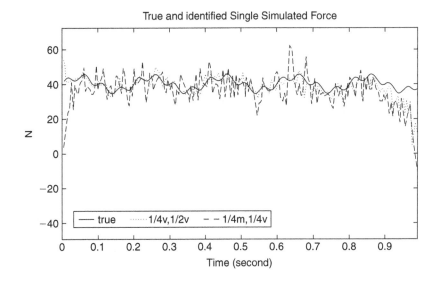

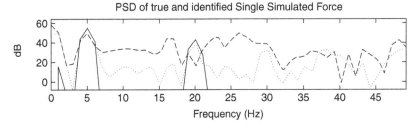

Figure 7.1 Identified time histories and PSDs of the true and identified forces

The reconstructed bending moment and velocity at 1/4 span are compared with the measured ones in Figures 7.2 and 7.3. The time histories match closely with the true ones. The PSDs of the responses indicate that errors exist in the upper frequency range which is again suspected due to the simulated noise effect.

7.2.5.2 Two-Forces Identification

Bending moment and/or velocity responses at 1/4, 1/2 and 3/4 spans in 20 combinations as shown in Table 7.2 are used to identify the two forces. The results are obtained in a manner similar to the single force identification.

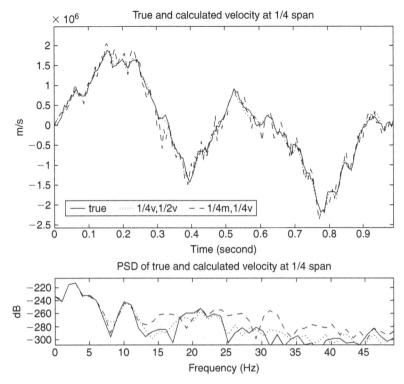

Figure 7.2 The reconstructed and measured velocities at 1/4 Span

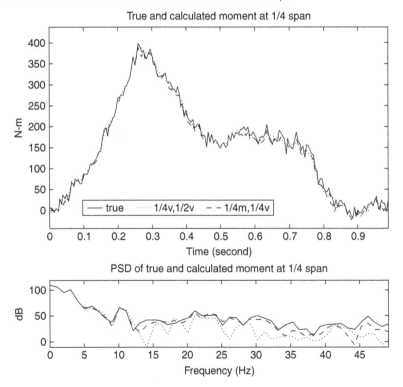

Figure 7.3 Reconstructed and measured bending moments at 1/4 span

Table 7.2 Error in two-forces identification (in percent) (with 5 percent noise in response)

Sensor location and response type	First Force	Second Force	Total Force
1/4m, 1/2m	35.6 (+)	37.4 (+)	24.3
1/4v, 1/2v	47.5 (222)	48.7 (78.6)	33.5
1/2m, 1/2v	36.3 (+)	38.4 (+)	24.5
1/4m, 1/4v	38.7 (+)	40.6 (+)	27.3
1/2m, 1/4v	36.4 (+)	38.4 (+)	24.1
1/4m, 1/2m, 3/4m	32.1 (+)	33.1 (+)	20.6
1/4v, 1/2v, 3/4v	38.9 (10.7)	39.8 (10.7)	27.5
1/4m, 1/2m, 1/4v	35.6 (+)	37.4 (+)	24.3
1/2m, 3/4m, 1/4v	33.5	34.1	21.5
1/2v, 3/4m, 3/4v	35.5	37.7	25.8
1/4m, 1/2m, 1/2v	35.6 (+)	37.4 (474)	24.3
1/4m, 1/4v, 1/2v	38.7 (206)	40.6 (156)	27.3
1/2m, 1/4v, 1/2v	36.3 (193)	38.4 (46.8)	24.5
1/2m, 1/4v, 3/4v	36.4	38.4	24.1
1/4m, 1/2m, 1/4v, 1/2v	35.6 (201)	37.4 (49.7)	24.0
1/4m, 1/2m, 1/2v, 3/4v	32.1	33.2	20.7
1/2m, 1/4v, 1/2v, 3/4v	36.4	38.4	24.1
1/4m, 1/2m, 3/4m, 1/4v, 1/2v	32.1	33.2	20.7
1/4m, 1/2m, 1/4v, 1/2v, 3/4v	35.6	37.3	24.0
1/4m, 1/2m, 3/4m, 1/4v, 1/2v, 3/4v	32.1	33.1	20.6

Note: + indicates error larger than 1000 percent.

The errors between the calculated and the true forces are shown in Table 7.2 for different sensor combinations. The error calculated by the Time Domain Method (TDM) (Law et al., 1997), presented in Chapter 6, is included in brackets for comparison. Note that the TDM uses acceleration measurements instead of velocity measurements in the identification. The proposed method gives far smaller error, in general, than the Time Domain Method. Only in the case of using three acceleration measurements does the TDM gives better results than the present case using three velocities. The results shown in Table 7.2 indicate that (a) many sensor combinations would give similar errors in the identified forces; (b) the use of more than two sensors may not improve the result; and (c) both velocity and bending moment give approximately the same accuracy in the identified forces.

The time histories and PSDs of the identified forces from using (1/4v, 1/2v, 3/4v) and (1/4m, 1/2m, 3/4m) sensor combinations are shown in Figures 7.4 to 7.6. All the identified forces vary close to the true force in the middle length of the time duration between 0.25 seconds and 0.75 seconds. The results have a lot of variations due to the simulated noise effect. Other results, not shown here, from using different sensor combinations, also exhibit similar patterns in the identified forces. Note that the discrepancies at the start and end of the time histories contribute greatly to the overall error.

The above results show that accuracy in the two-forces identification is poorer than that in single force identification. This is due to a component in the simulated individual forces with same amplitude and opposite phase. This results in large errors in the time histories. The combined force shown in Figure 7.6 indicates that most of these errors

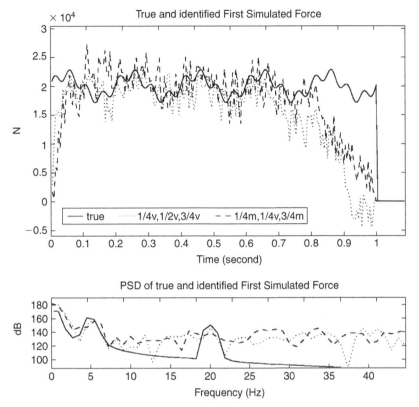

Figure 7.4 Time histories and PSDs of the identified first forces from using different sensor combinations

are cancelled out leading to a greatly improved time history and reduced error in the combined identified force as shown in Table 7.2.

7.2.6 Experiment and Results

The experimental setup with a model vehicle moving on top of a simply supported steel beam described in Section 5.4 is used for the following studies of moving forces identification. The parameters of the beam and model vehicle are the same as those described in Section 5.4.

The experimental procedures have been described in Section 5.4. The time histories between two successive measured data points of the responses were sub-divided into ten sub-divisions such that the time difference between two time steps is 7.8125×10^{-4} second. A smaller sampling time interval is important to the accuracy of the iterative computation. The measured acceleration records were integrated to velocity records with an algorithm developed by Petrovski and Naumovski (1979) with a low pass filter at 0.0625 Hz.

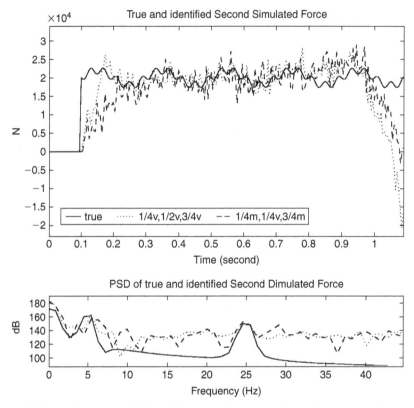

Figure 7.5 Time histories and PSDs of the identified second forces from using different sensor combinations

In this practical case when f_{true} is not known, the semi-norm of the solution against the regularization parameter B is plotted. The semi-norm of the estimated forces is:

$$E1 = \|f_{j+1}^{(identify)} - f_j^{(identify)}\| \tag{7.29}$$

where $f_j^{(identify)}$, $f_{j+1}^{(identify)}$ are the identified forces with B_j and $B_j + \Delta B$. The value of B corresponds to the smallest semi-norm is the optimal value.

7.2.6.1   Single-Force Identification

The model vehicle with two axles moves on top of the beam and the combined axle loads are identified as a single moving force in this study. Nine combinations of the measured responses at 1/4 and 1/2 span are used for the single force identification. Correlation coefficients between the reconstructed responses using the identified forces and measured responses at 3/4 span are calculated to evaluate the accuracy of the identified force. Clearly a larger coefficient means that the identified forces are more accurate than that with a smaller coefficient, but the larger coefficient does not mean

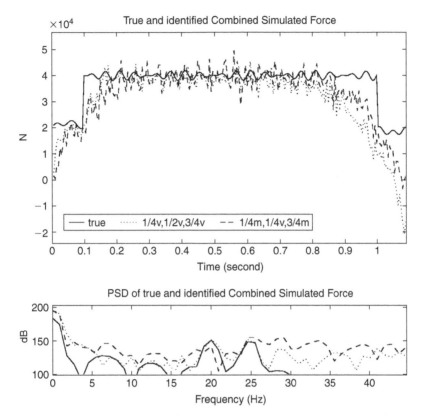

Figure 7.6 Time histories and PSDs of the identified combined force from using different sensor combinations

Table 7.3 Correlation coefficients between measured and reconstructed responses – Single force identification

Combinations of the responses	Comparing moment at 3/4 span	Comparing velocity at 3/4 span
1/2m	0.349	0.155
1/4m	0.801	0.600
1/4v	0.369	0.869
1/2v	0.200	0.254
1/2m, 1/4m	0.939	0.744
1/2v, 1/4v	0.364	0.946
1/4m, 1/2v	0.814	0.629
1/4m, 1/4v	0.796	0.594
1/2m, 1/4v	0.520	0.142

the identified force is accurate enough because this indirect method of checking on the identified results is not fully sufficient. The coefficients are shown in Table 7.3, and some of the identified results are shown in Figures 7.7 to 7.9. It is not possible to identify the static component of the forces using only velocity measurements, and

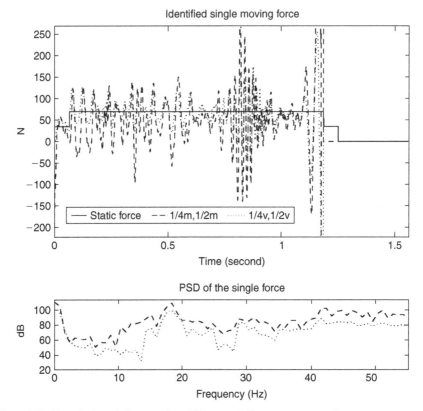

Figure 7.7 Identified single force and its PSD using different sensor combinations

the static forces are added to the identified forces in the figures for convenience of comparison.

The results in Table 7.3 show that the use of bending moments at both 1/2 span and at 1/4 span is the most suitable for single moving force identification. Figures 7.8 and 7.9 show that the PSDs of the responses match closely, although the time histories are not very close. It is suspected that the calibration errors lead to these differences in the time histories.

7.2.6.2 Two-Forces Identification

Twelve combinations of the measured responses are used to identify the two axle loads of the model vehicle. The velocity of the main beam at the 3/8 span is reconstructed using the identified forces, and it is compared with the measured response.

Some of the identified results are shown in Figures 7.10 to 7.13. The component with the same amplitude, and opposite phase in the two identified forces in Figures 7.10 and 7.11, is due to the pitching motion of the model car. The two identified forces are added to obtain a resultant force as shown in the figures. This resulting force can be used as a good estimate of the total equivalent static load of the vehicle.

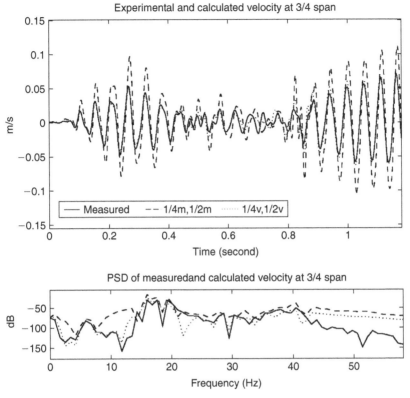

Figure 7.8 Measured and reconstructed velocities at 3/4 span

Figures 7.12 and 7.13 show that those curves obtained from using velocity responses closely match the measured responses, while those obtained from using bending moment responses has a large difference in the amplitude. This again leads to the suspicion of the existence of calibration error in the strain measurements.

Correlation coefficients are calculated between the reconstructed and measured responses and they are shown in Table 7.4. They show that (a) velocity response alone gives better results for the two-forces identification than the bending moment responses; and (b) 1/4 span responses are better than the other responses. Comparison of the correlation coefficients in Tables 7.3 and 7.4 and comparing the responses in Figures 7.8 to 7.11, show results from the identification as two moving forces are more accurate than those from the identification as a single force.

7.2.7 Discussions on the Performance of Method I

The ill-conditioned identified forces have large fluctuations in their time histories. Their magnitudes are, however, bounded with the use of non-linear least-squares minimization, as shown in Equation (7.21), and the errors computed for the identified forces are much smaller than those obtained from TDM (Law et al., 1997) without

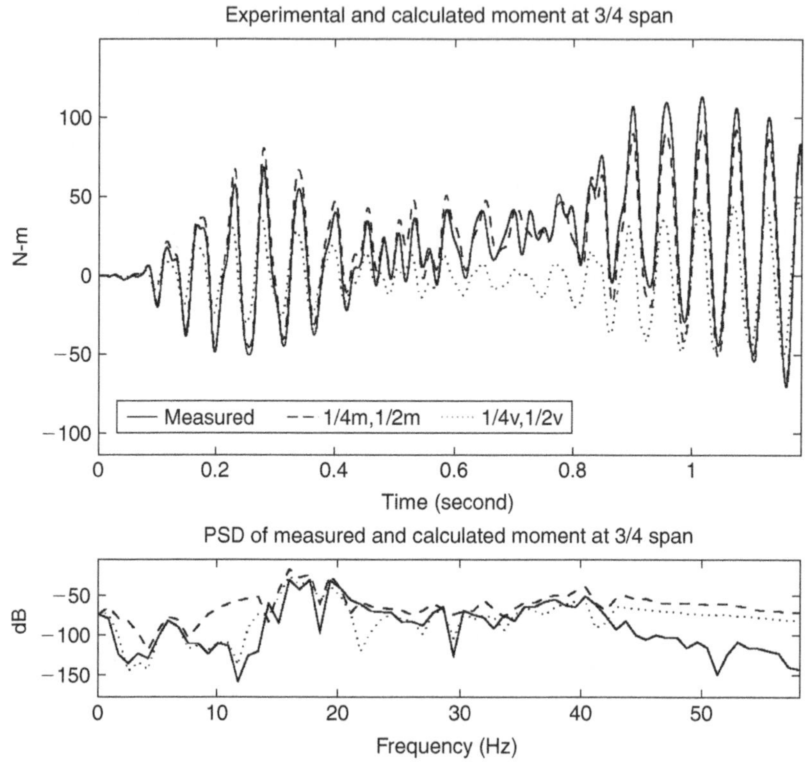

Figure 7.9 Measured and calculated moments at 3/4 span

regularization. However, the short periods at the start and end of the time histories still contribute greatly to the total error as computed from Equation (7.28). This error can be reduced, as reported by Choi and Chang (1996), by using different smaller **B** matrix at these two time durations. The forces identified from the experimental data are fairly smooth, varying around the static force, and the identified combined force can be useful as a reliable estimate of the static weight of the vehicle crossing the bridge.

7.3   Method II – Solution based on Regularization Algorithm

7.3.1   Discrete Time State–Space Model

If the response of the orthotropic plate is represented by N_s output quantities in the output vector $v(t)$ from sensors such as accelerometers, velocity transducers, displacement transducers or strain gauges, an output equation can be expressed as:

$$v = R_a \ddot{u} + R_v \dot{u} + R_d u \tag{7.30}$$

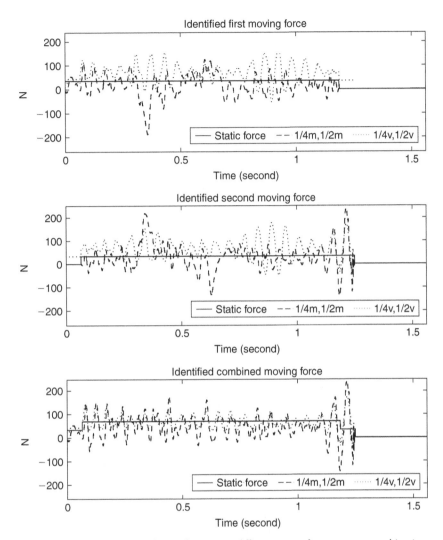

Figure 7.10 Identified moving forces from using different sets of two sensor combinations

where R_a, R_v and R_d are output influence matrices for acceleration, velocity, strain/displacement measurements, respectively. Solving for $\ddot{u}$, $\dot{u}$, u from Equation (7.1) and substitute into Equation (7.30) to yield:

$$v = RX + D\phi f \tag{7.31}$$

$$R = [R_d - R_a M^{-1} K \quad R_v - R_a M^{-1} C]; \quad D = R_a M^{-1} \tag{7.32}$$

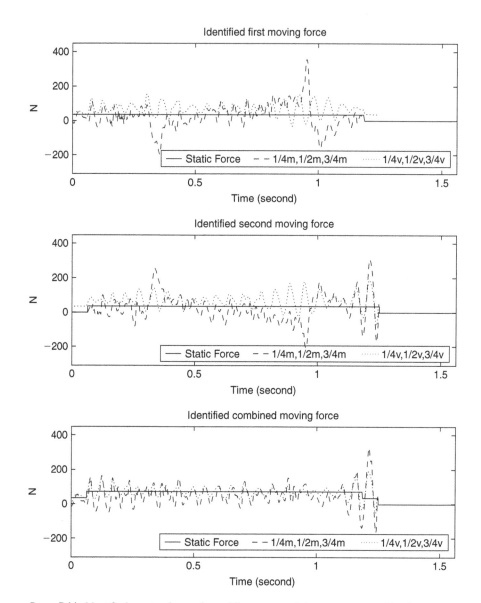

Figure 7.11 Identified moving forces from different sets of three sensor combinations

where ϕ is the vector of the location of the applied force f. Equations (7.4) and (7.31) are converted into discrete equations using the exponential matrix, and the final discrete model is:

$$X_{j+1} = AX_j + B\phi_j f_j$$
$$v_j = RX_j + D\phi_j f_j \quad (j = 1, 2, \ldots, N) \tag{7.33}$$

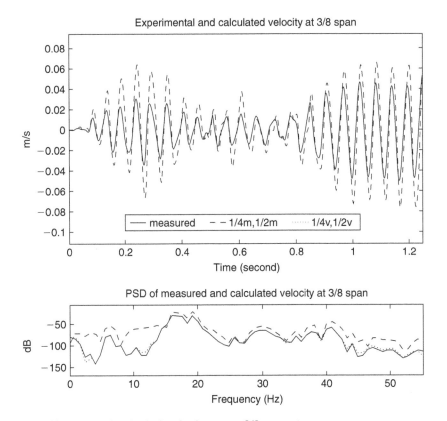

Figure 7.12 Measured and calculated velocities at 3/8 span using two sensors

where N is the total number of sampling points. τ is the time step between the variable states X_{j+1} and X_j, and

$$A = \exp(\overline{K}\tau); \quad \overline{B} = \overline{K}(A - I)B$$

Solving for the output with zero initial conditions from Equation (7.33) in terms of the previous inputs and ϕ_j and f_i $(i = 1, 2, \ldots, j)$ yields:

$$v_j = \sum_{i=0}^{j} H_i \phi_{j-i} f_{j-i} \tag{7.34}$$

where

$$H_0 = D; \quad H_i = RA^{(i-1)}\overline{B} \tag{7.35}$$

The constant matrices in the series are known as system Markov parameters. The Markov parameters are commonly used as the basis for identifying mathematical models in linear dynamic systems. The Markov parameters represent the response of the discrete system to unit impulse, and they must be unique for the system (Juang, 1994).

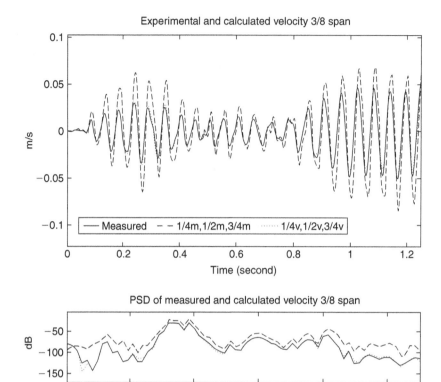

Figure 7.13 Measured and calculated velocity at 3/8 span using three sensors

7.3.2 Moving Load Identification

Rewrite Equation (7.34) to give the matrix convolution equation as (Zhu et al., 2006):

$$\overline{H}P = V \tag{7.36}$$

where

$$\overline{H} = \begin{bmatrix} H_0\phi_0 & 0 & \dots & 0 \\ H_1\phi_0 & H_0\phi_1 & \dots & 0 \\ \vdots & \vdots & \dots & 0 \\ H_{N-1}\phi_0 & H_{N-2}\phi_1 & \dots & H_0\phi_{N-1} \end{bmatrix}$$

$$P = \{f_0^T, f_1^T, \dots, f_{N-1}^T\}^T$$

$$V = \{v_0^T, v_1^T, \dots, v_{N-1}^T\}^T$$

Table 7.4 Correlation coefficients between measured and reconstructed velocity at 3/8 span – Two-forces identification

Response combinations location and type	Correlation coefficient
1/2m,1/4m	0.742
1/2m,1/4m,3/4m	0.748
1/2v,1/4v	0.989
1/2v,1/4v,3/4v	0.988
1/2m,1/2v	0.018
1/2m,1/4m,1/2v	0.743
1/2m,1/4m,1/2v,1/4v	0.744
1/4m,1/4v	0.739
1/4m,1/4v,1/2v	0.747
1/2m,1/4v	0.236
1/2m,1/4v,1/4m	0.743
1/4v,1/2v,1/2m	0.245

This is an ill-posed problem due to the lack of continuous dependence of the solution on the data when the loads are on the supports. A straightforward least-squares solution produces an unbounded result at these locations. Regularization is used to circumvent the problem of the lack of continuous dependence. One approach to regularization proposed by Tikhonov and Arsenin (1977) is to replace Equation (7.36) with the associated equation:

$$[\overline{H}^T \overline{H} + \lambda \overline{L}]P = \overline{H}^T V \tag{7.37}$$

where $\overline{L}$ is typically a discrete approximation to the ith-order derivative operator. For $\overline{L} = I$, it is zeroth-order Tikhonov regularization. For $\lambda > 0$, the matrix operator $[\overline{H}^T \overline{H} + \lambda \overline{L}]$ is unique, and therefore, its inverse is continuous. Solving Equation (7.37) is equivalent to solve the following unique problem of:

$$\min J(P,\lambda) = \|\overline{H}P - V\|^2 + \lambda \|\overline{L}P\|^2 \tag{7.38}$$

It is clear from the second term that the non-negative regularization parameter has the effect of forcing a bounded solution. If the measured displacements, velocities, strains and accelerations (or their combinations) are used together to identify the moving loads, each response component in the vector $v(t)$ in Equation (7.30) should be scaled by their respective norms similar to Equation (5.28) to have dimensionless units.

7.3.3 Numerical Studies

An orthotropic bridge deck simply supported along $x = 0$ and $x = a$, with the other two edges free, is studied. The structure consists of five I-section steel beams and a concrete slab as shown in Figure 7.14, it is similar to the one studied in Section 4.4 but has different properties. The natural frequencies and physical parameters of the structure are listed in Tables 7.5 and 7.6 respectively. There are five guide rails on

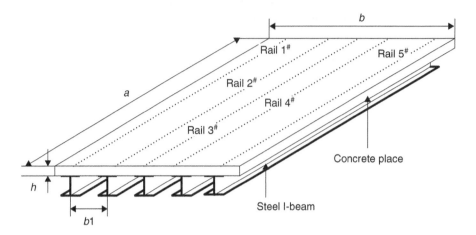

Figure 7.14 Typical single span bridge deck

Table 7.5 Natural frequency of the orthotropic bridge deck (Hz)

m \ n	1	2	3	4	5	6	7
1	4.960*	6.310	10.015	16.074	24.814	33.650	48.264
2	19.842*	21.285	25.412	32.171	41.508	45.910	59.720
3	44.645*	46.067	50.325	57.329	67.059	68.449	81.124
4	79.369*	80.805	85.071	92.171	102.059	101.604	113.232
5	124.016*	124.024	125.424	125.443	125.463	145.215	146.056
6	178.583*	179.316	181.572	185.484	191.117	191.165	191.454
7	243.074*	243.576	243.720	243.768	243.784	243.801	243.833

Note: * denotes bending modes; m and n denote longitudinal and transverse mode number respectively.

Table 7.6 Physical parameters of bridge deck

Concrete slab	I-beam	Diaphragm
length $a = 24.325$ m	beam spacing $b_1 = 2.743$ m	diaphragms spacing $d = 4.865$ m
width $b = 13.715$ m	web thickness $w_t = 0.01111$ m	cross-sectional area $A = 0.001548$ m^2
height $h = 0.2$ m	web height $w_h = 1.490$ m	$I_y = 0.707 \times 10^{-6}$ m^4
$E_x = 4.1682 \times 10^{10}$ N/m^2	flange width $f_w = 0.405$ m	$I_z = 2 \times 10^{-6}$ m^4
$E_y = 2.9733 \times 10^{10}$ N/m^2	flange thickness $f_t = 0.018$ m	$J = 1.2 \times 10^{-7}$ m^4
$\rho = 3000$ kg/m^3	$\rho = 7800$ kg/m^3	$\rho = 7800$ kg/m^3
$v_{xy} = v_{yx} = 0.3$		

the bridge deck for vehicle movement. Rail $1^{\#}$, $2^{\#}$, $3^{\#}$, $4^{\#}$ and $5^{\#}$ correspond to $1/8b$, $3/8b$, $1/2b$, $5/8b$ and $7/8b$ respectively measured from the left edge of the deck. The rigidities of the equivalent orthotropic plate can be calculated (Zhu and Law, 2003c) as $D_x = 2.415 \times 10^9$ Nm, $D_y = 2.1813 \times 10^7$ Nm, $D_k = 2.2195 \times 10^8$ Nm. Sensors are located at the bottom of the I-beams.

White noise is added to the calculated responses to simulate the polluted measurement as:

$$v = v_{\text{calculated}} \times (1 + E_p \, N_{\text{oise}})$$

where v is the matrix of measured responses used in the identification; E_P is the noise level; N_{oise} is a standard normal distribution vector with zero mean and unit standard deviation. $v_{\text{calculated}}$ is the set of calculated responses. The Relative Percentage Error in the identified results are calculated by Equation (7.39), where $\| \cdot \|$ is the norm of matrix, $P_{\text{identified}}$ and P_{true} are the identified and the true force time histories respectively.

$$RPE = \frac{\| P_{\text{identified}} - P_{\text{true}} \|}{\| P_{\text{true}} \|} \times 100\% \tag{7.39}$$

7.3.3.1 Validation of Method II

A vehicle is modeled with a two-axle model with 4.26 m axle spacing. The axle loads are:

$$P_1(t) = 6268(1.0 + 0.1\sin(10\tau t) + 0.05\sin(40t)) \, \text{kg};$$
$$P_2(t) = 12332(1.0 - 0.1\sin(10t) + 0.05\sin(50t)) \, \text{kg}.$$

The sampling frequency is 200 Hz and the vehicle moves at 20 m/s along Rail $3^{\#}$ at zero eccentricity. Nine vibration modes are used in both the identification and the response analysis ($m = 3$, $n = 3$). Figure 7.15 shows the identified results using nine accelerations, nine velocities and nine strains with 3 percent noise. Figure 7.16 shows the sensor patterns correspond to Sensor Arrangement SA9, 9-3 and 10 respectively with the sensors placed at the middle and quarter spans. The five beams are drawn as solid dark lines and Rails $1^{\#}$ to $5^{\#}$ are named in sequence from left to right of the deck. Thirty-five sensor arrangements are listed in Figure 7.16, and 28 of them will be used for the study in Section 7.3.3.2 below. The relative percentage errors of the identified results for different noise level from nine accelerations in SA9, from a combination of six accelerations and three strains (SA8) and nine strains (SA10) are listed in Table 7.7. The following observations are made from Table 7.7 and Figure 7.15:

- The relative percentage errors increase with the noise level. But the identified results from accelerations alone exhibits smaller errors than those from strains alone at the beginning and end of the time histories. This is because the acceleration responses capture the higher frequency responses of the structure from the excitation of the force in the form of an impulsive force at entry of the bridge deck, while the strain responses retain only the lower frequency responses, and thus causing larger error at these two time instances of the force time histories.
- The first half of the identified force time histories from nine accelerations match the true time histories very well, but they deviate from the true curves afterwards. This is due to computation error in the solution process since the solution in a latter time step is dependent on that in the previous time step, and computation error would be accumulating over the complete time history leading to large error at the latter time steps.

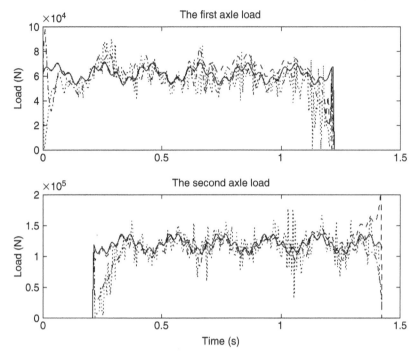

Figure 7.15 Identified axle loads with 3 percent noise (—— true force; - - - 9 accelerations; -.-.- 9 velocities; · · · · 9 strains)

- Error in the identification using accelerations comes from mainly the large fluctuation at the end of the time histories. Error from using velocities comes from mainly the fluctuations at both the start and end of the time histories, while error from using strains comes from these fluctuations as well as large fluctuations at higher frequencies throughout the duration of the time history. An inspection of the curves shows that a large proportional of them is contributed from the large fluctuations at both the start and end of the time history.
- The nine accelerations give better results than that from nine strains for a noise level up to 5 percent. There is a drastic increase in the error when the noise level increases up to 10 percent.
- The combination of six accelerations and three strains gives very good results for a noise level up to 5 percent.

7.3.3.2 Study on the Effects of Sensor Type and Location

The above vehicle model is again used for this study. The axle loads move at a constant speed of 20 m/s along Rails 3#, 4# and 5# of the deck in turn. To study the effect of different sensor arrangement and type of measured information on the identified results, 28 sensor arrangements (SA1 to 28) as illustrated in Figure 7.16 are used in

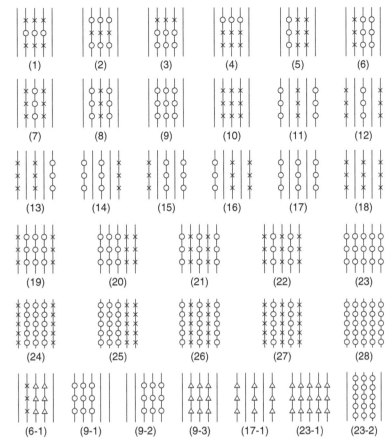

Figure 7.16 Arrangement of strain gauges and accelerometers for 9, 15 and 25 measured points
(○ – accelerometer; △ – velocity sensor; × – strain gauge)

Table 7.7 The Relative Percentage Error (%) of identification with different noise levels

Noise level (%)	From 9 accelerations (SA9)		From 6 accelerations and 3 strains (SA8)		From 9 strains (SA10)	
	Axle-1	Axle-2	Axle-1	Axle-2	Axle-1	Axle-2
0	0.969	1.383	1.429	1.110	16.674	17.752
1	1.653	1.768	1.487	2.376	17.826	18.363
3	8.955	5.127	4.996	4.534	20.667	22.523
5	14.826	11.019	7.468	8.834	22.634	24.377
10	27.318	23.532	18.664	15.240	26.902	27.714

the simulation. Only the strain and acceleration measurements are studied. There are three types of comparison on the effect of sensor arrangements.

(a)   Different type of sensors on the beam (SA1 to 4) versus same types of sensors on the beam (SA5 to 10).
(b)   Sensors on interior beams (SA5 to 10) versus sensors on edge beams (SA11 to 18).
(c)   Three sensors on all the beams (SA19 to 23) versus five sensors on all the beams (SA24 to 28).

The accelerometers and strain gauges are placed at $a/4, a/2$ or $3/4a$ along the beams in SA 1 to 23, and they are evenly distributed along the beams in SA 24 to 28. The number of modes for the identification and response analysis are nine $(m=3, n=3)$ for nine sensors, 15 $(m=3, n=5)$ for 15 sensors and 25 $(m=5, n=5)$ for 25 sensors. No noise is included in the analysis. The other conditions are the same as for Section 7.3.3.1. Table 7.8 shows the relative percentage errors of the identified results for different sensor arrangements and with eccentricities at 0, $b/8$ and $3/8b$. The identified results from SA17 with *zero* and $3/8b$ eccentricities are plotted in Figure 7.17.

The following observations are made from Table 7.8 and Figure 7.17:

- The results identified from accelerations alone are better than those from strains alone, as well as many cases of combination of strain and acceleration. This observation is very prominent when the loads are moving at a large eccentricity. This leads to the suspicion that the measured strain is not very sensitive to eccentric loads.
- Table 7.8 shows that when the same type of sensor is placed on the beam, the identified results would be more accurate compared with those having different sensors placed on the beam. This indicates that information of the same type should be obtained from all selected locations in the same beam for the identification.
- Table 7.8 also shows that sensors on the edge beams do not give distinctly more accurate results compared with sensors on the interior beams.
- When the number of sensors is equal to the number of vibration modes for the identification, increasing the number of sensors does not significantly improve the identified result in general, except for the case with large load eccentricity.
- This study demonstrates that optimal sensor type and location can be selected for an optimal identified result.

7.3.3.3   *Further Studies on the Sensor Location Effect and Velocity Measurement*

Three more studies are conducted basing on observations in Section 7.3.3.2.

(a)   Study in Section 7.3.3.2 is based on equal number of vibration modes in both the longitudinal and lateral directions, and thus limiting the responses from only a single pattern of vibration modes for the identification. Sensor arrangement SA23 is studied again with comparison to a new arrangement SA23-2 as shown in Figure 7.16. Both of them have 15 sensors while the former has five sensors in the lateral direction at $a/4, a/2$ or $3/4a$ along the beams, and the latter has

Table 7.8 The Relative Percentage Error (%) of identification from different sensor arrangements and eccentricities

Number of measured points	Sensor arrangement	Eccentricity (Rail number)					
		0 (Rail 3#)		b/8 (Rail 4#)		3/8b (Rail 5#)	
		Axle-1	Axle-2	Axle-1	Axle-2	Axle-1	Axle-2
9	(1)	14.325	13.655	14.867	14.016	16.354	15.966
	(2)	6.354	5.985	6.890	6.372	9.087	8.653
	(3)	6.890	6.324	7.660	6.816	9.568	9.001
	(4)	13.695	13.243	14.358	13.654	15.001	14.760
	(5)	2.610	2.467	2.960	2.786	6.358	4.987
	(6)	1.102	1.206	1.246	1.405	2.726	2.502
	(7)	1.484	1.377	1.809	1.782	4.251	3.940
	(8)	1.429	1.110	1.954	1.552	2.814	2.324
	(9)	0.969	1.383	1.090	1.473	1.330	1.852
	(9-1)	0.934	1.314	2.045	2.385	18.631	19.055
	(9-2)	0.932	1.315	2.475	2.318	1.049	1.475
	(9-3)	15.372	16.410	15.916	17.058	16.754	18.890
	(10)	16.674	17.752	18.034	20.086	22.487	23.584
	(11)	0.662	0.781	2.883	2.870	10.815	11.203
	(12)	1.002	1.321	2.780	2.901	8.486	9.816
	(13)	1.966	2.110	3.784	3.975	4.886	5.219
	(14)	0.550	0.863	2.976	3.666	14.712	13.109
	(15)	0.542	0.796	1.421	1.285	3.724	3.330
	(16)	1.954	2.076	2.801	3.217	11.275	12.179
	(17)	0.480	0.565	0.741	1.252	2.510	2.902
	(17-1)	14.904	15.263	17.670	18.321	21.911	26.072
	(18)	16.180	16.783	17.867	19.036	29.214	30.505
15	(19)	1.320	0.938	1.572	1.206	1.965	1.516
	(20)	0.805	0.686	1.339	1.524	2.653	1.912
	(21)	0.742	0.625	1.089	1.207	2.917	2.098
	(22)	1.022	1.118	1.160	1.316	2.886	3.021
	(23)	0.543	0.825	1.680	1.116	2.904	2.464
	(23-1)	14.011	15.004	14.237	14.652	15.739	16.008
25	(24)	0.647	0.555	1.020	1.202	1.610	1.301
	(25)	0.673	0.587	1.188	1.541	1.580	1.579
	(26)	0.976	1.024	1.878	2.128	2.798	3.055
	(27)	0.760	0.649	2.867	3.022	1.682	1.740
	(28)	0.626	0.523	1.249	1.049	1.551	1.507

five sensors evenly distributed on the beams in the longitudinal direction. Both arrangements have the sensors placed closed to the middle of the bridge deck. The effect from different combinations of vibration modes in the longitudinal and lateral directions of the bridge deck is studied. The same two-axle vehicle model and conditions for the previous studies are used, and the vehicle runs along Rail 3#. The relative percentage errors of identification for different load eccentricities are listed in Table 7.9.

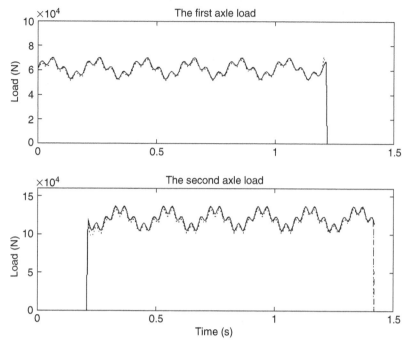

Figure 7.17   Identified axle loads from nine accelerations (SA 17) (— true force; --- with zero eccentricity; ···· with 3/8b eccentricity)

Results from Table 7.9 show that:

- The number of sensors in the lateral direction should be equal or greater than the number of lateral vibration modes used for the identification (Case 2). The same conclusion is drawn for sensors in the longitudinal direction (Case 5).
- When the load eccentricity is large, the torsional response of the structure is larger. A reduction in the number of lateral vibration modes for the identification from Case 2 to Case 1 would significantly affect the accuracy of results from SA23-2 but with little effect on SA23. Contrarily when the load eccentricity is small, the torsional response of the structure is smaller. A reduction in the number of longitudinal vibration mode from Case 5 to Case 4 does not have any significant effect on the results from the two sensor arrangements.

(b)   Two other arrangements of sensors, SA9-1 and SA9-2, are further studied for the effect of asymmetrically placed sensors on the identified results. All conditions are the same as for Section 7.3.3.2. The relative percentage errors of each identified axle loads are shown in Table 7.8 and $(m = 3, n = 3)$ mode combination is used. These results, as well as those from SA13 and 14, show that the identified results would be more accurate when the sensors are closer to the moving loads. They also confirm again that acceleration is more useful than strain measurements for the identification.

Table 7.9 The Relative Percentage Error (percent) of identification from different mode combinations and load eccentricities

Case	m	n	Total number of modes	Eccentricity (rail)					
				0 (Rail 3#)		$b/8$ (Rail 4#)		$3/8b$ (Rail 5#)	
				Axle-1	Axle-2	Axle-1	Axle-2	Axle-1	Axle-2
1	3	4	12	0.992/0.974	1.342/1.314	1.537/1.384	1.866/1.614	6.043/1.664	8.290/2.316
2	3	5	15	0.544/0.543	0.869/0.825	1.241/1.080	1.181/1.116	3.171/2.904	3.352/1.106
3	3	6	18	0.474/0.475	0.741/0.723	2.750/2.407	2.988/3.233	3.054/3.560	3.642/3.543
4	4	3	12	1.015/1.709	1.318/1.984	2.038/2.820	1.426/2.524	2.262/4.108	2.272/4.282
5	5	3	15	0.959/1.670	1.257/1.896	1.894/2.671	1.191/2.354	2.237/3.568	2.270/3.954
6	6	3	18	0.950/1.660	1.246/1.875	1.766/2.774	1.084/2.391	2.222/3.825	2.245/3.963

Note: •/• denotes results from sensor arrangement SA23-2 and SA23 respectively.

(c) Section 7.3.3.2 only presents the effectiveness of strain and acceleration measurements. The effectiveness of velocity measurement is further studied here for comparison. Sensor arrangements SA9-3, 17-1 and 23-1 are studied, and the results are compared with those from using strain and acceleration measurements. Errors in the identified loads in Table 7.8 clearly show that velocity is slightly better than strain measurements with comparable errors. Results from Section 7.3.3.1 also support this conclusion.

7.3.3.4 *Effect of the Aspect Ratio of the Bridge Deck*

To study the effect of the dimensions of the bridge deck on the identification, the width of the deck is kept constant at 13.715 m, and the length takes up different values of 6.868 m, 13.715 m, 20.573 m, 27.430 m, 34.288 m, 68.58 m and 137.15 m in turn, corresponding to an aspect ratio of 0.5, 1.0, 1.5, 2.0, 2.5, 5.0 and 10.0 respectively. The two-axle vehicle used for previous studies moves along Rail 3#. The number of modes used in identification is taken the same as for the response analysis but with different mode number in the longitudinal and lateral directions. Other conditions remain the same as for last study. Table 7.10 shows the relative percentage errors of the identified results for SA9. The following observations are made:

- The use of only one longitudinal vibration modes for the identification does not give any meaning results for all the cases studied since three sensors in the lateral direction could not capture the nine lateral modes.
- Acceptable results can be obtained over a wide range of the aspect ratio from 1.0 to 5.0 so long as when the number of longitudinal vibration mode is more than two.
- For the cases when the number of vibration mode is larger than the number of sensors, the accuracy of the identified result does not improve significantly.

Table 7.10 The Relative Percentage Error (percent) of identification for different aspect ratio of the bridge deck

M			1	2	3	4
N			9	5	3	4
Total No. of modes			9	10	9	16
Aspect Ratio	0.5	Axle-1	14.110	3.084	3.966	3.530
		Axle-2	14.353	2.059	2.383	2.195
	1.0	Axle-1	29.560	0.930	0.552	0.437
		Axle-2	23.920	1.107	0.445	0.331
	1.5	Axle-1	32.199	0.842	0.923	0.944
		Axle-2	25.321	1.307	1.185	1.211
	2.0	Axle-1	36.079	0.445	0.500	0.461
		Axle-2	25.955	0.650	0.143	0.141
	2.5	Axle-1	44.415	1.078	0.934	0.909
		Axle-2	44.773	1.574	0.849	0.814
	5.0	Axle-1	41.900	1.760	3.268	3.267
		Axle-2	25.800	0.967	0.919	0.919
	10.0	Axle-1	44.835	3.394	3.303	3.304
		Axle-2	25.212	3.386	3.342	3.343

Table 7.11 The Relative Percentage Error (percent) of identification with 10 percent noise for different types of measured information

	SA-5	SA-6	SA-6-1	SA-9	SA-9-3	SA-10
Axle-1	18.942	20.736	25.272	27.318	28.605	26.902
Axle-2	15.015	19.581	27.110	23.532	30.251	27.714

- In practice, the vibration mode combination $(m = 3, n = 3)$ gives accurate identified results, and this corresponds to the case with the least error in the study when the aspect ratio varies between 1.0 to 2.5 which is a practical range for ordinary bridge decks.

7.3.3.5 Further Study on the Effect of Noise in Different Types of Measurements

Section 7.3.3.2 has included limited study on the noise effect on the load identification. Further study is made in this section to compare the effect of 10 percent noise on different types of response measurements from different patterns of sensor arrangement, i.e. SA 5, 6, 6-1, 9, 9-3 and 10 as listed in Figure 7.16, and the relative percentage errors are listed in Table 7.11. This study makes use of the same two-axle vehicle for previous studies, and it moves along Rail $3^{\#}$. The $(m = 3, n = 3)$ mode combination is used both in the identification and the response analysis, and other conditions remain the same for Section 7.3.3.2. The following observations are made:

- The errors in the identified forces are large for all the cases when only one type of measured information is used (SA9 and 10).

Table 7.12 Sensor sets for moving load identification in experiment

Sensor Set	Total number of sensors	Sensor type and number	Sensor location Transverse	Longitudinal
I	9	6 strains	Beams 2[#] and 3[#]	¼, ½, ¾ spans
		3 accelerations	Beam 4[#]	- ditto -
2	9	9 strains	Beams 2[#], 3[#] and 4[#]	¼, ½, ¾ spans
3	9	6 strains	Beams I[#] and 3[#]	¼, ½, ¾ spans
		3 accelerations	Beam 5[#]	- ditto -
4	9	9 strains	Beams I[#], 3[#] and 5[#]	¼, ½, ¾ spans
5	12	6 strains	Beams I[#] and 2[#]	¼, ½, ¾ spans
		6 accelerations	Beams 4[#] and 5[#]	- ditto -
6	12	12 strains	Beams I[#], 2[#], 4[#] and 5[#]	¼, ½, ¾ spans
7	15	9 strains	Beams I[#], 2[#] and 3[#]	¼, ½, ¾ spans
		6 accelerations	Beams 4[#] and 5[#]	- ditto -
8	15	15 strains	Beams I[#], 2[#], 3[#], 4[#] and 5[#]	¼, ½, ¾ spans
9	20	20 strains	Beams I[#], 2[#] 3[#], 4[#] and 5[#]	¼, 3/8, ½, ¾ spans

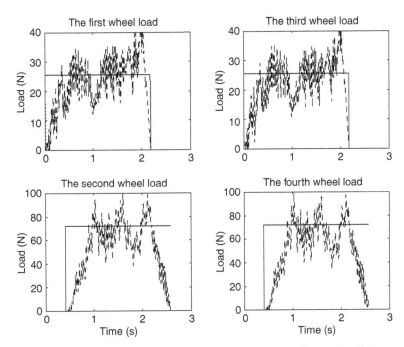

Figure 7.18 Identified wheel loads from the measured responses (Sensor Set 7) (— static load; - - - proposed method)

- When a mixture of measured information is used, acceleration performs better than velocity measurement.
- From the limited comparison in this study, the combination with more strain than acceleration measurements, i.e. six strains and three accelerations give the smallest error in the identified forces.

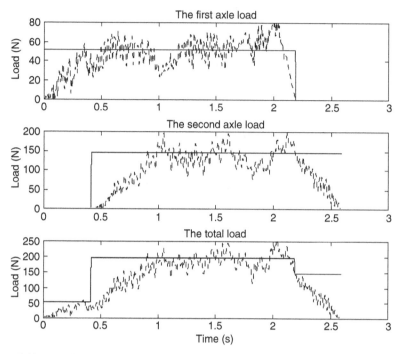

Figure 7.19 Identified experimental axle and total loads from the measured responses (sensor set 7) (— static load; - - - proposed method)

7.3.4 Experimental Studies

7.3.4.1 Experimental Set-up

The model vehicle and steel bridge deck fabricated in the laboratory, as described in Section 6.3.4.2, are used for this study. The measured strains and acceleration responses obtained from the tests are adopted in the following analysis. The average speed of the vehicle on the whole bridge deck is used for the identification of the moving loads. The force is not known in this practical case, and the Relative Percentage Error of the estimated forces is defined as:

$$\text{Relative Percentage Error} = \frac{\left\| f_{j+1}^{identify} - f_j^{identify} \right\|}{\left\| f_j^{identify} \right\|} \times 100\% \tag{7.40}$$

where $f_j^{identify}$, $f_{j+1}^{identify}$ are the identified forces with λ_j and $\lambda_j + \Delta\lambda$. The value of λ that corresponds to the smallest Relative Percentage Error is the optimal value.

7.3.4.2 Axle Loads and Wheel Loads Identification

The measured strains are re-sampled to have a time interval of 0.004 s. The model vehicle is moving at 1.11 m/s along rails $1^\#$, $2^\#$ and $3^\#$ in turn. The sensor sets adopted

Table 7.13 Correlation coefficients of measured and reconstructed strains from different sensor sets and load eccentricities

Eccentricity (Rail no.)	Sensor Sets	Beam 1[#]	Beam 2[#]	Beam 3[#]	Beam 4[#]	Beam 5[#]
0 (Rail 3[#])	1	–	0.978	0.968	0.979	–
	2	–	0.922	0.921	0.923	–
	3	0.970	–	0.971	–	0.964
	4	0.925	–	0.927	–	0.925
	5	0.970	0.981	–	0.981	0.965
	6	0.931	0.933	–	0.933	0.930
	7	0.971	0.983	0.985	0.983	0.970
	8	0.930	0.931	0.933	0.931	0.931
	9	0.932	0.932	0.935	0.932	0.930
1/8b (Rail 2[#])	1	–	0.932	0.930	0.931	–
	2	–	0.900	0.919	0.910	–
	3	0.902	–	0.923	–	0.930
	4	0.886	–	0.919	–	0.916
	5	0.911	0.926	–	0.931	0.933
	6	0.889	0.910	–	0.918	0.919
	7	0.904	0.928	0.932	0.935	0.930
	8	0.885	0.907	0.916	0.920	0.918
	9	0.884	0.906	0.908	0.916	0.912
3/8b (Rail 1[#])	1	–	0.903	0.909	0.944	–
	2	–	0.880	0.890	0.911	–
	3	0.300	–	0.909	–	0.944
	4	0.199	–	0.897	–	0.925
	5	0.311	0.908	–	0.954	0.956
	6	0.205	0.886	–	0.913	0.927
	7	0.303	0.902	0.922	0.961	0.967
	8	0.208	0.887	0.896	0.916	0.929
	9	0.211	0.890	0.899	0.919	0.931

Note: – denotes case not studied.

in the identification are shown in Table 7.12, and the number of modes employed in the identification is equal to the number of sensors, i.e. nine modes ($m = 3, n = 3$) for nine sensors, 12 modes ($m = 3, n = 4$) for 12 sensors, 15 modes ($m = 5, n = 3$) for 15 sensors and 20 modes ($m = 4, n = 5$) for 20 sensors.

Figures 7.18 and 7.19 show the identified wheel loads, axle loads and the total vehicle load compared with the corresponding static loads when the model vehicle moves along rail 3[#] from sensor set 7. The correlation coefficients between the measured and the reconstructed strain responses at 3/8a of each beam are tabulated in Table 7.13.

The following observations are made from Table 7.13, Figures 7.18 and 7.19:

• The combination of strains and accelerations give better results than the same number of strains alone, and the accuracy of identified axle loads is higher than that for the wheel loads.
• When the number of sensors is equal to the number of vibration modes in the identification, the use of more sensors does not improve the accuracy of the results significantly.

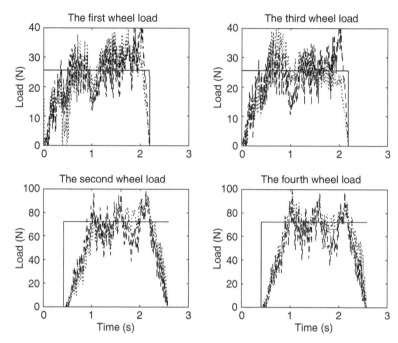

Figure 7.20 Identified experimental wheel loads from the measured responses (sensor set 8)
(— static load; - - - Method II; · · · · Dynamic programming method)

- When the vehicle moves along rail 1#, the correlation coefficients between the measured and the reconstructed strain responses at $3/8a$ of beam 1# is very poor, the others are all above 0.88. This is due to the low signal-to-noise ratio in the sensors further away from the moving loads giving less accurate results.

7.3.5 Comparison of the Two State–Space Approaches

The moving load identification problem has been solved using the dynamic programming method in Section 7.2 making use of velocity and strain measurements. In the second method described in Section 7.3, acceleration, velocity and strain measurements can be used as input, and is more flexible for practical use. This section compares the effectiveness and accuracy of load identification using both state space approaches. The measured strains in Section 7.3.4 are re-sampled to have a time interval of 0.005 s. The model vehicle is moving at 1.11 m/s along Rail 3#. Fifteen vibration modes are employed in the identification ($m = 3, n = 5$). The identified results calculated by the second method from Equation (7.40) and by the dynamic programming method from sensor set 8 listed in Table 7.12 are plotted in Figures 7.20 and 7.21. The identified results from the two methods are almost the same throughout the time histories. The second method takes 185 seconds to complete while the dynamic programming method takes 235 seconds to complete using MATLAB with a 2.4 GHz Pentiunm-4 PC and 512M RAM.

Another numerical comparison has been made with the EST method described in Section 6.3 (Zhu and Law, 2001b) which is also a time domain approach in the solution.

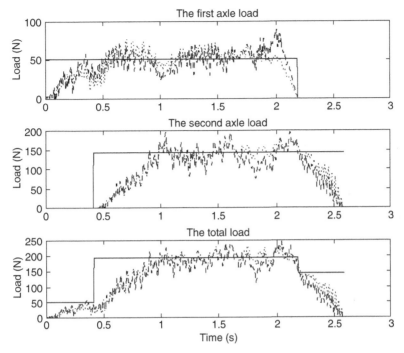

Figure 7.21 Identified experimental axle and total loads from the measured responses (sensor set 8) (— static load; - - - Method II; · · · · Dynamic programming method)

Table 7.14 Percentage errors in the identified results by Method II and that by Zhu and Law (2001b)

	Vibration modes		1% noise	3% noise	5% noise
Zhu and Law (2001b)	$m = 3, n = 3$	Axle-1	10.59	24.37	38.48
		Axle-2	5.48	14	22.54
		Time(s)	148	155	152
	$m = 3, n = 5$	Axle-1	6.39	12.79	17.92
		Axle-2	3.53	7.96	10.65
		Time(s)	290	301	234
Method II	$m = 3, n = 3$	Axle-1	2.11	5.92	10.38
		Axle-2	1.82	4.99	8.42
		Time(s)	122	99	76
	$m = 3, n = 5$	Axle-1	1.62	2.84	4.17
		Axle-2	1.41	2.15	3.16
		Time(s)	179	172	176

The bridge deck shown in Figure 7.14 is used in the comparison. The length of the bridge deck is 24.325 m. Other parameters are the same as those in previous numerical study in Section 7.3.3. Acceleration obtained from sensor arrangement SA9 in Figure 7.16 is used in the simulation and the moving loads are moving along Rail 2#.

Time interval between adjacent data point is 0.005 s. Table 7.14 shows the error of identification in the moving loads using Method II and the EST method. Nine ($m = 3, n = 3$) and 15 ($m = 3, n = 5$) vibration modes are used in the comparison study, and three types of noise levels are included. Results show that Method II gives much better results than Zhu and Law (2001b) with 10–100 percent less computation time.

7.4 Summary

Two moving load identification algorithms in state–space have been presented in this chapter. These methods provide bounds to the identified forces in the ill-conditioned problem. The first solution algorithm based on dynamic programming has a natural smoothing capability in the recurrence computations while Method II provides the smoothing effect via the damped least-squares solution or regularization. The errors of identification from Method II are much smaller than those obtained from the Exact Solution Technique (EST) from using different combinations of measured responses in both simulation and laboratory studies. Method II based on regularization is also more suitable for general application than the first method due to its flexibility to accept different types of dynamic response as input. The state–space formulation of the dynamic system can be extended to include a more complicated finite element model of a structure under multiple forces excitation.

Chapter 8

Moving Force Identification with Generalized Orthogonal Function Expansion

8.1 Introduction

Existing methods of moving load identification can be broadly classified into two categories, with one based on a continuous bridge model and modal superposition technique to decouple the equation of motion and the subsequent solution using optimization schemes as in Chapters 5 and 6. The second category is based on discrete bridge model with the finite element method to decouple the equation of motion, such as the state space approach in Chapter 7 and the finite element method (FEM) in Chapter 9. The modal superposition technique has good accuracy for identification but it demands extensive computation when multiple vehicles cross a multi-span bridge structure. The FEM approach is flexible when dealing with vehicular axle-loads moving on top of a bridge system with complex boundary conditions. However, a great deal of care must be used in transforming the displacement or strain to velocities and accelerations (O'Connor and Chan, 1988). Numerical differentiation may lead to large error in the identified results.

This chapter introduces the moving load identification through the generalized orthogonal function expansion to overcome the above computation problem. The method has efficient computational performance and good identification accuracy, especially with the orthogonal function smoothing technique to obtain the velocities and accelerations from the measured strains (Zhu and Law, 2001a). Orthogonal functions, such as the generalized orthogonal functions and wavelets, are introduced in Section 8.2. The moving load identification with these functions is presented in Section 8.3. The applications by numerical simulations and laboratory experiments are discussed in Sections 8.4 and 8.5.

8.2 Orthogonal Functions

8.2.1 Series Expansion

In order to improve the accuracy of the computed velocities and accelerations from strains or displacement measurements, the measurement is approximated with an analytical function. The velocity and acceleration are then obtained by differentiation of the function. The function is represented as a series expansion:

$$f(t) = \sum_{i=1}^{m} a_i \phi_i(t) \tag{8.1}$$

where $\phi_i(t)$ is the set of functions. There are many series expansions, such as polynomial functions, Fourier series and orthogonal functions, such as the Legendre polynomials, Chebyshev polynomials, Bessel functions and B-spline functions. Generalized orthogonal functions and wavelets will be discussed in the following sections.

8.2.2 Generalized Orthogonal Function

The Chebyshev polynomial has been used widely in numerical analysis. This section introduces the generalized orthogonal function from the first kind of Chebyshev polynomial $T_n(x)$. It is a polynomial in x of degree n, defined by the relation (Mason and Handscomb, 2003):

$$T_n(x) = \cos n\theta \quad \text{when } x = \cos \theta \tag{8.2}$$

If the range of the variable x is in the interval $[-1,1]$, then the range of the corresponding variable θ can be taken as $[0, \pi]$. These ranges are traversed in opposite directions, since $x = -1$ corresponds to $\theta = \pi$ and $x = 1$ corresponds to $\theta = 0$. We have the fundamental recurrence relation from Equation (8.2):

$$T_n(x) = 2xT_{n-1}(x) - T_{n-2}(x), \quad (n = 2, 3, \ldots) \tag{8.3}$$

with the initial conditions:

$$T_0(x) = 1, \quad T_1(x) = x \tag{8.4}$$

All the polynomials $\{T_n(x)\}$ can be generated recursively from Equation (8.3). The orthonormal polynomials can be obtained as follows, with scaling of the polynomials:

$$\frac{1}{\sqrt{\pi}}T_0(x), \left\{ \frac{2}{\sqrt{\pi}}T_i(x), \quad i = 1, 2, \ldots \right\} \tag{8.5}$$

For an independent variable t in a general range $[0, T]$, we can map the independent variable t to the variable x with the transformation

$$x = \frac{2t}{T} - 1 \tag{8.6}$$

and this leads to a shifted Chebyshev polynomial (of the first kind) $T_n^*(t)$ of degree n in variable t in the interval $[0, T]$ which is a generalized orthogonal function given as:

$$T_1^* = \frac{1}{\sqrt{\pi}}$$

$$T_2^* = \sqrt{\frac{2}{\pi}} \left(\frac{2}{T}t - 1 \right)$$

$$T_3^* = \sqrt{\frac{2}{\pi}} \left(2 \left(\frac{2}{T}t - 1 \right)^2 - 1 \right)$$

. . . .   (8.7)

$$T_{j+1}^* = 2 \left(\frac{2}{T}t - 1 \right) T_j^* - T_{j-1}^*$$

8.2.3 Wavelet Deconvolution

The Daubechies wavelets and associated scaling functions $\varphi_{j,k}(t)$ are obtained by translation and dilation of functions $\psi(t)$ and $\varphi(t)$ respectively (Law et al., 2008).

$$\psi_{J,k}(t) = 2^{J/2}\psi(2^J t - k) \quad J, k \in Z \tag{8.8}$$

$$\varphi_{J,k}(t) = 2^{J/2}\varphi(2^J t - k) \quad J, k \in Z \tag{8.9}$$

where J is the resolution. The scaling function $\varphi(t)$ and wavelet function $\psi(t)$ can be derived from the dilation equation as:

$$\varphi(t) = \sum_k a_k \varphi(2t - k) \tag{8.10}$$

$$\psi(t) = \sum_k (-1)^k a_{1-k} \varphi(2t - k) \tag{8.11}$$

where a_k, a_{1-k} are the filter coefficients and they are fixed for specific wavelet or scaling function basis. It is noted that only a finite number of a_k, a_{1-k} are nonzero for compactly supported wavelets.

The scaling function $\varphi(t)$ and wavelet function $\psi(t)$ have the following properties:

$$\int_{-\infty}^{\infty} \varphi(t)dt = 1 \tag{8.12}$$

$$\int_{-\infty}^{\infty} \varphi(t - j)\varphi(t - k)dt = \delta_{j,k}, \quad j, k \in Z \tag{8.13}$$

$$\int_{-\infty}^{\infty} t^m \psi(t)dt = 0, \quad (m = 0, 1, \ldots, L/2 - 1) \tag{8.14}$$

where m denotes the number of vanish moments and L is the order of Daubechies wavelet with $L = 2m$.

The translation of the scaling and wavelet functions on each fixed scale forms the orthogonal subspaces:

$$V_J = \{2^{J/2}\varphi(2^J t - k), \quad J \in Z\} \tag{8.15}$$

$$W_J = \{2^{J/2}\psi(2^J t - k), \quad J \in Z\} \tag{8.16}$$

such that V_J forms a sequence of embedded subspaces:

$$\{0\}, \ldots, \subset V_{-1} \subset V_0 \subset V_1, \ldots, \subset L^2(R) \quad \text{and} \quad V_{J+1} = V_J \oplus W_J \tag{8.17}$$

where $\oplus$ is the operator for the addition of two subspaces. At a certain resolution J, the approximation of a function $f(t)$ in $L^2(R)$ space using $\varphi_{J,k}(t)$ as basis can be denoted as:

$$P_J(f) = \sum_k \tilde{\alpha}_{J,k}\varphi_{J,k}(t), \quad J, k \in Z \tag{8.18}$$

where $P_J(f)$ is the approximation of $f(t)$ and $\tilde{\alpha}_{J,k}$ is the approximation coefficient.

Let $Q_J(f)$ be the detail of the function using $\psi_{J,k}(t)$ as basis at the same level J, and

$$Q_J(f) = \sum_k \tilde{\beta}_{J,k}\psi_{J,k}(t), \quad J, k \in Z \tag{8.19}$$

where $\tilde{\beta}_{J,k}$ is the detail coefficient. The approximation $P_{J+1}(f)$ of the next level $(J+1)$ of resolution is given by:

$$P_{J+1}(f) = P_J(f) + Q_J(f) \tag{8.20}$$

This forms the basis of multi-resolution analysis associated with wavelet approximation.

The Wavelet-Galerkin approximation to the signal $f(t)$ at a certain resolution J can be expressed as:

$$h(t) = \sum_k \tilde{\alpha}_{J,k}2^{J/2}\varphi(2^J t - k), \quad J, k \in Z \tag{8.21}$$

from Equations (8.8) and (8.18). Substituting $y = 2^J t$ into Equation (8.21), we obtain:

$$h(y) = \sum_k \alpha_{J,k}\varphi(y - k); \quad \alpha_{J,k} = 2^{J/2}\tilde{\alpha}_{J,k}, \quad J, k \in Z \tag{8.22}$$

If y takes up only integer values, the approximation is discretized at all dyadic points with $t = 2^{-J}y$ as:

$$h(i) = h(i\Delta y) = h_i, \quad (i = 0, 1, 2, \ldots, N_T) \tag{8.23}$$

where N_T is the number of time instances. Equation (8.22) can be rewritten as:

$$h_i = \sum_k \alpha_k \varphi_{i-k} = \sum_k \alpha_{i-k}\varphi_k \tag{8.24}$$

with $\varphi_k = \varphi(k)$. In matrix form this becomes:

$$
\begin{Bmatrix} b_1 \\ b_2 \\ b_3 \\ \vdots \\ \vdots \\ \vdots \\ \vdots \\ b_{N_T-1} \end{Bmatrix} = \begin{bmatrix} 0 & 0 & 0 & \cdots & \varphi_{L-2} & \cdots & \varphi_2 & \varphi_1 \\ \varphi_1 & 0 & 0 & \cdots & 0 & \cdots & \varphi_3 & \varphi_2 \\ \varphi_2 & \varphi_1 & 0 & \cdots & 0 & \cdots & \varphi_4 & \varphi_3 \\ \vdots & \cdots & \cdots & \cdots & \cdots & \cdots & \cdots & \vdots \\ \varphi_{L-2} & \varphi_{L-3} & \varphi_{L-4} & \cdots & \cdots & \cdots & 0 & 0 \\ 0 & \varphi_{L-2} & \varphi_{L-3} & \cdots & \cdots & \cdots & 0 & 0 \\ \vdots & \cdots & \cdots & \cdots & \cdots & \cdots & \cdots & \vdots \\ 0 & 0 & 0 & \cdots & \varphi_{L-3} & \cdots & \varphi_1 & 0 \end{bmatrix} \begin{Bmatrix} \alpha_1 \\ \alpha_2 \\ \alpha_3 \\ \vdots \\ \vdots \\ \vdots \\ \vdots \\ \alpha_{N_T-1} \end{Bmatrix}
\tag{8.25}
$$

The periodic boundary condition has been included in Equation (8.25) for the finite domain analysis denoted as:

$$
\begin{Bmatrix} \alpha_{-1} = \alpha_{N_T-1} \\ \alpha_{-2} = \alpha_{N_T-3} \\ \vdots \\ \alpha_{-L+2} = \alpha_{N_T-L+2} \end{Bmatrix} \quad \text{and} \quad \begin{Bmatrix} \alpha_{N_T} = \alpha_0 \\ \alpha_{N_T+1} = \alpha_1 \\ \vdots \\ \alpha_{N_T+L-2} = \alpha_{L-2} \end{Bmatrix}
\tag{8.26}
$$

8.3 Moving Force Identification

8.3.1 Beam Model

8.3.1.1 Generalized Orthogonal Function Expansion

The strain in the Euler-Bernoulli beam at a point x and time t can be written as:

$$
\varepsilon(x,t) = -h \frac{\partial^2 w(x,t)}{\partial x^2}
\tag{8.27}
$$

where h is the distance between the lower surface and the neutral plane of bending of the beam. Substituting the transverse displacement $w(x,t)$ in Equation (2.7) of Chapter 2 into Equation (8.27), and assuming there are N modes in the responses, we have:

$$
\varepsilon(x,t) = \varphi Q
\tag{8.28}
$$

where

$$
\varphi = -\{h\phi_1''(x), h\phi_2''(x), \ldots, h\phi_N''(x)\}; \qquad Q = \{q_1(t), q_2(t), \ldots, q_N(t)\}^T.
$$

and $\phi_i''(x)$ is the second derivative of $\phi_i(x)$.

The strain is then approximated by a generalized orthogonal function $T(t)$ as shown in Equation (8.5) as:

$$
\varepsilon(x,t) = \sum_{i=1}^{N_f} T_i(t) C_i(x)
\tag{8.29}
$$

where $\{T_i(t), i = 1, 2, \ldots, N_f\}$ are the generalized orthogonal functions; $\{C_i(x), i = 1, 2, \ldots, N_f\}$ is the vector of coefficients in the expanded expression. Note that the wavelet form of the orthogonal function in Equation (8.21) can also be used. The strains at the N_s measuring points can be expressed as:

$$\boldsymbol{\varepsilon} = \boldsymbol{CT} \tag{8.30}$$

where

$$\boldsymbol{T} = \{T_0(t), T_1(t), \ldots, T_{N_f}(t)\}^T;$$
$$\boldsymbol{\varepsilon} = \{\varepsilon(x_1, t), \varepsilon(x_2, t), \ldots, \varepsilon(x_{N_s}, t)\}^T;$$
$$\boldsymbol{C} = \begin{bmatrix} C_{10}(x_1) & C_{11}(x_1) & \cdots & C_{1N_f}(x_1) \\ C_{20}(x_2) & C_{21}(x_2) & \cdots & C_{2N_f}(x_2) \\ \vdots & \vdots & \vdots & \vdots \\ C_{N_s0}(x_{N_s}) & C_{N_s1}(x_{N_s}) & \cdots & C_{N_sN_f}(x_{N_s}) \end{bmatrix}$$

and $\{x_1, x_2, \ldots, x_{N_s}\}$ is the vector of the location of the strain measurements. By the least-squares method, the coefficient matrix can be obtained as:

$$\boldsymbol{C} = \boldsymbol{\varepsilon T}^T(\boldsymbol{TT}^T)^{-1} \tag{8.31}$$

Substitute Equation (8.28) into Equation (8.30), we have:

$$\boldsymbol{Q} = (\boldsymbol{\Phi}^T\boldsymbol{\Phi})^{-1}\boldsymbol{\Phi}^T\boldsymbol{CT} \tag{8.32}$$

where

$$\boldsymbol{\Phi} = -\begin{bmatrix} h\phi_1''(x_1) & h\phi_2''(x_1) & \cdots & h\phi_N''(x_1) \\ h\phi_1''(x_2) & h\phi_2''(x_2) & \cdots & h\phi_N''(x_2) \\ \vdots & \vdots & \vdots & \vdots \\ h\phi_1''(x_{N_s}) & h\phi_2''(x_{N_s}) & \cdots & h\phi_N''(x_{N_s}) \end{bmatrix}$$

and it can be obtained from Equation (2.4) in Chapter 2.

8.3.1.2 Moving Force Identification Theory

The vector of generalized coordinates obtained from Equation (8.32) can be substituted into Equation (2.8) for the beam, and rewrite it in matrix form to become:

$$I\ddot{\boldsymbol{Q}} + C_d\dot{\boldsymbol{Q}} + K\boldsymbol{Q} = \boldsymbol{BP} \tag{8.33}$$

where

$$C_d = diag(2\xi_i\omega_i);$$
$$K = diag(\omega_i^2)$$

$$B = \begin{bmatrix} \phi_1(\hat{x}_1(t))/M_1 & \phi_1(\hat{x}_2(t))/M_1 & \cdots & \phi_1(\hat{x}_{N_p}(t))/M_1 \\ \phi_2(\hat{x}_1(t))/M_2 & \phi_2(\hat{x}_2(t))/M_2 & \cdots & \phi_2(\hat{x}_{N_p}(t))/M_2 \\ \vdots & \vdots & \vdots & \vdots \\ \phi_N(\hat{x}_1(t))/M_N & \phi_N(\hat{x}_2(t))/M_N & \cdots & \phi_N(\hat{x}_{N_p}(t))/M_N \end{bmatrix}$$

The required $\ddot{Q}$ and $\dot{Q}$ can be obtained by directly differentiating Equation (8.32) to have:

$$\ddot{Q} = (\mathbf{\Phi}^T \mathbf{\Phi})^{-1} \mathbf{\Phi}^T C \ddot{T}$$
$$\dot{Q} = (\mathbf{\Phi}^T \mathbf{\Phi})^{-1} \mathbf{\Phi}^T C \dot{T}$$

The moving forces obtained from Equation (8.33) using a straight forward least-squares solution would be unbound. Let the left-hand-side of Equation (8.33) be represented by U. Regularization technique is used to solve the ill-posed problem in the form of minimizing the function:

$$J(P, \lambda) = \|BP - U\|^2 + \lambda \|P\|^2 \tag{8.34}$$

where λ is the non-negative regularization parameter.

The success of solving Equation (8.34) lies in how to determine the regularization parameter λ. Two methods are used in this chapter. If the true forces are known, the parameter can be determined by minimizing the error between the true forces and the predicting values as:

$$S = \|\hat{P} - P\| \tag{8.35}$$

In the practical case when the true forces are not known, the method of generalized cross-validation (GCV) is used to determine the optimal regularization parameter. The GCV function to be minimized in this work is defined by (Golub et al., 1979):

$$g(\lambda) = \frac{\|B\hat{P} - U\|_2^2}{\{trace[I - B((B^T B + \lambda I)^{-1} B^T)^{-1}]\}^2} \tag{8.36}$$

where $\hat{P}$ is the vector of estimated forces.

8.3.2 Plate Model

The displacement $w(x_s, y_s, t)$ at location (x_s, y_s) and at time t is rewritten in matrix form from Equation (3.16) in Chapter 3 as:

$$w(x_s, y_s, t) = W_s Q \quad (s = 1, 2, \ldots, N_s) \tag{8.37}$$

where N_s is the number of measuring points, and Q is a matrix of $q_{ij}(t)$ from Equation (3.16).

$$W_s = \{W_{11}(x_s, y_s), W_{12}(x_s, y_s), \ldots, W_{1n}(x_s, y_s), W_{21}(x_s, y_s), \ldots, W_{mn}(x_s, y_s)\}$$

The modal strains in x-direction can be written as:

$$W_{ij}(x_s, y_s) = -z_t \left(\frac{i\pi}{a}\right)^2 \sin\left(\frac{i\pi}{a}x_s\right) Y_{ij}(y_s), \quad (i = 1, 2, \ldots, m; \ j = 1, 2, \ldots, n).$$

where z_t is the distance from the measuring point at the outer surface to the neutral surface of bending. For N_s measuring points:

$$w_{ns} = W_{ns}Q \tag{8.38}$$

where

$$w_{ns} = [w(x_1, y_1, t), w(x_2, y_2, t), \ldots, w(x_{N_s}, y_{N_s}, t)]^T$$

$$W_{ns} = \begin{bmatrix} W_{11}(x_1, y_1) & W_{12}(x_1, y_1) & \cdots & W_{mn}(x_1, y_1) \\ W_{11}(x_2, y_2) & W_{12}(x_2, y_2) & \cdots & W_{mn}(x_2, y_2) \\ \vdots & \vdots & \cdots & \vdots \\ W_{11}(x_{N_s}, y_{N_s}) & W_{12}(x_{N_s}, y_{N_s}) & \cdots & W_{mn}(x_{N_s}, y_{N_s}) \end{bmatrix}_{N_s \times m \cdot n}$$

The modal displacement can be obtained from Equation (8.38) by least-squares method as:

$$Q = (W_{ns}^T W_{ns})^{-1} W_{ns}^T w_{ns} \tag{8.39}$$

Since the displacements or strains are measured, the velocities and accelerations can be obtained by dynamic programming filter (Trujillo and Busby, 1983) or orthogonal polynomial method described in Section 8.2, and the modal velocities and accelerations are calculated by the least-squares method from Equation (8.39). They are then substituted into Equation (3.17) for the plate to form the matrix equation:

$$B = SP \tag{8.40}$$

where

$$S = \begin{bmatrix} \dfrac{2\sin\left(\frac{\pi}{a}\hat{x}_1(t)\right) Y_{11}(\hat{y}_1(t))}{\rho h a \int_0^b Y_{11}^2(y)dy} & \dfrac{2\sin\left(\frac{\pi}{a}\hat{x}_2(t)\right) Y_{11}(\hat{y}_2(t))}{\rho h a \int_0^b Y_{11}^2(y)dy} & \cdots & \dfrac{2\sin\left(\frac{\pi}{a}\hat{x}_{N_p}(t)\right) Y_{11}(\hat{y}_{N_p}(t))}{\rho h a \int_0^b Y_{11}^2(y)dy} \\[3ex] \dfrac{2\sin\left(\frac{\pi}{a}\hat{x}_1(t)\right) Y_{12}(\hat{y}_l(t))}{\rho h a \int_0^b Y_{12}^2(y)dy} & \dfrac{2\sin\left(\frac{\pi}{a}\hat{x}_2(t)\right) Y_{12}(\hat{y}_2(t))}{\rho h a \int_0^b Y_{12}^2(y)dy} & \cdots & \dfrac{2\sin\left(\frac{\pi}{a}\hat{x}_{N_p}(t)\right) Y_{12}(\hat{y}_{N_p}(t))}{\rho h a \int_0^b Y_{12}^2(y)dy} \\[3ex] \vdots & \vdots & \cdots & \vdots \\[3ex] \dfrac{2\sin\left(\frac{m\pi}{a}\hat{x}_1(t)\right) Y_{mn}(\hat{y}_l(t))}{\rho h a \int_0^b Y_{mn}^2(y)dy} & \dfrac{2\sin\left(\frac{m\pi}{a}\hat{x}_2(t)\right) Y_{mn}(\hat{y}_2(t))}{\rho h a \int_0^b Y_{mn}^2(y)dy} & \cdots & \dfrac{2\sin\left(\frac{m\pi}{a}\hat{x}_{N_p}(t)\right) Y_{mn}(\hat{y}_{N_p}(t))}{\rho h a \int_0^b Y_{mn}^2(y)dy} \end{bmatrix}$$

$$B = \left\{ \begin{array}{c} \ddot{q}_{11}(t) + 2\zeta_{11}\omega_{11}\dot{q}_{11}(t) + \omega_{11}^2 q_{11}(t) \\ \ddot{q}_{12}(t) + 2\zeta_{12}\omega_{12}\dot{q}_{12}(t) + \omega_{12}^2 q_{12}(t) \\ \vdots \\ \ddot{q}_{mn}(t) + 2\zeta_{mn}\omega_{mn}\dot{q}_{mn}(t) + \omega_{mn}^2 q_{mn}(t) \end{array} \right\}; \quad P = \left\{ p_1(t), p_2(t), \ldots, p_{N_p}(t) \right\}^T$$

(8.41)

The moving load P can be obtained by the straightforward least-squares method from Equation (8.40). But the solutions are frequently unstable in the sense that small noises in the responses would result in large changes in the predicted moving force. Regularization technique is utilized to improve the conditioning. The load identification is formulated as a nonlinear least-squares problem.

$$\min J(P, \lambda) = (B - SP, R(B - SP)) + \lambda(P, P)$$

(8.42)

where λ is an optimal regularization parameter or a vector. R is a weight matrix and it can be determined from the measured information (Santantamarina and Fratta, 1998). Generalized cross-validation method (Golub et al., 1979) and L-Curve method (Hansen, 1992) are then used to determine the optimal regularization parameter in this study.

8.4 Applications

8.4.1 Identification with a Beam Model

The method, described in previous sections, is illustrated in the following simulation studies. The effect of discarding some of the information contained in the measured responses on the error of identification is studied.

8.4.1.1 Single-Span Beam

A single span simply supported beam is studied with two varying forces moving on top at a constant spacing of 4.27 m.

$$\begin{aligned} f_1(t) &= 9.9152 \times 10^4 [1 + 0.1\sin(10\pi t) + 0.05\sin(40\pi t)] \text{ N}; \\ f_2(t) &= 9.9152 \times 10^4 [1 - 0.1\sin(10\pi t) + 0.05\sin(50\pi t)] \text{ N}. \end{aligned}$$

(8.43)

The parameters of the beam are as follow: $EI = 2.5 \times 10^{10} \text{ Nm}^2$, $\rho A = 5000 \text{ kg/m}$, $L = 30 \text{ m}$, $h = 1 \text{ m}$. The first eight natural frequencies of the beam are 3.9, 15.61, 35.13, 62.48, 97.58, 140.51, 191.25 and 249.8 Hz, and they are used in the computation of the analytical mode shapes from Equation (2.4) in Chapter 2. The forces are moving at a speed of 30 m/s. Random noise is added to the calculated strains to simulate the polluted measurement and 1, 5 and 10 percent noise levels are studied with:

$$\varepsilon = \varepsilon_{calculated} + E_p \cdot N_{iose} \cdot \text{var}(\varepsilon_{calculated})$$

(8.44)

where ε is the vector of strains; E_p is the noise level; N_{iose} is a standard normal distribution vector with zero mean and unit standard deviation; $\varepsilon_{calculated}$ is the vector of

Table 8.1 Error of Identification for single span beam

Number of mode shapes	Noise level					
	1%		5%		10%	
	First force	Second force	First force	Second force	First force	Second force
2	31.51	31.53	45.11	45.05	174.27	112.50
3	11.56	12.00	13.04	13.24	16.31	16.08
4	6.78	6.18	7.05	6.95	8.47	8.69
5	5.16	3.62	5.07	3.89	5.46	4.80
6	3.90	3.10	4.02	3.28	4.14	3.66
7	3.43	2.99	3.45	3.13	3.66	3.44
8	3.15	2.86	3.19	3.01	3.42	3.29
9	9.52	8.96	9.48	8.94	9.48	8.94
10	18.02	17.30	18.51	17.99	18.42	17.88

calculated strains; $var(\varepsilon_{calculated})$ is the standard deviation of $\varepsilon_{calculated}$. The errors in the identified forces are calculated as:

$$Error = \frac{\|\hat{P} - P_{True}\|}{\|P_{True}\|} \times 100\% \qquad (8.45)$$

Table 8.1 shows the errors of identification from using different number of mode shapes in the identification. The time step is 0.001 s in the calculation. The strain consists of responses from the first eight mode shapes polluted with 5 percent noise level. Ten measuring points are available in the identification and they are evenly distributed along the beam length. The different combination of number of mode shapes used in the identification and the number of measuring points are studied. Figure 8.1 shows the identified results using three and six mode shapes. The following observations were made:

1. Results in Table 8.1 show that the errors in the identified forces are insensitive to the noise level in the responses. This is because orthogonal functions have been used to approximate the strains in the identification, and this approximation suppresses the errors due to high frequency measurement noise.
2. When the number of mode shapes used in the identification is the same as the number of mode shapes in the responses, i.e. eight mode shapes, the errors of identification are the smallest. The errors become large when the number of mode shapes used in identification is either larger or smaller than the number of mode shapes in the responses. This indicates that the pairing of the number of mode shapes in both the responses and the identified forces has a large effect on the errors in the identification. The correct pairing can be determined from an inspection of the frequency content in the measured responses.
3. Figure 8.1 shows that there are large discrepancies in the identified forces near the beginning and the end of the moving forces when only three modes are used in the identification. These discrepancies are much less when six modes are used. This is because of the sudden appearance and disappearance of the forces at these points,

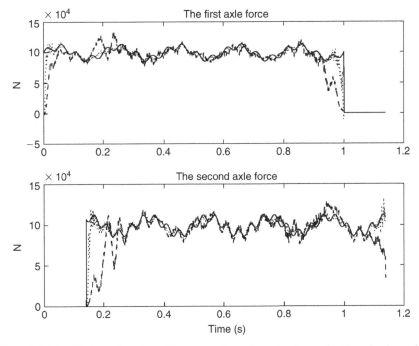

Figure 8.1 Identified results with different number of mode shapes (— True loads; -- from 3 modes; ···· from 6 modes)

which can be represented by an equivalent impulsive force. These impulsive forces excite the beam with a broad-band vibration that covers a large number of modal frequencies. Therefore, more mode shapes should be used in the identification to take advantage of the information of the forces at higher modal frequencies in the responses at the beginning and the end of the time histories.

8.4.1.2 Two-Span Continuous Beam

Table 8.2 shows the errors in the identified moving forces on a two-span continuous beam with different numbers of mode shapes and numbers of measuring points. The parameters of the beam are the same as for the single-span beam except that each span measures 30 m long. The first eight natural frequencies of the beam are 3.9, 6.1, 15.61, 19.75, 35.12, 41.22, 62.43 and 70.48 Hz. Figure 8.2 shows the identified forces from using strains polluted with 5 percent noise level at six measuring points. Inspection of the results in Table 8.2 and Figure 8.2 gives the following observations:

1. Results in Table 8.2 show that the errors increase as the noise level in the response increase. The errors are more than twice of that under similar conditions for the single-span beam. Therefore, moving load identification in a multi-span beam would be less accurate than that in a single-span beam.
2. When the number of mode shapes used in the identification equals to that in the responses as shown in the first two rows and the lower part of Table 8.2, the

Table 8.2 Identified Errors for Two-span Beam

No. of mode shape in responses N_1	No. of mode shapes in identification N_2	No. of measuring points N_s	Noise level					
			1%		5%		10%	
			First force	Second force	First force	Second force	First force	Second force
10	10	14	7.85	9.07	18.48	19.44	27.64	28.99
10	10	10	7.78	8.98	16.67	17.98	26.28	27.42
10	9	10	8.90	10.61	15.95	17.23	24.24	25.12
10	8	10	11.54	13.71	23.05	24.12	31.37	32.59
10	7	10	13.81	16.19	17.02	19.01	21.99	23.70
10	6	10	17.05	19.55	20.76	22.99	26.61	28.54
6	6	6	15.99	18.15	20.15	21.91	26.26	27.78
6	6	8	16.06	18.21	20.93	22.69	27.74	29.24
6	6	10	15.98	18.15	19.98	21.77	26.11	27.69
6	6	12	15.90	18.08	18.90	20.73	24.02	25.65
6	6	14	15.85	18.04	18.19	20.06	22.47	24.14

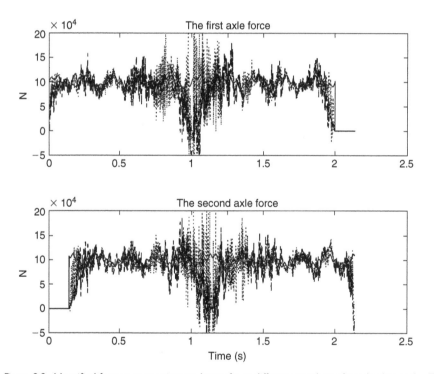

Figure 8.2 Identified forces on continuous beam from different number of mode shapes (— True loads; - - from with 3 modes; · · · · from with 6 modes)

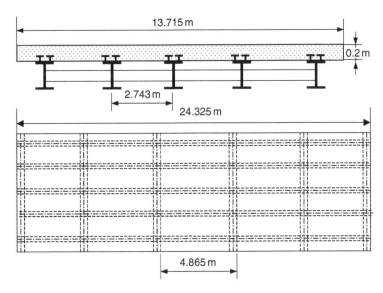

Figure 8.3 A typical single span bridge deck

errors in the identified forces varies only slightly with more measuring points. The number of the measuring points is best selected to be equal to the number of mode shapes.

3. Results from the upper part of Table 8.2 also show that the errors would be smallest when the number of mode shapes in the identification is the same as that in the responses. This confirms the observation made in the case of the single span beam.

4. The identified forces in Figure 8.2 have large fluctuations close to the intermediate support at 1.0 s. This is due to the presence of the small responses generated at this time instance with subsequently a small signal to noise ratio in the measured data.

8.4.2 *Identification with a Plate Model*

A simply supported prototype bridge composed of five I-section steel girders and a concrete deck as shown in Figure 8.3. It is noted that the model is similar to the one used by Fafard and Mallikarjuna (1993) in their study of bridge–vehicle interaction. It is also similar to the continuous bridge deck shown in Figure 4.12 of this book. It is wide enough to accommodate four lanes of traffic. The parameters of the bridge deck are listed as follow: $a = 24.325$ m, $b = 13.715$ m, $h = 0.2$ m, $E_x = 4.1682 \times 10^{10}$ N/m^2 $E_y = 2.9733 \times 10^{10}$ N/m^2, $\rho = 3000$ kg/m^3, $\nu_{xy} = 0.3$. For the steel I-beam: web thickness $= 0.01111$ m, web height $= 1.490$ m, flange width $= 0.405$ m, flange thickness $= 0.018$ m. For the diaphragms, the distance between two diaphragms is 4.865 m, cross-sectional area $= 0.001548$ m^2, $I_y = 0.707 \times 10^{-6}$ m^4, $I_z = 2 \times 10^{-6}$ m^4, $J = 1.2 \times 10^{-7}$ m^4. The rigidities in the x-direction of the equivalent orthotropic plate can be calculated according to Bakht and Jaeger (1985), as $D_x = 2.415 \times 10^9$ Nm,

Table 8.3 Natural Frequencies of the equivalent orthotropic plate (Hz)

m \ n	1	2	3	4	5	6	7	8
1	4.96*	6.31	10.01	16.07	24.81	36.46	51.08	68.70
2	19.84*	21.26	25.41	32.17	41.51	53.49	68.22	85.82
3	44.65*	46.07	50.32	57.33	67.06	79.48	94.60	112.49
4	79.37*	80.80	85.07	92.17	102.06	102.50	114.74	115.58
5	124.01*	124.02	125.42	125.44	125.46	129.40	129.43	129.51

* Longitudinal bending modes.

$D_y = 2.1813 \times 10^7$ Nm, $D_{xy} = 2.2195 \times 10^8$ Nm. The natural frequencies of the bridge deck are listed in Table 8.3. It should be noted that this structure is similar to, but not the same as, that in Chapter 4.

A two-axle vehicle model is used in the simulation. The axle spacing and wheel spacing are 4.26 m and 1.829 m respectively. The four wheel loads are listed as follows:

$$\begin{cases} P_1(t) = 3134.* \, (1 + 0.1 \sin(10\pi t) - 0.1 \sin(20\pi t) + 0.05 \sin(40\pi t)) \, \text{kg} \\ P_2(t) = 6166.* \, (1 - 0.1 \sin(10\pi t) - 0.1 \sin(20\pi t) + 0.05 \sin(40\pi t)) \, \text{kg} \\ P_3(t) = 3134.* \, (1 + 0.1 \sin(10\pi t) + 0.1 \sin(20\pi t) + 0.05 \sin(40\pi t)) \, \text{kg} \\ P_4(t) = 6166.* \, (1 + 0.1 \sin(10\pi t) + 0.1 \sin(20\pi t) + 0.05 \sin(40\pi t)) \, \text{kg} \end{cases}$$

(8.46)

where P_1 and P_3 are the front wheels and P_2 and P_4 are the rear wheels with P_2 and P_4 after P_1 and P_3 respectively. The total vehicle load is 18.6 Tonnes and the proportion of axle loads follows the pattern of vehicle type H20-44 from AASHTO (2002). The vehicle moving speed is 20 m/s, and the time step of analysis is 0.001 s in the simulation. White noise is added to the calculated displacements or strains to simulate the polluted measurements.

8.4.2.1 Study on the Noise Effect

The vehicle is moving along the centerline of the deck. The measured responses from 25 modes ($m = 5$, $n = 5$) in Table 8.3 are used in the calculation, and the number of modes in the identification Equation (8.40) is the same as that in the responses. According to discussions in Chapter 6, the number of measuring points should not be less than the number of vibration modes in the measured information. And therefore 25 measuring points are selected evenly distributed on the five I-beams. The identified individual wheel loads from using displacement responses with 1 percent and 5 percent noise levels are shown in Figure 8.4. The following observations are made:

1. The beginning or end of the identified results is under-estimated when there is noise in the responses. This is due to the small responses at the beginning or the end of the time duration, and the fact that the regularization parameter has been optimized over the total time duration of the event.
2. Errors in the identified results increase with the noise level. Hence when the noise level is high, a data treatment process (such as filtering or smoothing) should be used to reduce the noise in the responses before the computation.

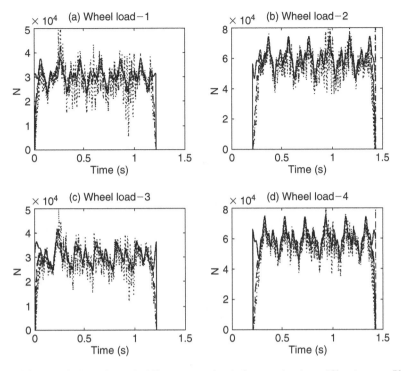

Figure 8.4 Identified results with different noise levels (— true load; - - - 1% noise; · · · · 5% noise)

8.4.2.2 Identification with Incomplete Modal Information

In practice, the vibration modes selected for identification are not the same as that in the responses. In general the lower modes of the structure are dominating the measured responses, and they are used in the identification. The moving loads are identified again with this incomplete modal information with fewer modes in the identification than those in the responses. The parameters of the system used in the simulation are the same as those in the last study. The vehicle is moving along the centerline of the bridge deck. Table 8.4 shows the errors in the identified results, and Figure 8.5 shows the identified results using 20 modes ($m = 4$, $n = 5$) or ($m = 5$, $n = 4$) with 1 percent noise in the responses. The following observations are made from the results:

1. When $m \geq 4, n \geq 4$, an acceptable result can be obtained with most of the errors less than 10 percent at 1 percent noise level. This is because the natural frequencies of these modes (shown in Table 8.3) have covered most of the excitation frequency range of the car as shown in Equation (8.46). In practice, the frequency range required in the identification can be obtained from the spectrum of the responses.
2. The more vibration modes used in the identification, the fewer errors are found in the identified results (Table 8.4). However, large errors still exist at the beginning and the end of the load time histories as seen in Figure 8.5. This is due to the fact that impulses are generated by the moving loads at the beginning and the end of

Table 8.4 Errors (percent) in the identified loads from different measured information

Noise Level			1%		5%		10%	
m	n	Total mode no.	Axle-1	Axle-2	Axle-1	Axle-2	Axle-1	Axle-2
5	5	25	6.97	6.00	17.90	14.46	28.68	22.16
			6.51	5.98	15.58	13.38	24.77	20.28
5	4	20	9.05	7.74	17.95	14.80	28.07	21.94
			9.23	6.60	16.50	12.78	26.31	20.41
5	3	15	30.56	18.62	32.25	20.32	36.11	23.99
			31.27	18.29	32.64	19.75	36.02	22.91
4	5	20	7.06	6.10	18.01	14.59	28.79	22.33
			6.59	6.06	15.69	13.51	24.89	20.44
4	4	16	9.13	7.82	18.05	14.93	28.17	22.10
			9.34	6.68	16.63	12.91	26.44	20.57
4	3	12	30.86	18.82	32.53	20.51	36.32	24.15
			31.57	18.43	32.90	19.88	36.23	23.04
3	5	15	10.82	9.83	23.71	21.93	34.64	32.22
			9.90	9.37	20.68	20.12	30.20	29.67
3	4	12	12.57	12.03	23.23	21.97	33.66	31.59
			14.67	10.29	23.12	20.03	32.58	30.45
3	3	9	34.69	23.48	36.19	25.91	40.10	32.10
			38.30	23.86	38.86	25.21	40.69	30.02
2	5	10	22.91	18.95	35.41	37.72	44.28	49.50
			19.71	17.13	32.53	35.17	40.23	46.76

Note: The errors in table correspond to each wheel load as $\dfrac{wheel\ 1\,|\,wheel\ 2}{wheel\ 3\,|\,wheel\ 4}$

the time duration, and a lot of higher modes of the structure are excited which are not covered by the selected vibration modes in the identification.

3. There are large errors in the identified results with $m = 5$, $n = 3$. This shows that the torsional modes are also very important in the moving load identification on bridge decks even if the vehicle is moving along the centerline.

8.4.2.3 *Effects of Travel Path Eccentricity*

There are four lanes on the bridge deck. Normally the vehicle is not moving exactly along the centerline. Table 8.5 shows the errors in the identified results with the car moving at different eccentricities and using different number of vibration modes in the identification. The identified results for different eccentricities with 25 modes ($m = 5$, $n = 5$) are shown in Figure 8.6. The parameters are the same as for the above studies, and the responses are calculated with 25 modes ($m = 5$, $n = 5$). The following intermediate conclusions can be drawn from Table 8.5 and Figure 8.6.

1. When $m \geq 3$, $n = 5$, an acceptable result can be obtained with most of the errors less than 10 percent at 1 percent noise level. This shows that the method proposed in the chapter is also effective to identify the eccentric moving loads on the bridge deck.

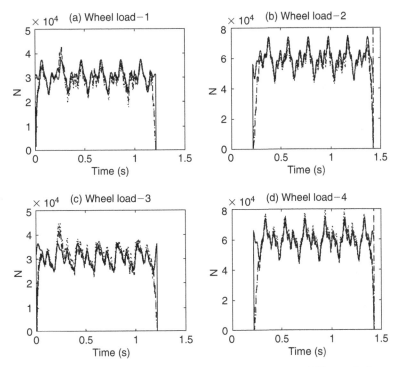

Figure 8.5 Identified results with different mode combinations (1% noise) (— true load; --- $m = 4, n = 5$; ···· $m = 5, n = 4$)

Table 8.5 Errors (percent) in the identified load moving at different eccentricities

Eccentricity	m	n	1%		5%		10%	
			Axle-1	Axle-2	Axle-1	Axle-2	Axle-1	Axle-2
3/8b	5	5	7.20	6.35	14.33	11.33	23.16	17.93
			6.16	4.37	19.06	12.26	34.71	21.75
	4	5	6.72	6.41	15.25	13.26	24.41	20.99
			10.16	9.95	25.20	21.58	41.72	33.84
	3	5	11.31	11.56	21.06	22.25	29.91	33.50
			13.56	13.25	29.46	27.37	45.00	40.97
	5	4	33.78	33.35	38.48	37.06	46.14	42.97
			16.83	20.04	26.33	21.33	41.11	26.60
1/8b	5	5	6.46	5.56	15.88	12.86	25.22	19.60
			7.73	7.07	15.42	12.35	22.31	20.31
	4	5	6.54	5.64	15.96	12.97	25.31	19.72
			7.80	7.16	15.51	14.48	22.41	20.47
	3	5	10.18	9.17	21.23	19.25	30.81	27.89
			11.22	10.93	20.08	21.19	27.42	29.28
	5	4	21.16	20.38	23.83	22.02	30.45	26.05
			22.03	18.21	23.46	17.44	26.87	18.61

Note: The errors in table correspond to each wheel load as $\dfrac{wheel\ 1 | wheel\ 2}{wheel\ 3 | wheel\ 4}$

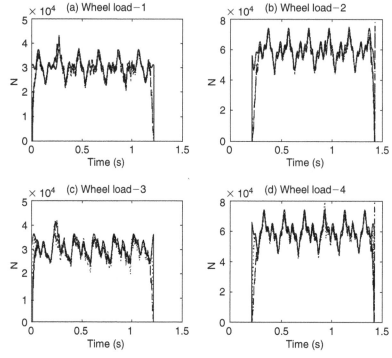

Figure 8.6 Identified results for different eccentricities (1 percent noise) (— true load; -.-.- e = 0; --- e = 1/8b; ⋯⋯ e = 3/8b)

2.   In the cases with $n < 5$, there are large errors in the identified eccentric load. This shows that the torsional modes are more important in the eccentric load identification than that for the case with the car moving along the centerline.

3.   When the eccentricity increases, the errors in the identified results also increase as seen in Table 8.5. Since the same measured information is used in all sets of identification, eccentric moving loads are more difficult to be identified accurately. There is a need for an optimum selection of measuring locations for different moving load configuration, and it is a subject of further research.

8.5 Laboratory Studies

8.5.1 *Beam Model*

8.5.1.1 *Experimental Setup and Measurements*

The measurements from the experimental system described in Section 6.3.4.1 are used for this study.

8.5.1.2 *Force Identification*

Strains at 1/8L, 1/4L, 3/8L, 1/2L, 3/4L, 7/8L are used in the identification. Table 8.6 shows the correlation coefficients between the measured and the reconstructed strains

Table 8.6 The correlation coefficients between measured and reconstructed responses at 5/8L

Case	Number of Mode Shapes	Measuring locations	Correlation Coefficient
A	3	1/4L, 1/2L, 3/4L	0.9809
B	4	1/8L, 1/4L, 1/2L, 3/4L	0.9470
C	5	1/8L, 1/4L, 1/2L, 3/4L, 7/8L	0.9752
D	3	1/8L, 1/4L, 3/8L, 1/2L, 3/4L, 7/8L	0.9853
E	4	1/8L, 1/4L, 3/8L, 1/2L, 3/4L, 7/8L	0.9837
F	5	1/8L, 1/4L, 3/8L, 1/2L, 3/4L, 7/8L	0.9822
G	6	1/8L, 1/4L, 3/8L, 1/2L, 3/4L, 7/8L	0.9716

at 5/8L obtained from the identified forces with different number of mode shapes in the identification. The number of measuring points is taken equal to the number of mode shapes in the identification. Figure 8.7 shows that the identified forces from Cases (A) and (G) of the study using 3 and 6 sensors respectively. The combined force is also presented in Figure 8.7(c). The following observations are made:

1. Table 8.6 shows that the correlation coefficients are all larger than 0.9 for different combination of modes and measuring points. It shows that the method based on generalized orthogonal function is effective to identify the moving forces in practice.
2. There is a low frequency component in the identified individual forces in Figure 8.7. This is the pitching motion of the moving car.
3. The identified forces from using six modes are closer to the static forces at the beginning and the end of the time histories than those obtained from using three modes. This gives experimental evidence that more mode shapes in the computation should be used to identify the moving forces near these locations.

8.5.2   Plate Model

8.5.2.1   Experimental Set-up

The model vehicle–bridge system fabricated in the laboratory as described in Section 6.3.4.2 is used for this study. The same model car as described in Section 6.3.4.2 is used in the experiment. Details on the experimental setup are referred to Figure 6.28. The layout plan of the sensors is reproduced in Figure 8.8 for easy reference in this study.

Twenty-five strain gauges were located at the bottom of the ribs to measure the strain of the bridge deck as shown in Figure 8.8. Six B&K model 4370 accelerometers were placed at the bottom of beams 4 and 5 at 1/4, 1/2 and 3/4 span for the acceleration measurements. $p_1(t)$ and $p_3(t)$ are the left and right wheel loads at the front looking in the direction of the traveling path; $p_2(t)$ and $p_4(t)$ are the left and right wheel loads at the back following $p_1(t)$ and $p_3(t)$.

The rigidities of the equivalent orthotropic plate are calculated as $D_x = 7.3677 \times 10^4$ Nm, $D_y = 4.2696 \times 10^3$ Nm and $D_k = 8.6018 \times 10^3$ Nm. The first ten measured

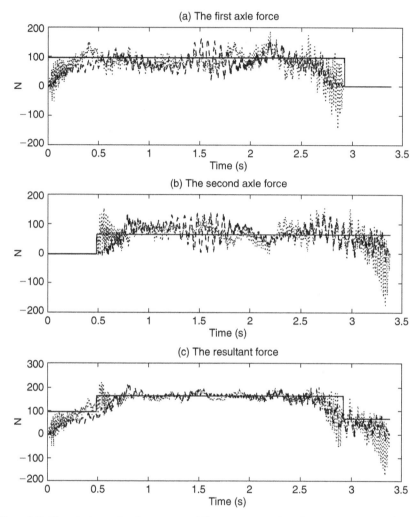

Figure 8.7 Identified results from using different modes (— Static load; -- with 3 modes (Case (A)); · · · with 6 modes (Case (G))

and calculated natural frequencies of the bridge deck are listed in Table 6.14 in Chapter 6. The response is sampled at 1000 Hz, and the number of data in each record segment is 7680. The average speed of the vehicle on the whole bridge deck is used for the identification of the moving loads in this study.

8.5.2.2 Wheel Load Identification

The measured strains are re-sampled to have a time interval of 0.003 s. When the model car is moving along the centerline (Rail 3), Rail 1 or Rail 2 in turn, the strains at 1/4a, 1/2a and 3/4a of each beam are measured to identify the moving wheel loads. The average speed of the model car is 0.54 m/s. Figures 8.9–8.11 show the identified

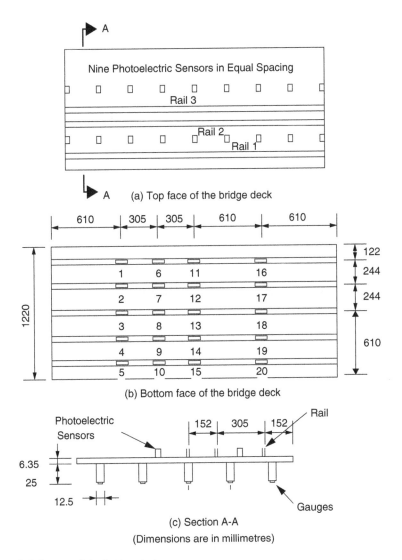

A

Nine Photoelectric Sensors in Equal Spacing

Rail 3

Rail 2
Rail 1

A   (a) Top face of the bridge deck

(b) Bottom face of the bridge deck

(c) Section A-A
(Dimensions are in millimetres)

Figure 8.8 Layout of the bridge deck

wheel loads, axle loads and the combined load from different vibration modes used in the identification with different number of measuring points when the model car moves along the centerline. Table 8.7 shows the correlation coefficients between the reconstructed and measured strains at 3/8a of each beam for different moving paths of the car. The following observations are made from the Figures and Table 8.7.

1. The method based on generalized orthogonal function is effective to identify individual moving wheel loads and acceptable results can be obtained.
2. The correlation coefficients between the reconstructed and measured strains on the beams adjacent to the moving path of the car are larger than 0.8. This shows

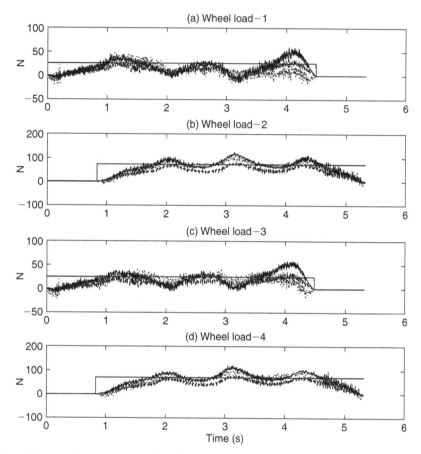

Figure 8.9 Identified wheel loads for different combinations of measured information (— static loads; -- $m = 3, n = 2(9)$; --- $m = 3, n = 3(15)$; ···· $m = 3, n = 2(15)$)

that the method is effective to identify the wheel loads moving with or without an eccentricity.

3. When the distance between the measuring point and the path of moving car is large, the correlation coefficient is small, as seen from Beam #1 for $e = 3/8b$. This is because of the small responses at the measuring points, and the reconstructed response is very sensitive to error in the identified loads.

4. The identified loads from the case with 15 vibration modes ($m = 3$, $n = 3$) is always smaller in all the results shown in Figures 8.9–8.11. The reason is due to an unequal number of modes used in the responses and in the identification, and it will be discussed in next section.

8.5.2.3 Effect of Unequal Number of Modes in the Response and in the Identification

Figures 8.9–8.11 show that the identified loads from ($m = 3$ and $n = 3$) is less than the loads identified from ($m = 3$ and $n = 2$). This difference cannot be the result of

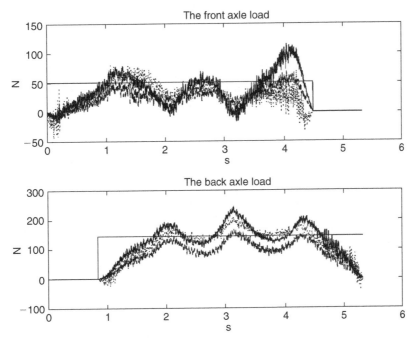

Figure 8.10 Identified axle loads for different combinations of measured information (— static loads; - - $m = 3, n = 2(9)$; - - - $m = 3, n = 3(15)$; · · · · $m = 3, n = 2(15)$)

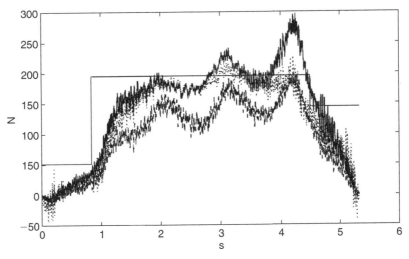

Figure 8.11 Identified total loads from using different modes (— static loads; - - $m = 3, n = 2(9)$; - - - $m = 3, n = 3(15)$; · · · · $m = 3, n = 2(15)$)

Table 8.7 Correlation coefficient between reconstructed and measured strains at 3/8a

Eccentricity	Modes	Beam 1	Beam 2	Beam 3	Beam 4	Beam 5
0	$m=3; n=4(15)$	0.783	0.897	0.931	0.909	0.799
(Rail 3)	$m=3; n=3(15)$	0.935	0.944	0.931	0.951	0.941
	$m=3; n=2(15)$	0.901	0.935	0.929	0.944	0.922
	$m=3; n=2(9)$	0.932	0.949	0.933	0.953	0.947
3/8b	$m=3; n=4(15)$	0.112	0.772	0.897	0.939	0.936
(Rail 1)	$m=3; n=3(15)$	0.166	0.793	0.915	0.951	0.948
	$m=3; n=2(15)$	0.039	0.813	0.914	0.948	0.947
	$m=3; n=2(9)$	0.044	0.692	0.839	0.849	0.837
1/8b	$m=3; n=4(15)$	0.550	0.794	0.848	0.790	0.758
(Rail 2)	$m=3; n=3(15)$	0.859	0.937	0.945	0.949	0.974
	$m=3; n=2(15)$	0.896	0.947	0.953	0.948	0.966
	$m=3; n=2(9)$	0.880	0.922	0.920	0.929	0.945

Note: (15) denotes 15 measuring points located evenly on the five beams; (9) denotes nine measuring points located evenly on the three beams near the moving path of the car.

any calibration error. An inspection of Equations (8.38) to (8.41) gives the following reasons for the existence of this error.

Equation (8.38) is valid for both the measured responses and for the identification. Let $N_R = m_R \times n_R$ and $N_I = m_I \times n_I$ be the number of the modes in the responses and in identification respectively. Equation (8.38) can be rewritten as follows:

$$w_{ns} = W_{N_s N_R} Q_{N_R} \tag{8.47}$$

where

$$W_{N_s N_R} = \begin{bmatrix} W_1(x_1, y_1) & W_2(x_1, y_1) & \cdots & W_{N_R}(x_1, y_1) \\ W_1(x_2, y_2) & W_2(x_2, y_2) & \cdots & W_{N_R}(x_2, y_2) \\ \vdots & \vdots & \cdots & \vdots \\ W_1(x_{N_s}, y_{N_s}) & W_2(x_{N_s}, y_{N_s}) & \cdots & W_{N_R}(x_{N_s}, y_{N_s}) \end{bmatrix}_{N_s \times N_R}$$

$$Q_{N_R} = \{q_1(t), q_2(t), \ldots, q_{N_R}(t)\}^T$$

We have two possible cases:

Case (a): When $N_I < N_R$, Q_{N_I} can be obtained from partitions of Equation (8.47) as

$$Q_{N_I} = (W_{N_s N_I}^T W_{N_s N_I})^{-1} W_{N_s N_I}^T (w_{ns} - W_{N_s N_R - N_I} Q_{N_R - N_I}) \tag{8.48}$$

with

$$W_{N_s N_R - N_I} = \begin{bmatrix} W_{N_I+1}(x_1, y_1) & W_{N_I+2}(x_1, y_1) & \cdots & W_{N_R}(x_1, y_1) \\ W_{N_I+1}(x_2, y_2) & W_{N_I+2}(x_2, y_2) & \cdots & W_{N_R}(x_2, y_2) \\ \vdots & \vdots & \cdots & \vdots \\ W_{N_I+1}(x_{N_s}, y_{N_s}) & W_{N_I+2}(x_{N_s}, y_{N_s}) & \cdots & W_{N_R}(x_{N_s}, y_{N_s}) \end{bmatrix}_{N_s \times (N_R - N_I)} \tag{8.49}$$

$$Q_{N_R - N_I} = \{q_{N_I+1}(t), q_{N_I+2}(t), \ldots, q_{N_R}(t)\}^T$$

The terms in last bracket in Equation (8.48) represents the responses from the lower sets of N_I modes in identification. But in practice, the total measured responses are used instead, leading to an over-estimation of the forces when substituting Q_{N_I} and its derivatives into Equation (3.17) in Chapter 3.

Case (b): When $N_R < N_I$, Q_{N_I} can be obtained in a similar way as:

$$Q_{N_I} = (W_{N_S N_I}^T W_{N_S N_I})^{-1} W_{N_S N_I}^T (w_{ns} + W_{N_S N_R + N_I} Q_{N_R + N_I}) \qquad (8.50)$$

where $W_{N_S N_R + N_I}$ and $Q_{N_R + N_I}$ are similarly defined as in Equation (8.49). The last term in bracket in Equation (8.50) represents the total responses corresponding to the modes used in identification. But, in practice, only the measured responses w_{ns} is used in the equation leading to under-estimation in the forces.

In the simulation studies, the measured responses are computed from 25 modes ($m = n = 5$). In the cases with $N_I < N_R$, the modes using in the identification covered the excitation frequency range of the moving loads. 200 terms in the orthogonal polynomial have been used in obtaining the derivatives of Q_{N_I}. The small magnitudes of modes higher than N_I are further reduced by the low pass filtering effect. The term $Q_{N_I - N_R}$ becomes insignificant small and the over-estimation from Equation (8.48) is therefore not noticeable.

In the experiment, the excitation or natural frequencies of the car are not small. But only 50 terms in the orthogonal polynomial have been used because of a very small signal to noise ratio in the higher mode responses. The number of vibration modes left after filtering is greatly reduced leading to $N_R < N_I$. The term $W_{N_S N_R + N_I} Q_{N_R + N_I}$ does not exist in the measured responses, and hence the modal coordinates Q_{N_I} and its derivatives are under-estimated giving smaller than true loads when substituting into Equation (3.17) in Chapter 3. There is a difference of three modes between the cases of ($m = 3, n = 3$) and ($m = 3, n = 2$), and yet the final results differs by a great percentage. This is because of the large responses in these three modes which should contribute greatly to the final identified results if they are included in the identification.

8.6 Summary

The moving load identification with the generalized orthogonal functions has been presented in this chapter. A generalized orthogonal function approach is proposed to obtain the modal velocity and acceleration from measured strain response. This reduces the error due to measurement noise, and the moving forces are identified with bounds in the errors using regularization method in the solution. The moving force identification method is illustrated with numerical simulations and experimental studies of a beam and plate structures. The significance of having enough structural vibration modes in the identification is shown with error studies in the structures.

Chapter 9

Moving Force Identification based on Finite Element Formulation

9.1 Introduction

The moving load identification techniques described in previous chapters have good accuracy for identification but they demand extensive computation with a vehicle crossing a multi-span bridge structure. The finite element approach is flexible when dealing with vehicle axle-loads moving on top of a bridge structure with complex boundary conditions. The method has efficient computational performance and good identification accuracy, especially with the orthogonal function smoothing technique to obtain the velocities and accelerations from the measured strains.

This chapter introduces the finite element model approach. The Interpretive Method I is discussed in Section 9.2.1, in which the bridge is modeled as an assembly of lumped masses interconnected by massless elastic beam elements. The Euler-Bernoulli beam model is used in the Interpretive Method II described in Section 9.2.2. The use of structural condensation technique to reduce the DOFs of the structure to have a determined set of identification equation is revisited in Section 9.2.3. Numerical simulation and experimental results in Sections 9.3 and 9.4 demonstrate the efficiency and accuracy of the method to identify the moving loads. A comparative study with the Exact Solution Technique Method is also presented in Section 9.5.

9.2 Moving Force Identification

9.2.1 Interpretive Method I

The Interpretive method I (IMI) is a moving load identification approach developed by O'Connor and Chan (1988), in which a bridge is modeled as an assembly of lumped masses interconnected by massless elastic beam elements. It consists of two basic components in which, the predictive analysis generates the theoretical bridge response, and the interpretive analysis identifies the original dynamic loads.

9.2.1.1 Predictive Analysis

A simply supported bridge can be modeled as a lumped mass system as shown in Figure 9.1. The nodal responses for displacements and/or bending moments at any instant are given by Equations (9.1) and (9.2) respectively:

$$\{Y\} = [Y_A]\{P\} - [Y_I][\Delta m]\{\ddot{Y}\} - [Y_I][C]\{\dot{Y}\} \qquad (9.1)$$

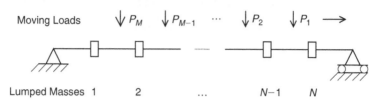

Figure 9.1 Beam-element model

$$\{M\} = [M_A]\{P\} - [M_I][\Delta m]\{\ddot{Y}\} - [M_I][C]\{\dot{Y}\} \tag{9.2}$$

where $\{P\}$ is the vector of wheel loads, $[\Delta m]$ is the diagonal matrix containing values of the lumped masses, $[C]$ is the damping matrix. $\{M\}, \{Y\}, \{\dot{Y}\}, \{\ddot{Y}\}$ are the nodal bending moment, displacement, velocity and acceleration vectors respectively. $[Y_A], [Y_I]$ are matrices for nodal forces to obtain nodal displacements, and $[M_A], [M_I]$ are matrices for nodal forces to obtain nodal bending moments. Here subscripts A and I represent the *A*cting load and *I*nertial force respectively.

9.2.1.2 Interpretive Analysis

The interpretive analysis predicts the dynamic loads from the measured response data. As stated in Section 9.2.1.1, solutions can be developed using the $\ddot{Y}$ (accelerations), Y (displacements) or M (bending moments). If Y is known at all times for all interior nodes, then $\dot{Y}$ and $\ddot{Y}$ can be obtained using numerical differentiation. Equation (9.1) becomes an over-determined set of linear simultaneous equations in which P can be solved. Similarly, if $\ddot{Y}$ is known, it can be integrated by an integration method to give $\dot{Y}$ and Y, and subsequently to get P. However, a particular difficulty arises if measured bending moments are used as input data. Remembering that the moving load P is not always at the nodes, and the relation between the nodal displacements and the nodal bending moments is:

$$\{Y\} = [Y_B]\{M\} + [Y_C]\{P\} \tag{9.3}$$

where $[Y_C]\{P\}$ allows for the deflections due to additional triangularly distributed bending moments that occur within elements carrying one or more point loads P. $[Y_C]$ can be calculated from the known locations of the loads. Both $[Y_B]$ and $\{M\}$ are known, but $\{Y\}$ can not be determined without a knowledge of $\{P\}$.

9.2.2 Interpretive Method II

Chan et al. (1999) used an Euler-Bernoulli beam to model the bridge deck in the interpretation of dynamic loads crossing the deck. If there are N_p moving loads on the single span beam with the mode shapes:

$$\phi_i(x) = \sin\left(\frac{i\pi x}{L}\right), \quad (i = 1, 2, \ldots, n)$$

and Equation (2.8) in Chapter 2 can be written as:

$$
\begin{Bmatrix} \ddot{q}_1 \\ \ddot{q}_2 \\ \vdots \\ \ddot{q}_n \end{Bmatrix}
+
\begin{Bmatrix} 2\xi_1\omega_1\dot{q}_1 \\ 2\xi_2\omega_2\dot{q}_2 \\ \vdots \\ 2\xi_n\omega_n\dot{q}_n \end{Bmatrix}
+
\begin{Bmatrix} \omega_1^2 q_1 \\ \omega_2^2 q_2 \\ \vdots \\ \omega_n^2 q_n \end{Bmatrix}
$$

$$
= \frac{2}{\rho L}
\begin{bmatrix}
\sin\dfrac{\pi(ct-\hat{x}_1)}{L} & \sin\dfrac{\pi(ct-\hat{x}_2)}{L} & \cdots & \sin\dfrac{\pi(ct-\hat{x}_{N_p})}{L} \\[2mm]
\sin\dfrac{2\pi(ct-\hat{x}_1)}{L} & \sin\dfrac{2\pi(ct-\hat{x}_2)}{L} & \cdots & \sin\dfrac{2\pi(ct-\hat{x}_{N_p})}{L} \\[2mm]
\vdots & \vdots & \vdots & \vdots \\[2mm]
\sin\dfrac{n\pi(ct-\hat{x}_1)}{L} & \sin\dfrac{n\pi(ct-\hat{x}_2)}{L} & \cdots & \sin\dfrac{n\pi(ct-\hat{x}_{N_p})}{L}
\end{bmatrix}
\begin{Bmatrix} P_1 \\ P_2 \\ \vdots \\ P_{N_p} \end{Bmatrix}
\tag{9.4}
$$

in which $\hat{x}_k$ is the distance between the kth load and the first load, and $\hat{x}_1 = 0$. Therefore, as mentioned above, the modal displacements at $x_1, x_2, \ldots, x_l$ can be obtained by solving Equation (9.4). If Equation (2.7) in Chapter 2 is expressed in a matrix form as:

$$
w = \left\{ \sin\dfrac{\pi x}{L} \quad \sin\dfrac{2\pi x}{L} \quad \cdots \quad \sin\dfrac{n\pi x}{L} \right\} \{ q_1 \quad q_2 \quad \cdots \quad q_n \}^T,
\tag{9.5}
$$

the displacements at $x_1, x_2, \ldots, x_l$ can then be calculated from Equation (9.4) as follows.

$$
\begin{Bmatrix} w_1 \\ w_2 \\ \vdots \\ w_l \end{Bmatrix}
=
\begin{bmatrix}
\sin\dfrac{\pi x_1}{L} & \sin\dfrac{2\pi x_1}{L} & \cdots & \sin\dfrac{n\pi x_1}{L} \\[2mm]
\sin\dfrac{\pi x_2}{L} & \sin\dfrac{2\pi x_2}{L} & \cdots & \sin\dfrac{n\pi x_2}{L} \\[2mm]
\vdots & \vdots & \ddots & \vdots \\[2mm]
\sin\dfrac{\pi x_l}{L} & \sin\dfrac{2\pi x_l}{L} & \cdots & \sin\dfrac{n\pi x_l}{L}
\end{bmatrix}
\begin{Bmatrix} q_1 \\ q_2 \\ \vdots \\ q_n \end{Bmatrix}
\tag{9.6}
$$

Furthermore, the accelerations on the beam at $x_1, x_2, \ldots, x_l$ can also be obtained from the second derivative of the displacements as

$$
\begin{Bmatrix} \ddot{w}_1 \\ \ddot{w}_2 \\ \vdots \\ \ddot{w}_l \end{Bmatrix}
=
\begin{bmatrix}
\sin\dfrac{\pi x_1}{L} & \sin\dfrac{2\pi x_1}{L} & \cdots & \sin\dfrac{n\pi x_1}{L} \\[2mm]
\sin\dfrac{\pi x_2}{L} & \sin\dfrac{2\pi x_2}{L} & \cdots & \sin\dfrac{n\pi x_2}{L} \\[2mm]
\vdots & \vdots & \ddots & \vdots \\[2mm]
\sin\dfrac{\pi x_l}{L} & \sin\dfrac{2\pi x_l}{L} & \cdots & \sin\dfrac{n\pi x_l}{L}
\end{bmatrix}
\begin{Bmatrix} \ddot{q}_1 \\ \ddot{q}_2 \\ \vdots \\ \ddot{q}_n \end{Bmatrix}
\tag{9.7}
$$

Similarly, the bending moments at the corresponding locations can be obtained from the relationship $M = -EI(\partial^2 v/\partial x^2)$.

$$
\begin{Bmatrix} M_1 \\ M_2 \\ \vdots \\ M_l \end{Bmatrix} = -EI\left(\frac{\pi}{L}\right)^2
\begin{bmatrix}
\sin\dfrac{\pi x_1}{L} & 2^2\sin\dfrac{2\pi x_1}{L} & \cdots & n^2\sin\dfrac{n\pi x_1}{L} \\[2mm]
\sin\dfrac{\pi x_2}{L} & 2^2\sin\dfrac{2\pi x_2}{L} & \cdots & n^2\sin\dfrac{n\pi x_2}{L} \\[2mm]
\vdots & \vdots & \ddots & \vdots \\[2mm]
\sin\dfrac{\pi x_l}{L} & 2^2\sin\dfrac{2\pi x_l}{L} & \cdots & n^2\sin\dfrac{n\pi x_l}{L}
\end{bmatrix}
\begin{Bmatrix} q_1 \\ q_2 \\ \vdots \\ q_n \end{Bmatrix}
\tag{9.8}
$$

If $P_1, P_2, \ldots, P_{N_p}$ are known constant moving loads and the effect of damping is ignored, the closed form solution of Equation (2.10) in Chapter 2 is given as:

$$
w(x,t) = \frac{L^3}{48EI}\sum_{i=1}^{N_p} P_i \sum_{n=1}^{\infty} \frac{1}{n^2(n^2-\alpha^2)}\sin\frac{n\pi x}{L}\left(\sin\frac{n\pi(ct-\hat{x}_i)}{L} - \frac{\alpha}{n}\sin\omega_n\left(t-\frac{\hat{x}_i}{c}\right)\right)
\tag{9.9}
$$

in which $\alpha = \pi c/L\omega_n$. Therefore, if the displacements of the beam at $x_1, x_2, \ldots, x_l$ caused by a set of constant moving loads are known, the magnitude of each moving load can be obtained by solving the following equation:

$$
\{w\} = [S_{vP}]\{P\}
\tag{9.10}
$$

or

$$
\begin{Bmatrix} w_1 \\ \vdots \\ w_i \\ \vdots \\ w_l \end{Bmatrix} =
\begin{bmatrix}
s_{11} & \cdots & s_{1j} & \cdots & s_{1N_p} \\
\vdots & \ddots & & & \vdots \\
s_{i1} & & s_{ij} & & s_{iN_p} \\
\vdots & & & \ddots & \vdots \\
s_{l1} & \cdots & s_{lj} & \cdots & s_{lN_p}
\end{bmatrix}
\begin{Bmatrix} P_1 \\ \vdots \\ P_j \\ \vdots \\ P_{N_p} \end{Bmatrix}
\tag{9.11}
$$

where,

$$
s_{ij} = \frac{L^3}{48EI}\sum_{n=1}^{\infty}\frac{1}{n^2(n^2-\alpha^2)}\sin\frac{n\pi x_i}{L}\left(\sin\frac{n\pi(ct-\hat{x}_j)}{L} - \frac{\alpha}{n}\sin\omega_n\left(t-\frac{\hat{x}_j}{c}\right)\right)
\tag{9.12}
$$

If $l \geq N_P$, which means that the number of displacement measuring stations is larger than or equal to the number of axle loads, $\{P\}$ can be obtained using the least-squares method

$$
\{P\} = ([S_{vP}]^T[S_{vP}])^{-1}[S_{vP}]^T\{w\}
\tag{9.13}
$$

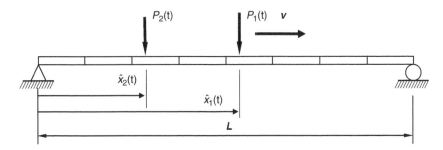

Figure 9.2 A continuous beam subject to moving loads

A similar equation can be obtained using bending moments instead of displacements as the bridge deck responses by considering the closed form solution in terms of bending moments.

$$M(x,t) = \frac{L}{4}\sum_{i=1}^{N_p} P_i \sum_{n=1}^{\infty} \frac{8n^2}{\pi^2}\sin\frac{n\pi x}{L}\frac{1}{n^2(n^2-\alpha^2)}\left(\sin\frac{n\pi(ct-\hat{x}_i)}{L} - \frac{\alpha}{n}\sin\omega_n\left(t-\frac{\hat{x}_i}{c}\right)\right)$$

(9.14)

It is noted that if the set of moving loads are time-varying, the method can still be applied to determine their static equivalent values as other traditional Weigh-In-Motion methods.

9.2.3 Regularization Method

9.2.3.1 Equation of Motion

Figure 9.2 shows two loads moving at a speed v over a bridge deck modeled with finite elements. The elemental mass and stiffness matrices are obtained using the Hermitian cubic interpolation shape functions. The supporting beam structure is discretized into $m-1$ beam element where m is the number of nodal points. The shape functions of the jth element in its local coordinate can be obtained as follows:

$$H_j = \left\{ 1 - 3\left(\frac{x}{l}\right)^2 + 2\left(\frac{x}{l}\right)^3 \quad x\left(\frac{x}{l}-1\right)^2 \quad 3\left(\frac{x}{l}\right)^2 - 2\left(\frac{x}{l}\right)^3 \quad x\left(\frac{x}{l}\right)^2 - \frac{x^2}{l} \right\}^T$$

(9.15)

where l is the length of the beam element. With the assumption of Rayleigh damping, the equation of motion for the bridge can be written as:

$$M_b\ddot{R} + C_b\dot{R} + K_bR = H_cP_{int}$$

(9.16)

where M_b, C_b, K_b are the mass, damping and stiffness matrices of the bridge structure respectively. $\ddot{R}, \dot{R}, R$ are the nodal acceleration, velocity and displacement vectors

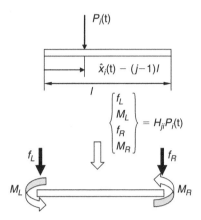

Figure 9.3 Equivalent nodal loads for a beam element loaded by the *i*th bridge-vehicle interaction force

of the bridge deck respectively, and $H_c P_{int}$ is the equivalent nodal load vector from the bridge–vehicle interaction force with:

$$H_c = \begin{Bmatrix} 0 & \cdots & 0 & \cdots & H_1 & \cdots & 0 \\ 0 & \cdots & H_2 & \cdots & 0 & \cdots & 0 \end{Bmatrix}^T$$

H_c is a $NN \times N_p$ matrix with zero entries except at the degrees-of-freedom corresponding to the nodal displacements of the beam elements on which the load is acting, and NN is the number of degrees-of-freedom of the bridge after considering the boundary condition. The components of the vector H_i $(i = 1, 2)$ evaluated for the *i*th interactive force on the *j*th finite element is given in Figure 9.3, and the shape function can be written in the global coordinates as:

$$H_j = \begin{Bmatrix} 1 - 3\left(\dfrac{\hat{x}_i(t) - (j-1)l}{l}\right)^2 + 2\left(\dfrac{\hat{x}_i(t) - (j-1)l}{l}\right)^3 \\[2mm] (\hat{x}_i(t) - (j-1)l)\left(\dfrac{\hat{x}_i(t) - (j-1)l}{l} - 1\right)^2 \\[2mm] 3\left(\dfrac{\hat{x}_i(t) - (j-1)l}{l}\right)^2 - 2\left(\dfrac{\hat{x}_i(t) - (j-1)l}{l}\right)^3 \\[2mm] (\hat{x}_i(t) - (j-1)l)\left(\left(\dfrac{\hat{x}_i(t) - (j-1)l}{l}\right)^2 - \left(\dfrac{\hat{x}_i(t) - (j-1)l}{l}\right)\right) \end{Bmatrix} \tag{9.17}$$

with $(j-1)l \le \hat{x}_i(t) \le jl$. $\hat{x}_1(t), \hat{x}_2(t)$ are the positions of the front axle and rear axle respectively at time t. To find the time response of the beam from Equation (9.16), a step-by-step solution can be obtained using the Newmark direct integration method. The deflection of the bridge at position x and time t can then be expressed as:

$$w(x,t) = H(x)R(t) \tag{9.18}$$

where $H(x) = \{0 \quad \cdots \quad H(x)_j^T \quad 0 \quad \cdots \quad 0\}$ with $(j-1)l \le x(t) \le jl$. $H(x)$ is a $1 \times NN$ vector with zero entries except at the degrees-of-freedom corresponding to the nodal displacements of the jth beam element on which the point x is located. The components of the vector $H(x)_j$ are calculated similar to Equation (9.17) with $x(t)$ replacing $\hat{x}_i(t)$.

9.2.3.2 Vehicle Axle Load Identification from Strain Measurements

According to Equation (9.18), the strain at a point x and at time t can be written as follow:

$$\varepsilon(x,t) = -z\frac{\partial^2 w(x,t)}{\partial x^2} = -z\frac{\partial^2 H(x)\,R(t)}{\partial x^2} \tag{9.19}$$

where z represents the distance from the neutral axis of the beam to the location of strain measurements at the bottom surface. Rewriting Equation (9.19), we have:

$$\varepsilon(x,t) = gR \tag{9.20}$$

where

$$g = -z\{g_1(x), g_2(x), \ldots, g_{NN}(x)\}$$

and $g_i(x)$ is the second derivative of H_i.

The strain can be approximated by a generalized orthogonal function $T(t)$ as:

$$\varepsilon(x,t) = \sum_{i=1}^{N_f} T_i(t)C_i(x) \tag{9.21}$$

where $\{T_i(t), i = 1, 2, \ldots, N_f\}$ is the generalized orthogonal function (Zhu and Law, 2001a), N_f is the number of terms in the generalized orthogonal function; $\{C_i(x), i = 1, 2, \ldots, N_f\}$ is the vector of coefficients in the expansion expression. The strains at the N_s measuring points can be expressed as:

$$\varepsilon = CT \tag{9.22}$$

where

$$T = \{T_0(t), T_1(t), \ldots, T_{N_f}(t)\}^T;$$

$$\varepsilon = \{\varepsilon(x_1,t), \varepsilon(x_2,t), \ldots, \varepsilon(x_{N_s},t)\}^T;$$

$$C = \begin{bmatrix} C_{10}(x_1) & C_{11}(x_1) & \cdots & C_{1N_f}(x_1) \\ C_{20}(x_2) & C_{21}(x_2) & \cdots & C_{2N_f}(x_2) \\ \vdots & \vdots & \vdots & \vdots \\ C_{N_s0}(x_{N_s}) & C_{N_s1}(x_{N_s}) & \cdots & C_{N_sN_f}(x_{N_s}) \end{bmatrix}$$

and $\{x_1, x_2, \ldots, x_{N_s}\}$ is the vector of location of the strain measurements. By the least-squares method, the coefficient matrix can be obtained as:

$$C = \varepsilon\, T^T\, (T\, T^T)^{-1} \qquad\qquad\qquad (9.23)$$

Putting Equation (9.20) into Equation (9.22) and using the least-squares method again, we have:

$$R = (G^T\, G)^{-1}\, G^T\, C\, T \qquad\qquad\qquad (9.24)$$

where

$$G = -z \begin{bmatrix} g_1(x_1) & g_2(x_1) & \cdots & g_{NN}(x_1) \\ g_1(x_2) & g_2(x_2) & \cdots & g_{NN}(x_2) \\ \vdots & \vdots & \vdots & \vdots \\ g_1(x_{N_s}) & g_2(x_{N_s}) & \cdots & g_{NN}(x_{N_s}) \end{bmatrix}$$

and matrices $\dot{R}$ and $\ddot{R}$ can be obtained by directly differentiating Equation (9.24) to have:

$$\dot{R} = (G^T\, G)^{-1}\, G^T\, C\, \dot{T},$$
$$\ddot{R} = (G^T\, G)^{-1}\, G^T\, C\, \ddot{T}.$$

Substituting the nodal responses $R, \dot{R}$ and $\ddot{R}$ into Equation (9.16), we have:

$$\cdot U = H_c P_{int} \qquad\qquad\qquad (9.25)$$

where $U = M_b \ddot{R} + C_b \dot{R} + K_b R$.

The moving loads can then be identified from Equation (9.25) by least-squares method. However, it is impractical to have measurements from all the degrees-of-freedom of the structure and Equation (9.25) would be under-determined. A structural condensation technique is required to reduce the unmeasured degrees-of-freedom to the measured degrees-of-freedom of the bridge deck such that Equation (9.25) is determinate.

The Improved Reduced System reduction scheme O'Callahan (1989) may be adopted to condense the unmeasured degrees-of-freedom (DOFs) to the measured degrees-of-freedom of the bridge deck. All the measured DOFs are designated as the master DOFs and denoted by $R_m(t)$. The remaining structural DOFs are called the slave DOFs, and are denoted by $R_s(t)$. The response vector of the bridge is then partitioned as:

$$R(t) = \begin{Bmatrix} R_m(t) \\ R_s(t) \end{Bmatrix} \qquad\qquad\qquad (9.26)$$

Accordingly, the bridge mass, damping, stiffness and shape function matrices are also partitioned as:

$$M_b = \begin{bmatrix} M_{mm} & M_{ms} \\ M_{sm} & M_{ss} \end{bmatrix}, \quad C_b = \begin{bmatrix} C_{mm} & C_{ms} \\ C_{sm} & C_{ss} \end{bmatrix}, \quad K_b = \begin{bmatrix} K_{mm} & K_{ms} \\ K_{sm} & K_{ss} \end{bmatrix},$$

$$H = \begin{bmatrix} H_{mm} \\ H_{ss} \end{bmatrix}^T, \quad H_c = \begin{bmatrix} H_{cmm} \\ H_{css} \end{bmatrix} \tag{9.27}$$

The total response matrix of the system $R(t)$ can then be represented by the partitioned master set of DOFs $R_m(t)$ multiplied by the transformation matrix as:

$$R(t) = W\, R_m(t), \tag{9.28}$$

where

$$W = W_s + W_i, \quad W_s = \begin{bmatrix} I \\ -K_{ss}^{-1} K_{sm} \end{bmatrix}$$

and

$$W_i = \begin{bmatrix} I \\ K_{ss}^{-1}(M_{sm} - M_{ss}K_{ss}^{-1}K_{sm})(W_s^T M_b W_s)^{-1}(W_s^T K_b W_s) \end{bmatrix}$$

where I is the identity matrix. Substituting Equations (9.27) and (9.28) into Equation (9.16) and pre-multiplying W^T to both sides yielding:

$$U = H_{cr} P_{int} \tag{9.29}$$

and

$$U = M_r \ddot{R}_m + C_r \dot{R}_m + K_r R_m$$

where $M_r = W^T * M_b * W, C_r = W^T * C_b * W, K_r = W^T * K_b * W$, and $H_{cr} = W^T * H_c$.

Substituting Equation (9.28) and $H = \begin{bmatrix} H_{mm} \\ H_{ss} \end{bmatrix}^T$ into Equation (9.19), matrices $R_m, \dot{R}_m$ and $\ddot{R}_m$ can be obtained by the generalized orthogonal function method. The identification with selected measuring points can then be performed using Equation (9.29). Since the proposed method is based on finite element method, it could be applied to complex structure with different boundary conditions, varying geometry and mass distribution.

9.2.3.3 Regularization Algorithm

The moving forces obtained from Equation (9.25) or (9.29) using a straight forward least-squares solution would be unbound. A regularization technique (Law et al., 2001) is used to solve the ill-posed problem in the form of minimizing the function.

$$J(P, \lambda) = \|B P_{int} - U\|^2 + \lambda \|P_{int}\|^2 \tag{9.30}$$

where B is H_c or H_{cr}; λ is the non-negative regularization parameter corresponding to the smallest relative percentage error calculated from Equation (9.32) or (9.33). The Generalized Cross Validation (Gorman and Heath, 1979) and L-curve method (Hansen, 1992) can be employed to determine the optimal regularization parameter. The solution of Equation (9.25) or (9.29) can be obtained by the damped least-squares method as:

$$P_{int} = (H_c^T * H_c + \lambda I)^{-1} H_c^T * U \qquad (9.31)$$

where I is the identity matrix, and singular-value decomposition is used in the pseudo-inverse calculation.

9.3 Numerical Examples

The effects of discretization on the structure, number of measuring points, sampling frequency, velocity of vehicle and noise level on the accuracy of the identified results are investigated in the following sections. The number of master DOFs is taken equal to the number of measuring points in all of the following studies.

The calculated responses are polluted with white noise to simulate the polluted measurement as:

$$\varepsilon = \varepsilon_{calculated}(1 + E_p * N_{oise})$$

where ε_j and $\varepsilon_{j\,calculated}$ are vectors of measured and calculated responses at the jth measuring point; E_p is the noise level; N_{oise} is a standard normal distribution vector with zero mean and unit standard deviation. The relative percentage error (RPE) in the identified results is calculated from Equation (9.32), where $\|\bullet\|$ is the norm of matrix, $P_{identified}$ and P_{true} are the identified and the true force time histories respectively.

$$RPE = \frac{\|P_{identified} - P_{true}\|}{\|P_{true}\|} \times 100\% \qquad (9.32)$$

The bridge deck is modeled as a 30 m long simply supported beam with the physical and material parameters same as those described in Section 8.4.1.1 with $\xi = 0.02$ for all modes. The two moving loads are at 4.26m spacing moving at a constant speed. They are:

$$P_1(t) = 6268(1.0 + 0.1 \sin(10\pi t) + 0.05 \sin(40\pi t)) \text{ kg};$$
$$P_2(t) = 12332(1.0 - 0.1 \sin(10\pi t) + 0.05 \sin(50\pi t)) \text{ kg}.$$

The measuring points are evenly distributed at the bottom of the beam and their locations are shown in Table 9.1 for different arrangements of sensors.

9.3.1 Effect of Discretization of the Structure and Sampling Rate

The loads move on top of the beam at a constant velocity of 15 m/s. Strain at the bottom of the beam is measured with three sensors, and the sampling frequency is taken to be 100, 200, 300, 400 and 500 Hz separately for the study. The simply supported beam is discretized into four, eight, 12 and 16 beam elements separately. No noise effect

Table 9.1 Sensor arrangements

Number of sensors	Location
3	1/4L, 1/2L, 3/4L
4	1/8L, 1/4L, 1/2L, 3/4L
5	1/8L, 1/4L, 1/2L, 3/4L, 7/8L
6	1/8L, 1/4L, 1/2L, 5/8L, 3/4L, 7/8L
7	1/8L, 1/4L, 3/8L, 1/2L, 5/8L, 3/4L, 7/8L

Table 9.2 The percentage error of the identified forces for different discretization scheme and sampling rate

Sampling frequency (Hz)	Number of elements							
	4		8		12		16	
	Axle-1	Axle-2	Axle-1	Axle-2	Axle-1	Axle-2	Axle-1	Axle-2
100	20.93	18.36	21.11	17.49	20.41	17.64	21.46	17.66
200	15.02	13.12	11.42	9.19	11.50	9.20	11.50	9.12
300	14.31	12.57	11.09	8.99	10.99	8.92	10.98	8.90
400	14.64	12.92	10.92	9.03	10.94	9.04	10.95	8.94
500	14.68	12.92	10.95	9.03	10.97	9.04	10.97	8.94

is included in this study. The relative percentage errors of the identified forces for different number of finite elements and sampling frequency are tabulated in Table 9.2. Figure 9.4 gives the identified results from the cases of four and eight elements.

The identified force time histories from both four and eight finite elements match the true forces very well in the middle half of the duration. The forces from four elements have large fluctuations after the entry of the second load and before the exit of the first load, while those from eight elements have slight fluctuations at these moments. Table 9.2 also shows that the relative percentage errors of the identified forces from four elements are much larger than those from the other discretization cases. These indicate discretizing the beam into eight elements would be sufficient for an accurate identification.

The sampling frequency is shown not to have any significant effect when it is larger or equal to 200 Hz. It is noted that the first five modes are included in the measured responses with this sampling frequency indicating that the higher modes do not have significant contribution to the identification accuracy.

9.3.2 Effect of Number of Sensors and Noise Level

The loads move on top of the beam at a constant speed of 15 m/s. The bridge is discretized into eight finite elements, and the force identification is studied with three, five and seven sensors separately and the sampling rate is 200 Hz. The measured data are polluted with 5 percent and 10 percent noise. The relative percentage errors of the identified results are listed in Table 9.3. Figure 9.5 shows the identified results from three and five sensors with 5 percent noise level.

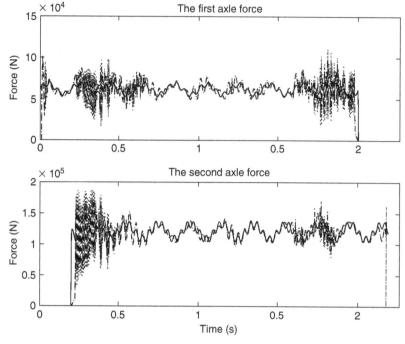

Figure 9.4 Identified results from three measured points without noise (—— true force, --- 8 elements, · · · · 4 elements)

Table 9.3 The percentage error of the identified forces for different sensor arrangements and noise level

Noise level (%)	Number of measuring points					
	3		5		7	
	Axle-1	Axle-2	Axle-1	Axle-2	Axle-1	Axle-2
0	11.09	9.99	6.33	3.28	4.22	2.64
5	13.27	11.28	8.12	5.81	6.06	5.11
10	16.59	14.81	11.27	9.85	10.34	9.97

The identified force time histories from five sensors agree with the true forces very well with only small differences at the start and end of the time histories. Identified results from three sensors are also acceptable with slight fluctuations around the true forces in the middle half of the duration. The relative percentage errors from three sensors are significantly larger than those from five and seven sensors only because of the large fluctuations in the time histories at the moments close to the entry of the second load and exit of the first load. The relative percentage error increases with the noise level, but it is not sensitive to the measurement noise.

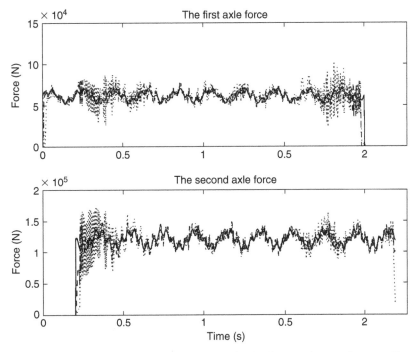

Figure 9.5   Identified results with 5 percent noise and eight elements (— true force, - - - 5 measured points, · · · · 3 measured points)

9.4   Laboratory Verification

9.4.1   *Experimental Set-up*

The experimental set-up of a model vehicle moving on top of a steel beam in the laboratory described in Section 6.3.4.1 is adopted for this study.

The model vehicle has two axles at a spacing 0.557 m and it runs on four steel wheels wrapped with rubber band on the outside. The mass of the model vehicle is 12.55 kg with the front axle load and rear axle load weigh 8.725 and 3.825 kg respectively. The first three measured natural frequencies of the beam are: 3.67, 16.83 and 37.83 Hz.

9.4.2   *Identification from Measured Strains*

The average velocity of the model vehicle at 0.787 m/s is used for the identification. The beam is discretized into eight Euler-Bernoulli beam elements. The measured strains are re-sampled at 100 Hz. The sensor arrangements are shown in Table 9.1.

Since the true force P_{true} is not known, both the norm of residuals E of the forces and the semi-norm $E1$ of the estimated forces are calculated as:

$$E = \| B\, P^{identify} - U \|$$

$$E1 = \| P^{identify}_{j+1} - P^{identify}_{j} \| \tag{9.33}$$

Table 9.4 Correlation coefficients of measured and reconstructed strains at 5/8L from different number of measured points

Number of measured points	Number of master degree-of-freedom				
	3	4	5	6	7
3	0.935	0.142	0.105	0.101	0.084
4	0.936	0.943	0.107	0.234	0.092
5	0.939	0.946	0.947	0.329	0.115
6	0.940	0.946	0.954	0.960	0.134
7	0.940	0.947	0.956	0.962	0.963

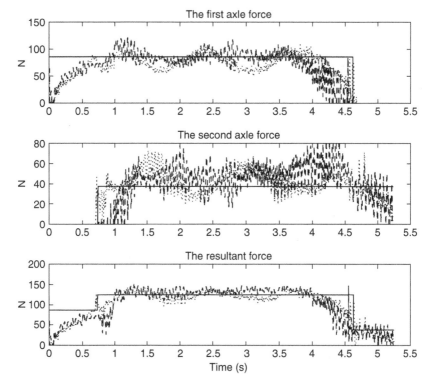

Figure 9.6 Experimental identified axle loads (— static loads; - - - from 3 measured strains; · · · · from 7 measured strains)

where $P_j^{identify}$, $P_{j+1}^{identify}$ are the identified forces with λ_j and $\lambda_j + \Delta\lambda$, and L-curve method (Hansen, 1992) is used to find the optimal regularization parameter λ_{opt}.

Since it is practically not possible to measure large number of responses, the number of master DOFs is varied to study its effect on the experimental identification. The dynamic strain at 5/8L is reconstructed by inputting the identified forces into the structure. The correlation coefficients of the measured strain and the reconstructed strain at 5L/8 for different number of master DOFs and measured points are tabulated in Table 9.4. Figure 9.6 shows the identified axle loads and the resultant loads from using three, five and seven master DOFs. The number of sensors is taken equal to the number of master DOFs in this study.

Table 9.4 shows that when the number of sensors is larger than or equal to the number of master DOFs, the correlation coefficients between the measured and reconstructed strains at 5L/8 are all larger than 0.935, even with as few as three master DOFs. The total load identified from three, five and seven master DOFs is relatively stable in the middle part of the time history, but those from three master DOFs only have much smaller value than those from five and seven master DOFs. This confirms the observations from the simulations that accepted results can be obtained from only a few measuring points by this method. The accuracy increases with the number of master DOFs particularly in the region towards the start and end of the time histories.

9.5 Comparative Studies

Methods have been developed in Chapters 5 to 8 to identify moving loads on top of a continuous bridge using measured vibration responses. A comparison is presented as follows to illustrate the robustness and accuracy of two time domain method, i.e. the Exact Solution Technique (EST) Method in Section 6.3 and the Finite Element Method based on regularization in Chapter 9. Numerical studies with a single-span bridge deck are presented in this section. Parameters that may influence the accuracy of moving load identification, such as sampling frequency, number of vibration modes and measuring points in the identification are discussed.

A single span simply supported beam with two forces $p_1(t)$ and $p_2(t)$ moving on top is studied.

$$\begin{cases} p_1(t) = 20000[1 + 0.1 \sin(10\pi t) + 0.05 \sin(40\pi t)] \, \text{N} \\ p_2(t) = 20000[1 - 0.01 \sin(10\pi t) + 0.05 \sin(50\pi t)] \, \text{N} \end{cases} \tag{9.34}$$

The parameters of the beam are: $EI = 2.5 \times 10^{10} \, \text{Nm}^2$, $\rho A = 5000 \, \text{kg/m}$, $L = 30 \, \text{m}$. The distance between the two moving forces is 4.27 m. The first six natural frequencies of the beam are 3.90 Hz, 15.61 Hz, 35.13 Hz, 62.45 Hz, 97.58 Hz and 140.51 Hz. White noise is added to the calculated responses of the beam to simulate the polluted measurements.

9.5.1 Effect of Noise Level

The first six modes are used in the simulation. The time interval between adjacent data points is 0.002 s. Six measuring points are evenly distributed on the beam at 1/7L spacing. The moving velocity of forces is 30 m/s, and 20 terms are used in the orthogonal function to approximate the measured responses. Monte Carlo method is used to simulate the noise in 20 sets of responses, and the noise level varies from 1 percent to 10 percent. Figure 9.7 shows the mean and standard deviation of the errors in the identified moving loads using the method based on the EST in Section 6.3, and Figure 9.8 shows those from using the method based on finite element formulation in Section 9.2.3.

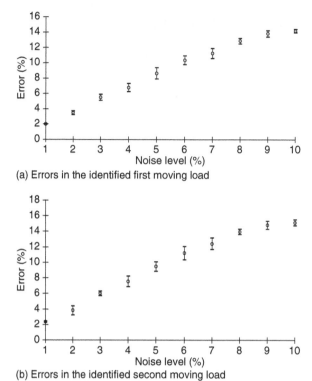

(a) Errors in the identified first moving load

(b) Errors in the identified second moving load

Figure 9.7 The mean (circles) and standard deviation (error bars) of the errors in the identified moving loads using EST Method

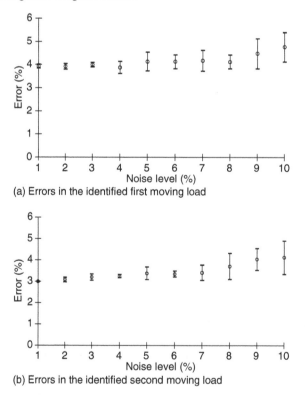

(a) Errors in the identified first moving load

(b) Errors in the identified second moving load

Figure 9.8 The mean (circles) and standard deviation (error bars) of the errors in the identified moving loads using FEM-based Method

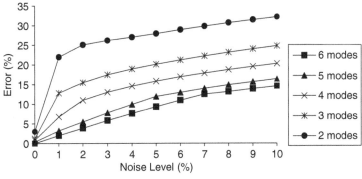

(a) Error in the identified first moving load

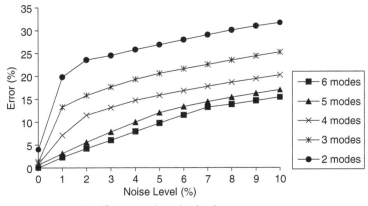

(b) Errors in the identified second moving load

Figure 9.9 Errors in the identified moving loads using EST Method

The errors from using EST method vary approximately linearly with the noise levels in the responses. The standard deviation in the errors is largest with 6 percent noise in the responses. The errors from using FEM-based Method exhibit little change with the noise level in the responses. This is because the orthogonal function approach in the identification reduces the effect of noise by its own filtering effect. When the noise level in the responses increases the standard deviation in the errors also increases. This indicates the EST method could give very accurate results at low noise level, but it may be badly influenced by the noise effect. On the other hand, the orthogonal function approximation in the FEM-based Method reduces consistently the noise effect to give accuracy results in all cases studied.

9.5.2 *Effect of Modal Truncation*

The first six modes are included in the responses, and six measuring points are evenly distributed on the beam at 1/7L spacing. The first two, three, four, five and six vibration modes are used in the identification in turn. Other parameters are the same as described in last section. Figures 9.9 and 9.10 show the errors in the identified results

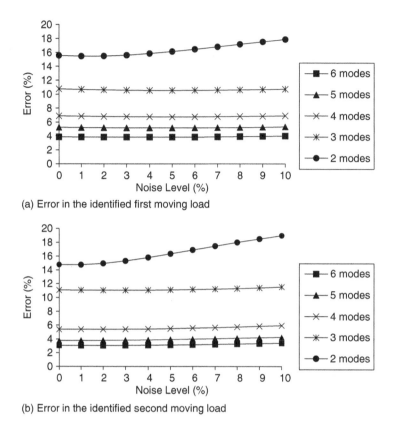

(a) Error in the identified first moving load

(b) Error in the identified second moving load

Figure 9.10 Errors in the identified moving loads using FEM-based Method

with different number of modes using EST Method and FEM formulation, respectively. Figure 9.11 shows the errors in the identified results with different number of terms in the orthogonal function in FEM-based Method when six modes are included in the responses.

The errors derived from EST Method increase roughly proportional to the noise level in the responses and with similar rate of change for different number of modes. The errors from using FEM-based Method exhibit little change with noise. This shows that the errors in the identified results using FEM-based Method are mainly governed by the efficiency of the filtering effect in the orthogonal function approach. FEM-based Method is, in general, much better than the EST Method in the identification.

The error shown in Figures 9.9 and 9.10 increases by a large extent when the number of the modes in the identification is less than three. This is because the first three natural frequencies of the beam cover the frequency range of the moving loads, and there is a loss of measured information in the identification when only two vibration modes are used. The errors in the identified forces in Figure 9.11 remain relatively constant for different noise levels when the number of terms in the orthogonal function in the FEM-based Method is less than 20. And the noise level would have a negative effect on the errors when there are more terms in the orthogonal function. This is because the frequency range in the orthogonal function increases with increasing number of terms,

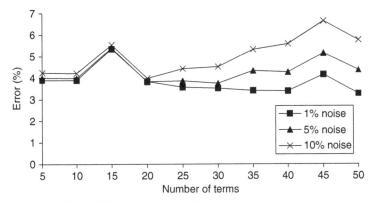

(a) Error in the identified first moving load

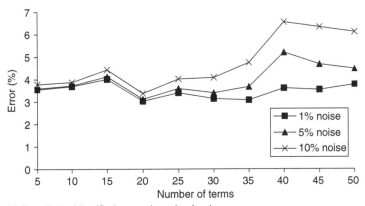

(b) Error in the identified second moving load

Figure 9.11 Error in the identified moving loads from FEM-based formulation with different number of terms in the orthogonal function

and the high frequency components in the noise would be retained in the calculation and thus affecting the final results.

9.5.3 Effect of Number of Measuring Points

Again the first six modes are used in the simulation. The number of measuring points is selected as six, seven, eight, nine, ten in turn. The measuring points are evenly distributed on the beam. Other parameters are the same as in the last section. Figures 9.12 and 9.13 show the errors in the identified results with different number of measuring points as the noise level in the responses is increased. The number of measuring points is shown insignificant to the identified results. It should be noted that the number of measuring points used are all larger than the number of the modes in the identification.

9.5.4 Effect of Sampling Frequency

The first six modes are used in the simulation. The responses are calculated with 0.001 s time interval between data points, and they are re-sampled with a time interval

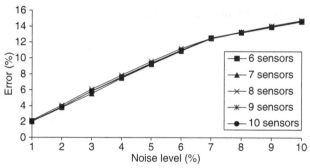

(a) Error in the identified first moving load

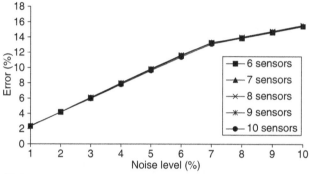

(b) Error in the identified second moving load

Figure 9.12 Errors in the identified moving loads using EST Method with different number of measuring points

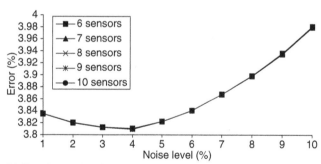

(a) Error in the identified moving load

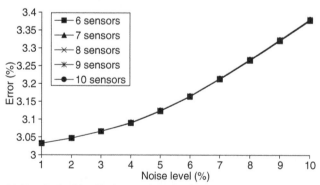

(b) Error in the identified second moving load

Figure 9.13 Errors in the identified moving loads using FEM-based formulation with different number of measuring points

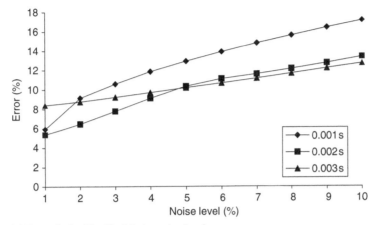

(a) Errors in the identified first moving load

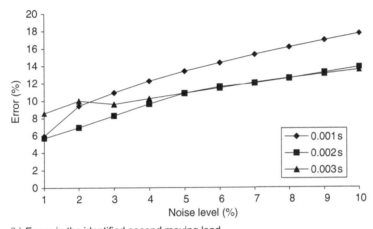

(b) Errors in the identified second moving load

Figure 9.14 Errors in the identified moving loads with different sampling frequencies using EST Method

of 0.002 s and 0.003 s in turn. The moving velocity of vehicle is 30 m/s. Figures 9.14 and 9.15 show the errors in the identified results with different sampling frequencies and noise levels using these two methods.

The errors in the identified results from EST Method are largest when the noise level is above 2 percent and the sampling time interval is 0.001 s. This is again due to the inclusion of the high frequency components of noise in the calculation with a higher sampling frequency. The errors from using FEM-based Method are smaller than those from using EST Method and with smaller variations.

9.6 Summary

A general moving load identification method basing on finite element formulation and the Improved Reduced System condensation technique has been presented in this

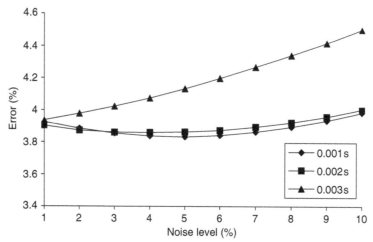

(a) Errors in the first identified moving load

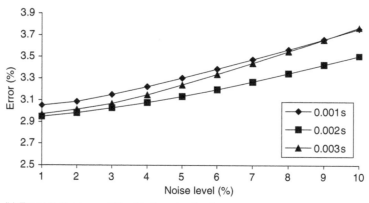

(b) Errors in the second identified moving load

Figure 9.15 Errors in the identified results with different sampling frequencies using FEM-based Method

chapter. The approximated finite element formulations, developed earlier, are also described. The measured displacement is expressed as the shape functions of the finite elements of the structure without the modal coordinate transformation as required in methods described in earlier chapters. Numerical simulation and experimental results demonstrate the efficiency and accuracy of the method to identify the moving loads or interaction forces between the vehicle and the bridge deck.

The number of master degrees-of-freedom of the system should be selected smaller than or equal to the number of measured points, and the identified results are relatively not sensitive to the sampling frequency, velocity of vehicle, measurement noise level and road surface roughness when a minimum of eight finite elements are used to model the simply supported bridge deck with measured information from at least three measuring points.

Chapter 10

Application of Vehicle–Bridge Interaction Force Identification

10.1 Merits and Disadvantages of Different Moving Force Identification Techniques

The four groups of methods presented in previous chapters of this book can be classified according to their solution techniques, which are the Frequency and Time Domain Method (FTDM), the Time Domain Method (TDM), the State Space Method (STM) and the Finite Element Method (FEM). If the method of formulation of the moving load problem is considered, they can be grouped into only two categories, namely those based on the Modal Superposition Technique (MST) and those based on the Finite Element Method (FEM). The capability of these methods in the moving force identification has been illustrated with numerical examples (Zhu and Law, 2002c) and experimental studies with the laboratory data (Zhu and Law, 2003a). A comparison between the FTDM and TDM is given in Chapters 5 and 6; of the two TDM methods in Chapter 6; the two types of STMs in Chapter 7 and comparison between the FEM and the EST method in Chapter 9.

The Modal Superposition Technique (MST), represented by the Exact Solution Technique (EST) Method in Chapter 6, has a good accuracy of identification but it demands heavy computation when multiple vehicles cross a multi-span bridge structure. The FEM approach is flexible when dealing with vehicle axle-loads moving on top of a bridge–vehicle system with complex boundary conditions. However a great deal of care must be used in transforming the displacements or strains into velocities and accelerations as the numerical differentiation may lead to large errors in the identified results. The FEM formulated Method, however, has acquired efficient computational capability and good identification accuracy when merged with the orthogonal function smoothing technique to obtain the velocities and accelerations from the measured strains as shown in Chapter 8.

Both the FEM formulation and the EST Method have been shown to be able to identify the bridge–vehicle interaction forces from strains with road surface roughness and vehicle braking on the bridge. The FEM gives consistently smaller error in the results for all noise levels while the accuracy of EST Method is greatly affected by noise. This indicates that the importance of having pre-processing of the measured data to remove the measurement noise before the identification is not over-emphasized. The orthogonal function approximation of the measured strains is also shown to be effective in filtering the high frequency noise components in the responses.

This chapter reviews on the industrial applications of the bridge Weigh-In-Motion method and the Moving Load Identification techniques in assessing the axle weights of travelling vehicles. Strategies on how to access a set of accurately measured data for the identification are discussed. The EST and FEM methods are then discussed further with more features in the practical problem of identifying vehicular interactive loads in numerical simulations and experimental tests in the laboratory.

10.2 Practical Issues on the Vehicle–Bridge Interaction Force Identification

10.2.1 Bridge Weigh-In-Motion

Information on the actual vehicular loads on highways is essential to the determination of maintenance programs of highways. One variation of the vehicle–bridge interaction force identification is the Bridge Weigh-In-Motion (B-WIM). A B-WIM system is based on the measurement of response of a bridge structure to estimate the attributes of passing traffic loads. The bridge structures are instrumented with strain gauges and, when necessary, with the axle detectors. Measurements during the entire period of the passage of vehicle over the bridge deck provide plenty of data, which facilitates the evaluation of axle loads. Karoumi et al. (2005) presented a cost-effective method for assessing the actual traffic loads on an instrumented railway bridge, and a B-WIM system was implemented with axle detection devices and four concrete strain transducers embedded in concrete. The longitudinal webs of an orthotropic box-girder bridge were instrumented to measure axle weights of trucks by Xiao et al. (2006). The Bridge Weigh-In-Motion was applied to a two-span continuous curved bridge with skew by Yamaguchi et al. (2009). Although the bridge is far from ideal for the Weigh-In-Motion, the accuracy of the B-WIM system was found satisfactory. Recently, the B-WIM system is combined with a structural health monitoring system where the data collected is used for traffic planning, pavement design, bridge rating and structural health monitoring (Cardini and Dewolf, 2009). A long-term B-WIM system using an existing structural health monitoring system and strain measurements from the steel girders was used to determine the truck weights.

Compared with the pavement WIM system (Znidaric, 2010), the B-WIM is particularly suitable for:

1. Short term measurements, as it can be easily installed and removed from the bridge. Unlike other WIM systems, the accuracy of results is not affected by the portability of installation of the instrument system.
2. Measurements on site, where cutting into the pavement is not allowed or is not feasible due to the heavy traffic.
3. Bridge assessments, as it provides supplementary structural information, such as the dynamic (impact) factors, the load distribution factor and the strain records.

Traditional B-WIM systems are commonly based on the static algorithm by Moses (1979), which assumes that a moving load will cause a bridge to bend in proportion to the product of the load magnitude and the influence line which is a reference curve representative of bridge behavior. The point of measurement is generally taken at midspan, where maximum stresses most often occur. The measured strain is the result of

Table 10.1 Bridge selection criteria

Criteria	Optimal	Acceptable
Bridge type	Steel girders, prestressed concrete girders, reinforced concrete girders, culvert, steel orthotropic decks	Concrete slab
Span length[1),2)] (m)	5–15	8–35
Traffic density	Free traffic-no congestion (traffic jam)	
Evenness of the pavement before and on the bridge	Class I or II	Class III
Skew (°)	≤10	≤25 ≤45[3)]

Notes: 1) this criterion applies for the length of the bridge part which influences the instrumentation;
2) except culverts;
3) after inspection of the calibration data.

all axle forces on the bridge deck. Based on Moses' theory, two prototype B-WIM systems were developed independently in Slovenia and in Ireland in the early 1990s (O'Brien et al., 1999). These B-WIM systems are suitable for short span bridges and basic criteria for the bridge selection have been recommended in COST323 (1999) and reproduced in Table 10.1. The response of the bridge to the moving vehicular loads is often influenced by the interaction between the vehicles and the road surface. Inaccurate assumed characteristics for the influence lines, bridge and vehicle dynamics and the presence of noise have been shown to be significant sources of error for the traditional static B-WIM algorithm (O'Brien et al., 1999).

10.2.2 *Moving Force Identification Techniques*

The moving force identification is an alternative B-WIM algorithm to reduce the dynamic uncertainty associated with the bridge measurements and to improve the accuracy of the B-WIM systems (Gonzalez et al., 2008). With the advances in computer power, data acquisition and sensor technologies, measurements with good resolution and high sampling rates are easily obtained. These advances make it possible to measure the vehicle axle forces continuously as the vehicle travels on the bridge deck. The identified vehicle axle forces will provide more reliable live load information for site-specific bridge fatigue assessment and performance evaluation (Liu et al., 2009; Law and Li, 2010).

The moving force identification techniques have been presented in Chapters 5 to 9 of this book and these algorithms could be used for the B-WIM systems. Field measurements have been carried out for a moving force identification study using an existing pre-stressed concrete bridge (Chan, Law and Yung, 2000). Seven strain gauges were installed on each of the five girders to acquire the dynamic bending moments of the bridge deck at the passage of a vehicle over the test span. Seven accelerometers were installed evenly distributed along the central girder alongside the strain gauges. Two axle sensors were installed on the road carriageway to indicate the entry and exit time of a vehicle passing over the test span such that the average speed, axle spacing and number of axles of the vehicles can be obtained. The time histories of the dynamic axle loads were then obtained from the strain measurements.

Gonzalez et al. (2008) presented the multiple-sensor B-WIM algorithm based on moving force identification theory, and the results showed that the Moving Force Identification algorithm is superior in giving more accurate results than the B-WIM method which is based on a simple static approach. An associated field test was carried out on the Vransko Bridge in Slovenia and an existing B-WIM system was applied to analyse the measured data whereby the static axle loads, impact factors and natural frequencies of the vehicle were obtained from the complete time histories of the identified moving forces (Rowley et al., 2009).

Deng and Cai (2010b) carried out a field test on a bridge deck carrying two-way traffic and dynamic axle loads are identified from both the deflection and strain time histories. Seven measurement locations were selected at the bottom of the seven girders. These measurement locations are 0.305m away from the mid-span of the girders to avoid stress concentration from incoming diaphragms at the mid-span of the girders. Strain gauges, accelerometers and cable extension transducers were placed at each of these locations.

Despite the above successful applications, there are still some difficulties in getting identified results with a very high accuracy. Some practical issues that should be considered to improve the accuracy of moving load identification are: the selection of the bridge structure, access of the vehicle information and structural responses, and accuracy of the available data. Other uncertainties arising from the effects of braking, jumping, suspension and tyre type and tyre pressure, may also affect the identification accuracy. The main issues of access to and accuracy of available data are discussed in the following sections.

10.2.2.1 Access to Available Data

Although there is no theoretical limit to the number of axles (vehicles) on top of the bridge deck during the measurements, the density of traffic limits the length of the instrumented bridge span which can be efficiently used for weighing. A larger number of axles on the bridge deck usually accompanies a lower accuracy in the identified results. In addition, the contributions of individual but closely spaced axles to the bridge responses are difficult to distinguish on long span bridges, which can further affect the quality of the identified forces. Also the type of the bridge (steel girder, prestressed concrete girders, reinforced concrete girders or concrete slab) may also have a minor influence on the accuracy.

Bridge Responses

Bridge responses such as strain, displacement and acceleration have all been used in the moving force identification. Strains are measured either by strain gauges or reusable strain transducers which are attached to or embedded in the main structural elements. The mid-span locations have been traditionally instrumented as they generally provide the highest strain values, but other locations can also be instrumented to provide adequate or even improved information on the structural behavior under traffic loads. Recently, fiber optical sensors have also been used to measure the structural responses, such as strains, deformations or accelerations. They could yield dynamic strains at a resolution of one micron and perform well under hostile environments.

To obtain the higher frequency component in the dynamic moving forces, accelerations are also used in the identification. The accelerometers are usually installed at the bottom of slab deck or at the level of neutral axis of the deck. The combination of strain and acceleration measurements could improve the accuracy of the identified results.

Vehicle Information

Accurate knowledge on the location of the vehicle/axle is vital for a successful identification of the moving loads with the algorithms described in this book. Most of the existing B-WIM systems use axle detectors, which can be either removable, such as tape switches, road hoses or similar pneumatic sensors, or permanent like piezoceramic or other similar built-in pavement sensors. Two detectors in each lane provide the travelling velocity of each axle and thus the dimensions, velocity and type of the vehicle. They are usually placed before the bridge entry or on the bridge deck depending on the requirement of subsequent data processing. Recently, the vehicle information, such as the vehicle speed and axle spacing, is determined (Cardini and Dewolf, 2009) by minimizing the difference between the measured and simulated responses of the bridge under the vehicular loads.

Knowledge on the natural frequencies of the vehicle may also be useful for an estimation of the frequency components of the interaction forces to be identified.

Data Acquisition System

The sensors can be part of a long-term monitoring system with the data logging system housed in a station at one end of the bridge deck. Or they can form part of a short-term monitoring system with the sensors installed immediately before the measurement. Use of a mobile measurement van is currently very popular for this purpose. Optimized locations for sensors can be determined with reference to the magnitude and the rate of change of the strain responses and to an acceptable level of error in the identified results.

10.2.2.2 Accuracy of Available Data

All the measured signals are pre-conditioned filtered with a low-pass filter before they are used for the analysis. The filter would be effective to reduce the high frequency noise in the measurements. The zero-shift in the measured data should be removed before the data is used for the identification. The time duration between two successive measured data points can be interpolated to have a higher sampling frequency to match the frequency components in the inverse problem.

System Calibration

All moving force identification techniques need the calibration of the measured data from a simple static test (Ashebo et al., 2007). A vehicle with a known gross weight, axle weight, axle spacing and distance between the left and the right wheels can be used for the calibration test. The vehicle is stopped at several specific locations along different lanes in the longitudinal direction of the bridge deck. All responses from strain gauges are recorded by the data acquisition system. The calibration factors and sensitivity for each strain gauge can then be determined by minimizing the difference between the theoretical responses and the measurements.

Modal Properties of Bridge Structure

The modal test is mainly required for the determination of the dynamic behavior of the bridge structure, such as the natural frequencies, mode shapes and the damping ratios. Experimental vibration modes of the deck can be obtained from its free vibration responses after the passage of a vehicle with appropriate sensor arrangement to collect responses at specified locations where mode shapes information is required. Detailed modal frequencies and vibration modes can then be obtained using commercial software, such as LMS Modal Analysis.

10.3 Further Comparison of the FEM Formulation and the EST Method in the Vehicle–Bridge Interaction Identification

The bridge–vehicle system is represented by a 30 m long simply supported beam subject to a moving vehicle with two axles and four degrees-of-freedom. The parameters of the beam are the same as those described in Section 8.4.1.1 with $\xi = 0.02$ for all modes.

A four DOFs vehicle model is presented in Figure 10.1. The characteristics of the vehicle model are adopted from Mulcahy (1983):

$$m_v = 17735 \, \text{kg}, \quad m_1 = 1500 \, \text{kg}, \quad m_2 = 1000 \, \text{kg}, \quad S = 4.27 \, \text{m}, \quad a_1 = 0.519,$$
$$a_2 = 0.481, \quad H = 1.80 \, \text{m}, \quad k_{s1} = 2.47 \times 10^6 \, \text{Nm}^{-1}, \quad k_{s2} = 4.23 \times 10^6 \, \text{Nm}^{-1},$$
$$k_{t1} = 3.74 \times 10^6 \, \text{Nm}^{-1}, \quad k_{t2} = 4.60 \times 10^6 \, \text{Nm}^{-1}, \quad c_{s1} = 3.00 \times 10^4 \, \text{Nm}^{-1}\text{s},$$
$$c_{s2} = 4.00 \times 10^4 \, \text{Nm}^{-1}\text{s}, \quad c_{t1} = 3.90 \times 10^3 \, \text{Nm}^{-1}\text{s}, \quad c_{t2} = 4.30 \times 10^3 \, \text{Nm}^{-1}\text{s},$$
$$I_v = 1.47 \times 10^5 \, \text{kgm}^2$$

The natural frequencies of the vehicle are 10.27 Hz, 14.44 Hz, 65.05 Hz and 94.90 Hz. The first six bridge modes are used in the calculation of the interaction

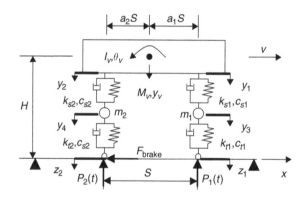

Figure 10.1 Bridge–Vehicle system model

forces by the method described in Chapter 4. Six measuring points are evenly located on the bridge. The mass ratio between the vehicle and bridge is 0.135. The time interval between two adjacent data points is 0.002 s.

10.3.1 Effect of Road Surface Roughness and Moving Speed

The road surface roughness has been described in Section 4.2.3 and in the ISO-8606 (1995) specification. Tables 10.2 and 10.3 show the errors in the identified moving loads with different moving speeds against the road surface roughness from using the FEM and EST Methods. Figure 10.2 shows the identified moving loads with Class B road surface roughness and 5 percent noise level in the responses. Only the strain responses are used.

The identified time histories are shown varying about the true time histories in Figure 10.2. These two methods can be used to identify the bridge–vehicle interaction forces from the bridge responses, and acceptable results can be obtained with different road surface roughness and moving speeds in the identification. The moving speed has little effect on the accuracy in the identified moving loads from both methods.

In the results from the FEM, the errors in the identified results increase as the road surface roughness increases but they change slightly for different noise levels. This is because the high frequency components caused by the road surface roughness are reduced by the fitting of the orthogonal function. In the EST Method, the errors in the identified results change slightly as the road surface condition deteriorates. However, they are sensitive to the noise level in the responses. It may be concluded from this study that the EST Method is good for measurements with low noise level and FEM is good for measurements with high noise level.

Table 10.2 Errors in the identified moving loads using FEM with different moving speeds and road surface roughness (in percent)

Speed (m/s)	Road Surface Roughness	1% noise		5% noise		10% noise	
		Error 1	Error 2	Error 1	Error 2	Error 1	Error 2
20	No	2.988	3.909	3.128	4.067	3.458	4.444
	A	3.412	3.416	3.562	3.530	3.890	3.844
	B	4.062	4.884	4.158	4.909	4.424	5.031
	C	11.312	15.747	11.316	15.742	11.391	15.807
	D	21.029	28.115	21.046	28.163	21.108	28.247
30	No	2.290	2.457	2.404	2.523	2.806	2.822
	A	3.126	3.108	3.182	3.181	3.430	3.446
	B	4.043	4.056	4.046	4.118	4.245	4.326
	C	11.691	13.279	11.654	13.344	11.676	13.487
	D	21.773	24.674	21.746	24.714	21.769	24.796
40	No	2.827	3.279	3.264	3.352	3.983	3.737
	A	3.203	4.104	3.509	4.054	4.124	4.207
	B	3.879	4.971	4.111	4.908	4.612	5.005
	C	10.714	13.110	10.770	13.022	10.932	13.059
	D	18.798	24.385	18.756	24.319	18.780	24.324

Note: Error 1 – error in the first axle load; Error 2 – error in the second axle load.

Table 10.3 Errors in the identified moving loads using EST with different moving speeds and road surface roughness (in percent)

Speed (m/s)	Road Surface Roughness	1% noise		5% noise		10% noise	
		Error 1	Error 2	Error 1	Error 2	Error 1	Error 2
20	No	1.997	2.235	8.631	9.009	14.206	14.463
	A	2.009	2.250	8.591	9.020	13.992	14.416
	B	2.026	2.269	8.610	9.072	13.904	14.434
	C	2.379	2.631	9.842	10.539	14.262	15.633
	D	3.199	3.557	13.263	14.240	18.527	19.098
30	No	2.091	2.246	9.238	9.619	14.435	14.843
	A	2.081	2.210	9.163	9.479	14.255	14.722
	B	2.083	2.193	9.150	9.419	14.169	14.696
	C	2.290	2.227	9.886	9.697	14.290	15.421
	D	2.858	2.648	12.247	11.623	18.523	17.901
40	No	2.008	2.349	8.843	9.534	14.623	15.466
	A	2.052	2.395	8.960	9.737	14.544	15.601
	B	2.091	2.435	9.095	9.919	14.544	15.752
	C	2.622	2.979	11.190	12.457	15.683	18.501
	D	3.372	4.031	14.336	17.443	19.652	22.532

Note: Error 1 – error in the first axle load; Error 2 – error in the second axle load.

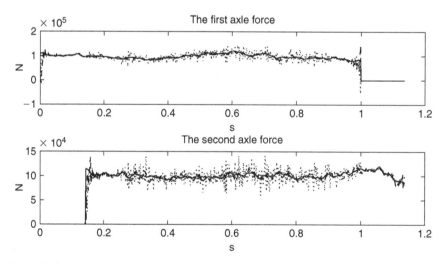

Figure 10.2 Identified forces on Class B road with 5 percent noise (— true forces; - - - using FEM; ···· using EST)

10.3.2   Identification of Moving Loads on a Bridge Deck with Varying Speeds

Normally a vehicle is moving on top of the bridge deck with a varying speed, and we shall discuss the moving load identification when the varying speed is known in this section. The first six modes are used in the simulation. Six measuring points are evenly

Table 10.4 Errors (in percent) in the moving loads identified with varying speeds using FEM

| Noise Level (%) | −1 m/s² (braking at entry) | | | | −3 m/s² (braking at 1/3L) | | | |
| | Instantaneous | | Average | | Instantaneous | | Average | |
	Error 1	Error 2	Error 1	Error 2	Error 1	Error 2	Error 1	Error 2
1	4.129	5.370	5.081	7.020	3.555	5.568	6.986	14.848
2	4.094	5.307	5.064	6.968	3.529	5.568	6.988	14.822
3	4.065	5.246	5.049	6.934	3.520	5.575	6.993	14.798
4	4.051	5.203	5.038	6.985	3.516	5.588	7.000	14.776
5	4.045	5.214	5.029	7.096	3.518	5.609	7.005	14.756
6	4.045	5.348	5.024	7.213	3.524	5.634	7.011	14.736
7	4.049	5.523	5.031	7.337	3.535	5.666	7.019	14.719
8	4.060	5.705	5.042	7.468	3.552	5.703	7.027	14.704
9	4.076	5.894	5.055	7.604	3.574	5.743	7.037	14.696
10	4.099	6.086	5.069	7.743	3.600	5.787	7.047	14.696

Note: Error1 – error in the first axle load; Error 2 – error in the second axle load.

Table 10.5 Errors (in percent) in the moving loads identified with varying speeds using EST Method

| Noise Level (%) | −1 m/s² (braking at entry) | | | | −3 m/s² (braking at 1/3L) | | | |
| | Instantaneous | | Average | | Instantaneous | | Average | |
	Error 1	Error 2	Error 1	Error 2	Error 1	Error 2	Error 1	Error 2
1	1.930	2.630	4.907	6.895	1.874	2.739	8.026	15.682
2	3.644	4.655	5.659	7.825	3.504	5.088	8.265	15.921
3	5.297	6.674	6.730	9.158	5.099	7.395	8.734	16.467
4	6.968	8.663	7.955	10.689	6.664	6.655	9.396	17.296
5	8.622	10.647	9.322	12.309	8.205	11.873	10.086	18.368
6	10.825	12.583	10.715	14.004	9.712	14.056	10.305	19.790
7	11.825	14.440	11.137	14.773	11.160	16.191	10.553	20.955
8	12.291	13.724	11.778	15.660	11.759	15.042	10.988	21.916
9	13.109	14.497	12.390	16.516	12.568	15.889	11.440	22.764
10	13.858	15.244	12.989	17.321	13.330	16.690	11.899	23.540

Note: Error 1 – error in the first axle load; Error 2 – error in the second axle load.

distributed on the bridge deck, and the time interval between data points is 0.002 s. Other parameters are the same as in previous discussions. The responses are calculated by the Modal Superposition Method (EST) in Chapter 2. Tables 10.4 and 10.5 show the errors in the identified results with different noise levels in the responses using the FEM and EST methods. Figure 10.3 shows the identified moving loads with the vehicle starts braking at the entry of the bridge deck with an acceleration of -1 m/s² and 5 percent noise in the responses using these two methods. Figure 10.4 shows the identified results with the vehicle starts braking at 1/3L and the acceleration is -3 m/s² and with 5 percent noise in the responses. The initial vehicle velocity is 30 m/s and the road roughness is Class *B* in both cases. The results are shown under the heading 'instantaneous' in the Tables.

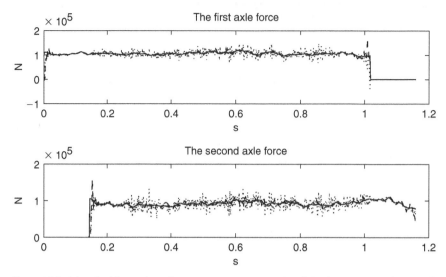

Figure 10.3 Identified forces based on instantaneous speed with braking at entry (— true forces; - - - using FEM; · · · · using EST Method)

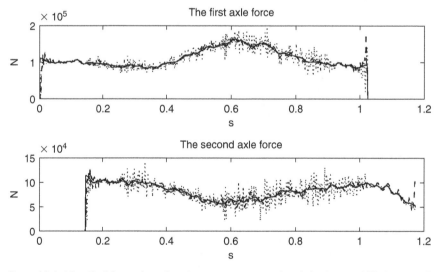

Figure 10.4 Identified forces based on instantaneous speed with braking at 1/3L (— true forces; - - - using FEM; · · · · using EST Method)

Since the instantaneous speed of the forces is known, both methods can be used to identify the moving loads from bridge strains, and acceptable results can be obtained from both methods at low noise level. Those from using FEM are consistently much better than those from using the EST Method for different noise levels in the study.

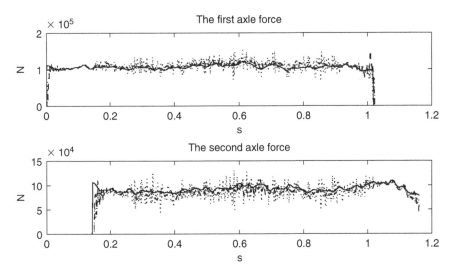

Figure 10.5 Identified forces based on averaged speed with braking at entry (— true forces; --- using FEM; ···· using EST Method)

10.3.3 Identification with Incomplete Vehicle Speed Information

In practice, the axle spacing, the number of axles of the vehicle, and the time that the vehicle enters or exits the bridge can be measured directly by axle sensors (Chan, Law and Yung, 2000). But the braking position and the acceleration are difficult to measure. The errors in result obtained from identifying using an average speed should be studied. Figure 10.5 shows the identified results with the vehicle starts braking at the entry with an acceleration of -1 m/s^2 using both methods. The average speed of 29.39 m/s is used. Figure 10.6 shows the identified results with the vehicle starts braking at $1/3L$ and the acceleration is -3 m/s^2. The average speed of 29.04 m/s is used. The initial vehicle velocity is 30 m/s and the road roughness is Class B in both cases. Tables 10.4 and 10.5 show the errors in the identified results with different noise levels in the responses. The results are shown under the heading 'average' in the Tables.

It is seen that the identified results from both methods using the average speed are acceptable when the acceleration is -1 m/s^2. But for the case with -3 m/s^2 acceleration, a large increase in the error for the second axle load is observed. This shows that the moving loads can be identified from strains using the average speed when the acceleration is not very large.

In the figures, the first moving load is seen over-estimated and the second moving load is under-estimated in both methods. This is because the moving loads are estimated by minimizing the error between the measured and reconstructed responses from the identified moving loads. The location of the resultant load using average speed lags behind that of the true resultant load and this difference is largest approximately at mid-span of the bridge deck in these two cases. The optimization, however, yields a location of the resultant load which is close and behind the true load. This leads to an

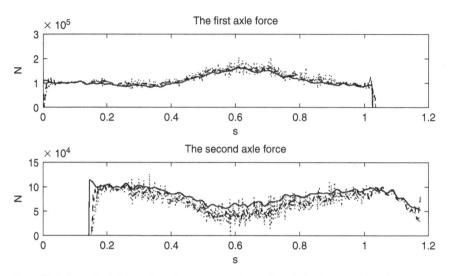

Figure 10.6 *Figure 10.6* Identified forces based on averaged speed with braking at 1/3L (— true forces; - - - using FEM; · · · · using EST Method)

over-estimated first axle load and an under-estimated second axle load. This behavior is opposite to the case of having the acceleration of the vehicle.

10.4 Dynamic Axle and Wheel Load Identification

10.4.1 Dynamic Axle Load Identification

The laboratory plate–vehicle system fabricated for the studies in Section 6.3.4.2 is used for the present studies. Measurements obtained from the test are used in the moving force identification.

Nine of the vibration modes ($m = 3, n = 3$), shown in Table 6.14 in Chapter 6, and the strains at nine or 15 measuring points are used in the identification. The measured signals, originally sampled at 1000 Hz, are re-sampled to have a time interval of 0.005s to reduce the computation time at the expense of accuracy. As the model car is moving along the central line (Rail 3), Rail 1 or Rail 2 in turn, the strains at 1/4a, 1/2a and 3/4a of each beam are used to identify the moving loads. Five sensor sets, shown in Figure 6.30 in Chapter 6, are used for a comparative study on the effect of sensor location selection.

10.4.1.1 Study 1: Effect of Number of Modes

Figures 10.7 and 10.8 show the identified results from different number of modes using the EST and FEM Methods respectively. Figure 10.7 is a reproduction of Figure 6.31 for easy reference in this study. The model car moves along rail 3 at a speed of 1.1079 m/s. Sensor set V from Figure 6.30 is used in the identification. Table 10.6 shows the correlation coefficients between the reconstructed and measured strains at 3/8a on the five beams. Figure 10.9 shows the identified axle loads from nine modes

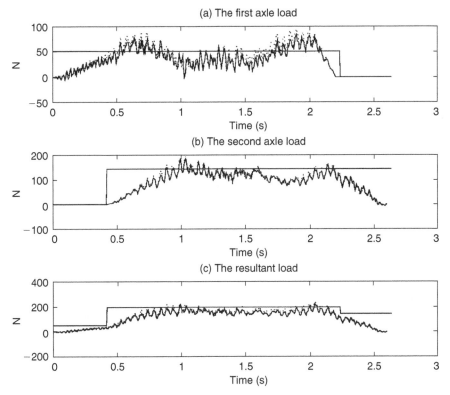

Figure 10.7 Identified axle loads along Rail 3 using different modes using EST Method (— static forces; - - [3;3;3]; - - - [4;3;2]; · · · · [1;1;1])

$(m = 3;\ n = 3)$ using both methods for comparison. In the FEM, 50 terms in the orthogonal functions are used to approximate the measured strains.

The identified axle loads from using the two methods are close to the static forces and the correlation coefficients between the reconstructed and measured strains at 3/8a on each beams are all over 0.9. This shows that both methods are effective to identify the moving vehicular axle loads from bridge strains and acceptable results can be obtained. Figures 10.7 and 10.8 show that the identified results from three longitudinal bending modes $(m = 3;\ n = 1)$ are close to that from using nine modes $(m = 3;\ n = 3)$. This shows that the effect of torsional vibration is small when the model car moves along Rail 3, and there is little difference in the results when the torsional modes are included in the calculation. The axle loads can be identified approximately using an equivalent beam model when the vehicle is moving along the central line of the bridge deck. Figure 10.9 also indicates the identified results using EST Method are close to that using FEM when sensor set V is used in the identification with FEM giving larger fluctuations than EST Method. This indicates that both methods are effective and accurate to identify the dynamic axle loads in practice using different number of vibration modes with 15 measuring points.

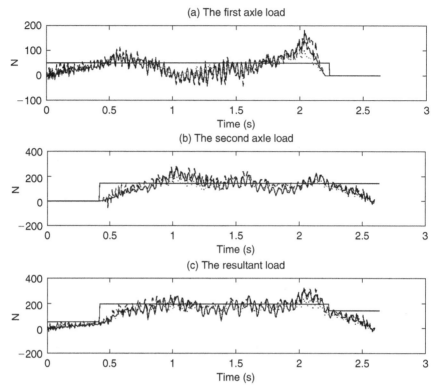

Figure 10.8 Identified axle loads along Rail 3 using different modes using FEM (— static forces; - - [3;3;3]; - - - [2;2;2]; · · · · [1;1;1])

Table 10.6 Correlation coefficients at 3/8a on the five beams

Method	Modes	Correlation Coefficients				
		Beam 1	Beam 2	Beam 3	Beam 4	Beam 5
EST	3;3;3	0.979	0.982	0.960	0.979	0.950
	4;3;2	0.982	0.983	0.961	0.981	0.956
	1;1;1	0.967	0.976	0.949	0.978	0.969
FEM	3;3;3	0.905	0.934	0.903	0.936	0.904
	2;2;2	0.923	0.927	0.904	0.931	0.921
	1;1;1	0.923	0.928	0.904	0.932	0.921

Note: 4;3;2 indicates four modes with $m = 1$, three modes with $m = 2$ and two modes with $m = 3$.

10.4.1.2 Study 2: Effect of Measuring Locations

The sensor sets and the parameters are the same as above. Figures 10.10 and 10.11 show the identified loads from strains at different measuring points using the EST and FEM Methods respectively as the model car moves along Rail 3. Figure 10.10 is a reproduction of Figure 6.32. The correlation coefficients between the reconstructed and measured strains at 3/8a of the beams are shown in Tables 10.7 and 10.8.

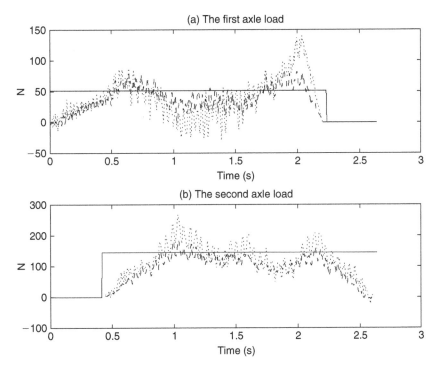

Figure 10.9 Identified axle loads along Rail 3 from different methods (— true forces; - - - using EST Method; · · · · using FEM)

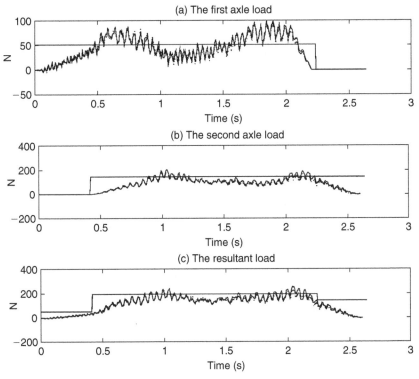

Figure 10.10 Identified axle loads along Rail 3 from EST Method (— static forces; - - identified with Set I; - - - identified with Set II; · · · · Identified with Set III)

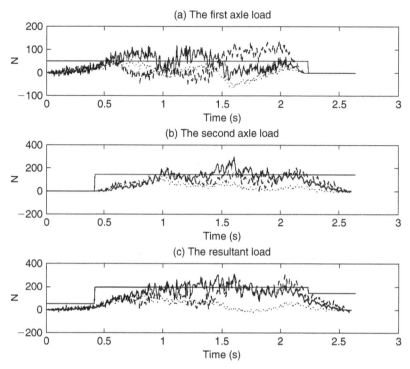

Figure 10.11 Identified axle loads along Rail 3 from FEM (— static forces; - - identified with Set I; - - - identified with Set II; · · · · Identified with Set III)

Table 10.7 Correlation coefficient between reconstructed and measured strains at 3/8a using EST Method

Sensor Set	Rail Number	Average Speed (m/s)	Correlation Coefficients				
			Beam 1	Beam 2	Beam 3	Beam 4	Beam 5
I	1	1.09	0.009	0.009	0.079	0.083	0.094
	2	1.09	0.938	0.975	0.957	0.947	0.976
	3	1.11	0.976	0.959	0.927	0.958	0.942
II	1	1.09	0.072	0.862	0.902	0.909	0.891
	2	1.09	0.955	0.980	0.960	0.950	0.975
	3	1.11	0.971	0.961	0.930	0.956	0.931
III	1	1.09	0.154	0.900	0.976	0.983	0.957
	2	1.09	0.958	0.977	0.955	0.946	0.970
	3	1.11	0.965	0.948	0.915	0.946	0.926
IV	1	1.09	0.071	0.871	0.969	0.977	0.955
	2	1.09	0.953	0.973	0.949	0.940	0.966
	3	1.11	0.966	0.948	0.915	0.946	0.926
V	1	1.09	0.001	0.104	0.289	0.0286	0.281
	2	1.09	0.911	0.976	0.969	0.961	0.980
	3	1.11	0.979	0.982	0.960	0.979	0.950

Table 10.8 Correlation coefficient between reconstructed and measured strains at 3/8a using FEM

Sensor Set	Rail Number	Average Speed (m/s)	Correlation Coefficients				
			Beam 1	Beam 2	Beam 3	Beam 4	Beam 5
I	1	1.09	0.067	0.002	0.036	0.036	0.013
	2	1.09	0.409	0.613	0.632	0.650	0.649
	3	1.11	0.887	0.936	0.914	0.931	0.881
II	1	1.09	0.036	0.372	0.812	0.900	0.896
	2	1.09	0.677	0.842	0.876	0.883	0.884
	3	1.11	0.725	0.753	0.715	0.764	0.718
III	1	1.09	0.016	0.253	0.494	0.604	0.625
	2	1.09	0.588	0.756	0.761	0.767	0.800
	3	1.11	0.621	0.743	0.742	0.739	0.631
IV	1	1.09	0.065	0.002	0.177	0.238	0.163
	2	1.09	0.700	0.867	0.890	0.895	0.907
	3	1.11	0.880	0.908	0.875	0.912	0.875
V	1	1.09	0.072	0.006	0.235	0.320	0.235
	2	1.09	0.741	0.886	0.909	0.911	0.921
	3	1.11	0.913	0.954	0.933	0.955	0.910

Figures 10.12 and 10.13 show the identified results with different sensor sets using EST Method as the model car moves along Rails 2 and 1 respectively. They are reproduced from Figures 6.33 and 6.34 for ease of comparison.

The EST Method gives correlation coefficients above 0.9 as the model car is moving along the central line (Rail 3) or Rail 2. But the correlation coefficients are very small when sensor Set I or II is used with the model car moving along Rail 1. This is because the measuring points are far away from the moving model car, and they cannot pick up the dominating bending modes, whereas the torsional modal responses are small, as found in Study 1 in Section 10.3.1.1 above. Figure 10.13 shows the extreme case from sensor Set I as the model car moved along Rail 1. It is concluded that the identified force time histories from EST Method are satisfactory from different sensor sets when the car moves along Rails 2 or 3.

The identified results are different from sensor Sets I, II and III using FEM as shown in Figure 10.11. They are also different with the identified results from sensor Set V shown in Figure 10.8. This shows that the performance of FEM is very dependent on the sensor locations and on the amount of measured information, and more sensors should be used to identify the moving loads using FEM. This observation is also supported by the correlation coefficients in Table 10.8 which are in general poorer than those from EST Method in Table 10.7. The results from FEM also exhibit unsatisfactory performance in identifying loads moving along Rails 1 and 2 from sensor Sets I to IV.

10.4.1.3 *Study 3: Effect of Load Eccentricities*

The following discussions refer to the results from EST Method shown in Figures 10.10, 10.12 and 10.13 and in Table 10.7. The EST Method is effective to identify the axle

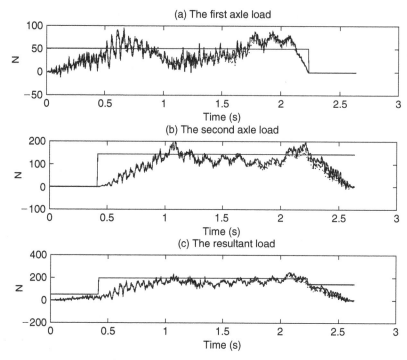

Figure 10.12 Identified axle loads along Rail 2 from EST Method (— static forces; - - identified with Set I; - - - identified with Set II; · · · · Identified with Set III)

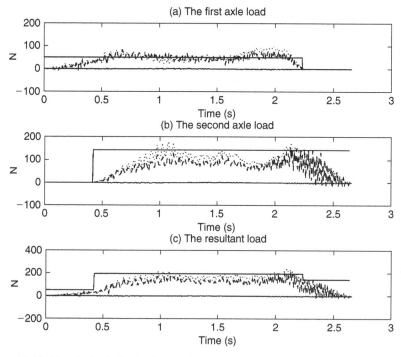

Figure 10.13 Identified axle loads along Rail 1 from EST Method (— static forces; - - identified with Set I; - - - identified with Set II; · · · · identified with Set III)

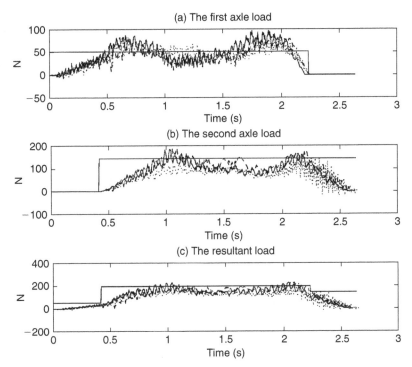

Figure 10.14 Identified axle loads along different rails from sensor Set II using EST Method (— static forces; - - Rail 3; - - - Rail 2; · · · · Rail 1)

loads from measured strains, and acceptable results can be obtained. When either one of sensor Set II, III or IV is used in the identification, almost all the correlation coefficients between the reconstructed and measured strains are over 0.9 even when the model car is moving on Rail 1 or Rail 2 at an eccentricity of 1/8b and 3/8b, respectively. The identified eccentric loads are close to that with no eccentricity as shown in Figure 10.14.

The identified results from sensor Set III in Figure 10.13 is the largest in the group and that from sensor Set II is larger than that from sensor Set I as the model car moves along Rail 1. This is because sensor Set III is close to the moving loads and is more subject to the bending modes of the bridge deck, and the signal to noise ratios of the measured strains are larger than that from sensor Sets I or II. Therefore the identified results from sensor Sets I and II are over-smoothed and are smaller than those from sensor Set III.

10.4.2 *Wheel Load Identification*

10.4.2.1 *Study 4: Effect of Measuring Locations*

The sensor sets and the parameters are the same as for the axle load identification in Section 10.3.1. The first and third wheels of the vehicle are on the front axle and the

Table 10.9 Correlation coefficients between reconstructed and measured strain at 3/8a for wheel load identification

Method	Sensor Set	Average Speed (m/s)	Rail	Correlation Coefficients				
				Beam 1	Beam 2	Beam 3	Beam 4	Beam 5
EST	I	1.11	3	0.977	0.960	0.926	0.958	0.944
	II			0.976	0.964	0.933	0.961	0.942
	III			0.970	0.959	0.930	0.957	0.932
	IV			0.969	0.951	0.917	0.948	0.929
	V			0.977	0.979	0.956	0.979	0.956
	II	1.09	2	0.959	0.980	0.959	0.950	0.975
	II	1.09	1	0.153	0.912	0.976	0.984	0.963
FEM	I	1.11	3	0.894	0.938	0.921	0.931	0.884
	II			0.818	0.827	0.786	0.836	0.811
	III			0.820	0.873	0.849	0.875	0.815
	IV			0.894	0.912	0.880	0.915	0.889
	V			0.911	0.926	0.895	0.930	0.907
	II	1.09	2	0.734	0.870	0.895	0.908	0.908
	II	1.09	1	0.029	0.406	0.824	0.913	0.906

second and fourth wheels are on the rear axle. The second and fourth wheels follow the first and third wheels respectively. Figures 10.15 and 10.16 show the identified wheel loads with different sensor sets using the EST and FEM Methods respectively as the model car moves along Rail 3. Table 10.9 shows the correlation coefficients between the reconstructed and measured strains at 3/8a. The sampling frequency is 200 Hz, and 50 terms of the orthogonal functions are used in the FEM. Figure 10.17 shows the comparative results from both methods from sensor Set V as the model car moves along the central line (Rail 3).

Both methods are effective to identify dynamic wheel loads along the central line from bridge strains, and acceptable results can be obtained as indicated by the correlation coefficients in Table 10.9. The identified force time histories from sensor Set II are nearly the same as from sensor Sets I or III using EST Method, but they are different from each other when using FEM. Table 10.9 also shows that more measured information is needed in FEM than in the EST Method in wheel load identification.

For results from the EST Method in Figure 10.15 (reproduced from Figure 6.35 for comparison), the left wheel loads (the first and second loads) from sensor Set I are larger than those from sensor Set III and the right wheel loads (the third and fourth loads) from sensor Set I are smaller than those from sensor Set III. No rolling motion of the vehicle is observed in the identified forces. Since there is no spring component in each wheel in the model car and the wheel spacing is very small, the four wheel loads behave similar to a single moving mass with some pitching effects. Therefore this difference in the identified forces can only be due to the proximity of the sensors to the loads, i.e. sensor Set I is close to the left wheel loads and sensor Set III is close to the right wheel loads.

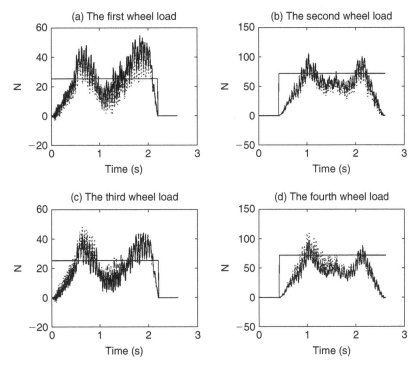

Figure 10.15 Identified wheel loads along Rail 3 using EST Method (— static forces; - - identified with Set I; - - - identified with Set II; · · · · identified with Set III)

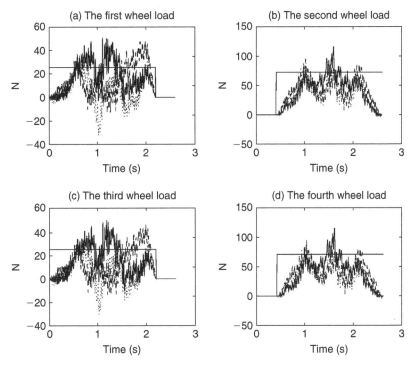

Figure 10.16 Identified wheel loads along Rail 3 using FEM (— static forces; - - identified with Set I; - - - identified with Set II; · · · · identified with Set III)

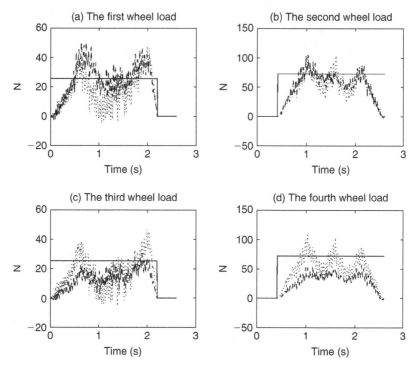

Figure 10.17 Identified wheel loads along Rail 3 using different methods (— static forces; - - - using EST Method; · · · · using FEM)

For the FEM results in Figure 10.16, the front wheel loads (the first and third loads) vary greatly from using different sensor sets whilst the rear wheel loads (the second and fourth loads) are relatively the same from using sensor Sets I, II and III. The identified force time histories are not consistent.

When all 15 sensors are used, the identified force time histories from both methods are similar with the FEM results showing larger fluctuations than results from the EST Method, as shown in Figure 10.17. This observation is similar to that in the axle load identification shown in Figure 10.9.

It may be concluded that the EST Method is effective to accurately identify loads moving along the central line of the bridge deck, while FEM does not give consistent results from using different sets of sensors.

10.4.2.2 Study 5: Effect of Load Eccentricities

The sensor sets and the parameters are the same as above. Figure 10.18 shows the identified results from sensor Set II using EST Method as the model car moves along different rails. Figure 10.18 is a reproduction of Figure 6.36 for comparison. Figure 10.19 shows the identified results from sensor Set V using FEM as the model car moves along different rails. All measured information is used as the accuracy of FEM is found from previous studies to be dependent on the amount of measured information.

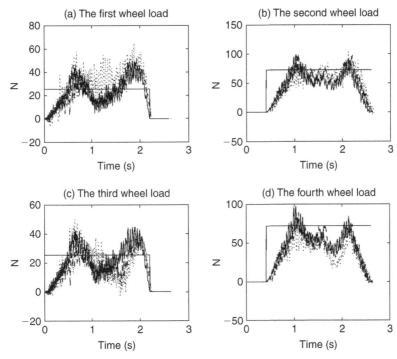

Figure 10.18 Identified wheel loads along different rails with sensor Set II using EST Method (— static forces; - - Rail 3; - - - Rail 2; ···· Rail 1)

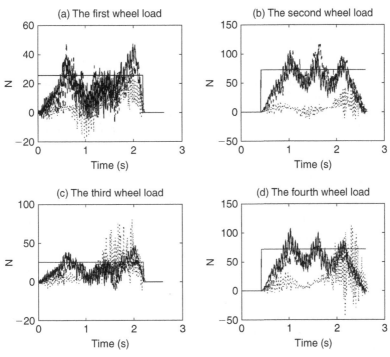

Figure 10.19 Identified wheel loads along different rails with sensor Set V using FEM (— static forces; - - Rail 3; - - - Rail 2; ···· Rail 1)

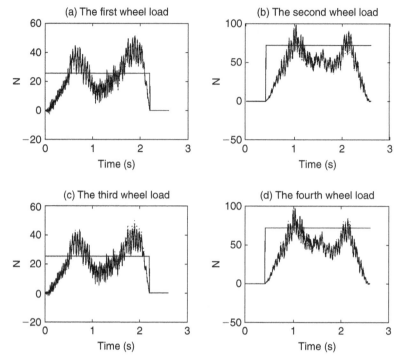

Figure 10.20 Identified wheel loads along Rail 3 using different modes using EST Method (— static forces; - - [3;3;3]; - - - [2;2;2]; · · · · [1;1;1])

The related correlation coefficients between the reconstructed and measured strain at 3/8a are also shown in Table 10.9. These two methods are found to be effective in identifying the dynamic wheel loads along Rails 2 and 3 based on sensor Set II for the EST Method and sensor Set V for the FEM. Both methods fail to identify forces along Rail 1 and the reason is given in Section 10.3.1.3 above.

10.4.2.3 Study 6: Effect of Number of Modes

The sensor sets and parameters are the same as above. Figure 10.20 shows the identified results from different number of modes using EST Method as the model car moves along the central line (Rail 3). Figure 10.20 is a reproduction of Figure 6.37 for comparison. Sensor Set II is used in the identification. Figure 10.21 shows the identified results using FEM and sensor Set V is used in the identification. The correlation coefficients between the reconstructed and measured strain at 3/8a are listed in Table 10.10. The correlation coefficients are all above 0.89 for both sets of results. The identified force time histories in Figures 10.20 and 10.21 are approximately the same when different numbers of the modes are used in the identification. This shows that the effect of torsional vibration is small when the model car moves along Rail 3, and there is little difference when the torsional modes are included in the calculation. This also supports the findings in Sections 10.3.1.1, 10.3.1.2 and 10.3.2.1 that the

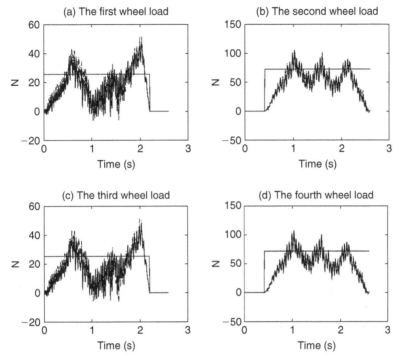

Figure 10.21 Identified wheel loads along Rail 3 using different modes using FEM (— static forces;
-- [3;3;3]; - - - [2;2;2]; · · · · [1;1;1])

Table 10.10 Correlation coefficients for wheel load identification with different number of modes

Method	Sensor Set	Modes	Correlation Coefficients				
			Beam 1	Beam 2	Beam 3	Beam 4	Beam 5
EST	II	$m=3; n=3$	0.976	0.964	0.933	0.961	0.942
		$m=2; n=2$	0.959	0.957	0.924	0.960	0.960
		$m=1; n=1$	0.958	0.957	0.924	0.960	0.960
FEM	V	$m=3; n=3$	0.911	0.926	0.895	0.930	0.907
		$m=2; n=2$	0.924	0.928	0.904	0.932	0.920
		$m=1; n=1$	0.923	0.918	0.904	0.932	0.921

sensors should be close to the moving loads to pick up the dominating bending modes
to have an accurate identified result.

10.5 Modifications and Special Topics on the Moving Load Identification Techniques

Recently, there has been an increase in the number of researchers working on moving
load identification. Many improvements and modifications have been developed for

these techniques. Since the bridge responses are composed of two components, which are the static (quasi-static) and the dynamic components, they theoretically require different values of optimal regularization parameters in the regularized solution. Pinkaew (2006) presented an updated static component technique. The bridge responses were decomposed into static and dynamic components. The static component was identified separately, while only the dynamic component remained in the regularization process. The dynamic programming algorithm described in Chapter 7 was used to identify the dynamic component. The obtained identified result was used to update the associated static component until convergent solution was achieved.

The Zeroth-Order Hold (ZOH) Discrete Method has been used universally in the discretization of measured data in which the sampled value remains constant within the time period between two samples. This leads to error when the sampling rate is not high with subsequent error in the inverse analysis. The First-Order Hold (FOH) Discrete Method improves the accuracy of computation with interpolation between two sampled data.

The Tiknohov regularization has been widely used in the moving load identification problem in previous chapters of this book and it has been shown to yield good identified results, particularly in the presence of measurement noise. The First-Order Regularization operates on the derivatives of the unknown forces and it has been shown to give good identified results in the whole force time history particularly at the beginning and end of the time duration where the Tikhonov regularization (Zeroth-Order Regularization) would fail to perform well in some cases. Both the First-Order Hold Discrete Method and the First-Order Regularization technique will be discussed in the following sections.

10.5.1 *First Order Hold Discrete versus Zeroth Order Hold Discrete*

10.5.1.1 *Zeroth-Order Hold Discrete Method in Response Analysis*

The equation of motion of a structural system can be expressed in the state space as following:

$$\dot{z} = A^C z + B^C L \cdot F \tag{10.1}$$

where

$$z = \begin{bmatrix} x \\ \dot{x} \end{bmatrix}, \quad A^C = \begin{bmatrix} 0 & I \\ -M^{-1}K & -M^{-1}C \end{bmatrix} \quad \text{and} \quad B^C = \begin{bmatrix} 0 \\ M^{-1} \end{bmatrix}.$$

where matrices M, C, and K are the mass, damping and stiffness matrixes of the structural system respectively. F is the vector of external excitation forces on the structure and L is the mapping matrix for the input excitation forces. $\ddot{x}$, $\dot{x}$ and x are vectors of acceleration, velocity and displacement of the structural system respectively. The superscript ^{C} denotes that the matrices are for the continuous system. Vector $y(t) \in \Re^{ns \times 1}$ is assumed to represent the output of the structural system and it is assembled from the measurements with:

$$y = R_a \ddot{x} + R_v \dot{x} + R_d x \tag{10.2}$$

where R_a, R_v and $R_d \in \mathfrak{R}^{m \times sdof}$ are the output influence matrices for the measured acceleration, velocity and displacement respectively, m is the dimension of the measured responses and $sdof$ is the number of DOFs of the structure. Equation (10.1) can be rewritten as:

$$y = R^C z + D^C \cdot L \cdot F \tag{10.3}$$

where $R^C = [R_d - R_a M^{-1} K \quad R_v - R_a M^{-1} C]$ and $D^C = R_a M^{-1}$. When the external force is known, or can be measured, the value of state variable z and y can be calculated accurately. However, in reality, the measurement data is discrete and the continuous state equation should be transformed into discrete equation.

Based on the ZOH discrete method (Franklin et al., 1998), Equations (10.1) and (10.3) can be converted into the following discrete equations as:

$$z(j+1) = A^D z(j) + B^D \cdot L \cdot F(j) \tag{10.4}$$

$$y(j) = Hz(j) + J \cdot L \cdot F(j) \quad (j = 1, 2, \ldots, N) \tag{10.5}$$

Superscript D denotes the matrices for the discrete structural system. N is the total number of sampling points, dt is the time step between the state variables $z(j)$ and $z(j+1)$, $A^D = \exp(A^C \cdot dt)$, $B^D = (A^C)^{-1}(A^D - I)B^C$, $H = R^C$ and $J = D^C$.

When the output $y(j)$ is solved from Equations (10.4) and (10.5) with zero initial conditions of responses in terms of the previous input $F(k)$, $(k = 0, 1, \ldots, j)$, we have:

$$y(j) = \sum_{k=0}^{j} H_k \cdot L \cdot F(j-k) \tag{10.6}$$

where $H_0 = J$ and $H_k = HA^D_{(k-1)}B^D$.

The constants in matrix H_k in Equation (10.6) are the system Markov parameters. Equation (10.6) can be rewritten to give the matrix convolution equation as:

$$Y = H_L F \tag{10.7}$$

where

$$H_L = \begin{bmatrix} H_0 & 0 & \cdots & 0 \\ H_1 & H_0 & \cdots & 0 \\ \vdots & \vdots & \ddots & 0 \\ H_{N-1} & H_{N-2} & \cdots & H_0 \end{bmatrix} L_S, \quad L_S = \begin{bmatrix} L & 0 & \cdots & 0 \\ 0 & L & \cdots & 0 \\ \vdots & \vdots & \ddots & 0 \\ 0 & 0 & \cdots & L \end{bmatrix}$$

$$Y = \{y(0)^T \quad y(1)^T \quad \cdots \quad y(N-1)^T\}^T, \quad F = \{F(0)^T \quad F(1)^T \quad \cdots \quad F(N-1)^T\}^T.$$

Matrix H_L is constant for a certain system, and the response vector Y can be derived from the measured responses. This is an ill-posed problem and the external force vector F can be identified from Equation (10.7) which is an inverse problem in structural mechanics.

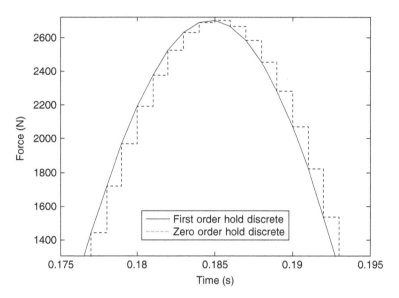

Figure 10.22 Discretization of applied force from ZOH and FOH discrete method

10.5.1.2 Triangle First-Order Hold Discrete Method

The response computation using the ZOH discrete method is subject to error and an inaccurate matrix B^D will result with large error in the state variable. This is because the force in a sampling period has been assumed to be constant as shown in Equation (10.4). Figure 10.22 shows a comparison of the two discrete methods in modeling the applied force data. This discretization within a sampling period is treated differently in the FOH discrete method, namely the triangle hold discrete method (Franklin et al., 1998), where the discretized data is interpolated as:

$$u(t) = u(i) + \frac{u(i+1) - u(i)}{T}(t - iT) \quad (iT \le t \le (i+1)T \quad i = 1, 2, \ldots, N-1) \quad (10.8)$$

where T is the sampling period.

$$u = LF \tag{10.9}$$

Set the impulse function δ as:

$$\delta(t) = \begin{cases} 1 & t = 0 \\ 0 & t \neq 0 \end{cases} \tag{10.10}$$

The impulse response and block diagram of the modified FOH are shown in Figure 10.23(a). The Laplace transformation of the extrapolation filter (Franklin et al., 1998) that follows the impulse sampling is:

$$H_{tri}(s) = \frac{e^{Ts} - 2 + e^{-Ts}}{Ts^2} \tag{10.11}$$

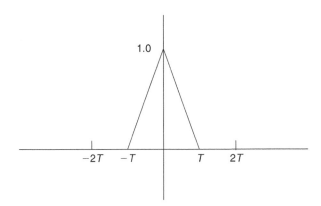

Figure 10.23(a) Impulse response of the extrapolation filter for the modified first order hold (triangle hold)

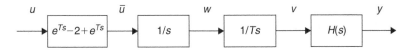

Figure 10.23(b) Block diagram of the triangle-hold equivalent

Based on the block diagram in Figure 10.23(b), the state variables (Franklin et al., 1998) v and w are defined as Equations (10.12) and (10.13):

$$v = w/T \tag{10.12}$$

$$\dot{w} = u(t + T)\delta(t + T) - 2u(t)\delta(t) + u(t - T)\delta(t - T) \tag{10.13}$$

where $\delta(t)$ is the unit impulse shown in Equation (10.10). It can be shown from the integration of Equations (10.12) and (10.13) that $v(i) = u(i)$ and $w(i) = u(i+1) - u(i)$, and a new state space equation can be derived as:

$$\begin{bmatrix} \dot{z} \\ \dot{v} \\ \dot{w} \end{bmatrix} = \begin{bmatrix} A^C & B^C & 0 \\ 0 & 0 & 1/T \\ 0 & 0 & 0 \end{bmatrix} \begin{bmatrix} z \\ v \\ w \end{bmatrix} + \begin{bmatrix} 0 \\ 0 \\ 1 \end{bmatrix} \bar{u} \tag{10.14}$$

where x, v, w and $\bar{u}$ as shown in Figure 10.23 can be taken as the input impulse function. The matrix on right-hand-side of the Equation (10.14) is defined as:

$$F_T = \begin{bmatrix} A_C & B_C & 0 \\ 0 & 0 & 1/T \\ 0 & 0 & 0 \end{bmatrix} \tag{10.15}$$

If the one step solution to Equation (10.14) is written as:

$$\zeta(iT + 1) = e^{F_T T}\zeta(iT) \tag{10.16}$$

then

$$\exp(F_T T) = \begin{bmatrix} \Phi & \Gamma_1 & \Gamma_2 \\ 0 & 1 & 0 \\ 0 & 0 & 1 \end{bmatrix} \qquad (10.17)$$

The equation in variable x can be written as:

$$x(i+1) = \Phi x(i) + \Gamma_1 v(i) + \Gamma_2 w(i) \qquad (10.18)$$

If a new state is defined as $z(i) = x(i) - \Gamma_2 u(i)$, Equation (10.18) for the modified FOH can be rewritten as:

$$z(i+1) = A^D z(i) + B^D u(i) \qquad (10.19)$$

The output equation is

$$y(i) = C^D z(i) + D^D u(i) \qquad (10.20)$$

The parameter for the state equation can then be represented as:

$$A^D = \Phi,$$
$$B^D = \Gamma_1 + \Phi \Gamma_2 - \Gamma_2,$$
$$C^D = H,$$
$$D^D = J + H \Gamma_2$$

Based on the above modified FOH discrete method, the force identification can be conducted following Equations (10.1) to (10.7) with more accurate results compared to the ZOH discrete method.

10.5.2 *First Order Regularization versus Zeroth Order Regularization*

10.5.2.1 *Tikhonov Regularization*

All the moving force identification methods aim at finding a value of the forcing term $\{f\}$ to best match the measured forces. In practice, it is not possible to measure all the structural responses and only certain combinations of the responses $\{\overline{X}\}$ are measured. The measured vector can be defined by:

$$\{Z\}_{2ns \times 1} = [Q]_{2ns \times 2nn} \{X\}_{2nn \times 1} \qquad (10.21)$$

where $[Q]$ is a selection matrix, and $\{X\}$ is a state vector including the responses from all DOFs. ns is the number of measurement points which is usually much less than the number of DOFs nn of the system, nf is the number of moving forces to be identified. The identification problem can be solved by setting up a non-linear least-squares minimization, called the Tikhonov regularization (Trujillo, 1978; Busby and Trujillo, 1997) as:

$$E(\overline{X}_k, f_k) = \sum_{k=1}^{N} [(\{Z\}_k - [Q]\{\overline{X}\}_k), [A]([Z]_k - [Q]\{\overline{X}\}_k) + (\{f\}_k, [B]\{f\}_k)] \quad (10.22)$$

in which (x, y) denotes the inner product of two vectors x and y. $[A]$ is usually a $ns \times ns$ identity matrix, and $[B]$ is a $nf \times nf$ diagonal matrix containing the optimal regularization parameter λ which has a smoothing effect on the solution. Matrices $[A]$ and $[B]$ are weighing matrices, $\{\overline{X}\}$ denotes a state vector of "measured" responses, N denotes the total time steps. The moving forces can be identified from Equation (7.15) in Chapter 7 if the system matrices are known. The above formulation of the inverse problems is also known as the Zeroth-Order Regularization.

10.5.2.2 First-Order Tikhonov Regularization

In the general formulation of the inverse problem, it is possible to replace the regularization term with the derivative of unknown forces instead of the forces themselves (Busby and Trujillo, 1997). Then Equation (7.15) can be rewritten as follows for such a first-order system

$$\{\tilde{X}\}_{j+1} = [J]_{j+1}\{\tilde{X}\}_j + [W]\{\dot{f}\}_j \quad (10.23)$$

in which $\{\tilde{X}\}_{j+1} = \left\{ \begin{array}{c} \{X\}_{j+1} \\ \{f\}_{j+1} \end{array} \right\}$, is a vector of state variables with dimension $(2nn + nf) \times 1$, $\{\dot{f}\}$ is the first order derivative of the unknown forces with dimension $(nf) \times 1$, and the system matrices are:

$$[J] = \begin{bmatrix} [FF] & [G] \\ 0 & [I] \end{bmatrix}_{(2nn+nf) \times (2nn+nf)}, \quad [W] = \begin{bmatrix} 0 \\ I \end{bmatrix}_{(2nn+nf) \times nf}$$

Accordingly Equation (10.21) can be rewritten as:

$$\{Z\}_{2ns \times 1} = [Q]_{2ns \times 2nn}\{\tilde{X}\}_{2nn \times 1} \quad (10.24)$$

With this condition, the problem can be solved correspondingly by the First-Order Tikhonov regularization

$$E(\overline{\overline{X}}_k, \dot{f}_k) = \sum_{k=1}^{N} [(\{Z\}_k - [Q]\{\overline{\overline{X}}\}_k), [A]([Z]_k - [Q]\{\overline{\overline{X}}\}_k) + (\{\dot{f}\}_k, [B]\{\dot{f}\}_k)] \quad (10.25)$$

where $\{\overline{\overline{X}}\}$ denotes a state vector including the 'measured' responses and the forces to be identified. An example of moving force identification results using the first-order regularization is shown in Figure 10.24.

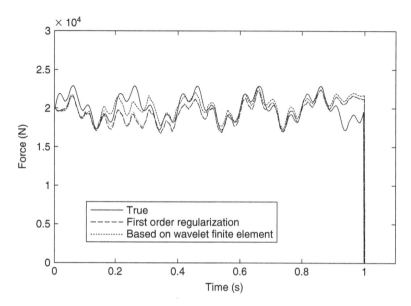

Figure 10.24 Identified moving force with 10 percent measurement noise

10.6 Summary

This chapter presented examples of industrial application of the bridge–vehicle interaction force identification with descriptions on the sensor placement, data analysis for the modal parameters of the bridge deck and calibration of the measured data from static test with discussions on the differences between the Moving Force Identification technique and the B-WIM technique.

The comprehensive experiences with a laboratory experiment indicate that a group of forces moving on top of the model bridge deck can be identified individually or, in terms of axle loads, with accuracy. The identified results for individual loads are poorer than those for axle loads. Both the EST Method and the FEM formulation can identify the moving loads with a small eccentricity, but FEM requires a lot more of measured information to have the same accuracy as the EST Method. Both methods fail to identify loads with a large eccentricity.

Many modifications and improvements have been made to the moving load identification techniques presented in this book, aiming at an enhanced accuracy and computation efficiency. The First-Order-Hold Discrete Method and the First-Order Regularization, which are the two more effective improvements to the techniques, are briefly introduced in this chapter for further reference of the readers.

Chapter 11

Concluding Remarks and Future Directions

11.1 State of the Art

Features of different moving force identification problems have been discussed in this book with descriptions on their implementation for field practices. Some important conclusions found in previous chapters are:

1. The dynamic behavior of a multi-span continuous bridge deck under moving vehicles has been presented. The bridge is modeled either as a multi-span continuous beam when the bridge is narrow compared with its length, or as an orthotropic rectangular plate if otherwise. The influence of different parameters such as the road-surface roughness of the bridge deck, the surface condition of the approach, multiple vehicles and their transverse positions, and braking or acceleration on the bridge deck are studied with computational simulations.

2. The algorithms to analyze the dynamic behavior of one-dimensional or two-dimensional structures subject to the effects of moving loads can be applied to three-dimensional structures, especially those based on finite element method.

3. A successful moving force identification algorithm needs to have good computational technique for the inverse solution. The precise integration method exhibits high precision and efficiency in computation when compared with the Newmark-Beta method. The First-Order Hold Discrete Method retains the gradient between two adjacent data points leading to an improved response computation. The generalized orthogonal function can be used to obtain the modal velocity and acceleration from measured strain response with the aim of reducing errors due to measurement noise. The regularization method applied to the inverse solution also gives greatly improved solution compared to the least-squares solution from using different combinations of measured responses.

4. The TDM is found better than FTDM in solving the ill-posed moving force identification problem. The Interpretive Method II (IMII) gives the least-squares solution on the forces while the TDM method gives highly accurate solution with the application of the regularization technique to provide bounds to the solution. The State Space Methods gives much smaller errors of identification than those obtained from TDM in comparing the identified forces from using different combinations of measured responses in both simulation and laboratory studies. They

can be conveniently extended to include a more complicated finite element model of a structure under multiple force excitations.

5. Both simulation and laboratory test results indicate that the total weight of a vehicle can be estimated indirectly using the moving force identification methods with acceptable accuracy. The methods presented in this book are effective and accurate to identify moving loads on continuous beam or plate from structural vibration responses, though all of them fail to identify moving loads at a large eccentricity. Other influencing factors affecting the accuracy in the identification, such as the sampling frequency, modal truncation in the responses, modeling error, non-uniform speed, sensor locations and the number of modes in the computation, have also been studied and discussed.

6. A general moving load identification method, based on finite element formulation and the Improved Reduced System Condensation technique, has been presented. The measured displacement is analytically expressed as function of the shape functions of the finite elements and the nodal responses of the structure without the usual modal coordinate transformation. The number of master degrees-of-freedom of the system should be smaller than, or equal to, the number of measured points. The identified results are found to be not sensitive to the sampling frequency, velocity of vehicle, measurement noise level and road surface roughness with measured information from at least three measuring points.

7. The bridge–vehicle interaction force identification is an application of the Moving Force Identification technique. Results obtained from a comprehensive laboratory experiment indicate that a group of forces moving on top of the model bridge deck can be identified individually or in terms of axle loads with accuracy. The identified results for individual loads are poorer than those for axle loads. Methods based on modal superposition technique, or based on a finite element model, can identify moving loads with a small eccentricity, but the latter category requires a lot more of measured information to have the same accuracy as the former category.

8. Methods described in this book can be further enhanced in the computation efficiency and accuracy with new advancements in the computational techniques such as the First-Order Hold Discrete Method and the First-Order Regularization to give greatly improved identification of the moving forces.

11.2 Future Directions

11.2.1 Effect of Uncertainties on Moving Force Identification

The dynamic interaction forces between the bridge and vehicle are essentially random in nature due to the effect of the random road surface roughness. Some researchers included the randomness in the excitation due to the road surface roughness in the analysis while the system parameters of the bridge and vehicle were treated as deterministic. An evolutionary spectral approach was developed by Ding et al. (2009) to evaluate the dynamic deflection and vehicle axle loads due to the passage of a vehicle moving along a rough bridge surface. The deterministic and random parts are calculated separately using the Runga-Kutta method. Wu and Law (2010a) presented a spectral stochastic finite element method for the bridge–vehicle problem with non-Gaussian uncertainties.

Karhunen-Loeve expansion was employed to represent both the random forces and system parameters which were assumed to be Gaussian distributed. Polynomial Chaos is also used to model responses with large uncertainties. The statistical relationship between the random moving force and the random structural responses is then established. A moving force identification technique based on this statistical system model was developed by Wu and Law (2010b) and further applied to the identification of the vehicle–bridge system (Wu and Law, 2010c).

11.2.2 *Moving Force Identification with Complex Structures*

The one-dimensional beam models and the two-dimensional orthotropic plate models are used to represent the physical bridge structures in this book. However, the moving force identification techniques can be extended to three-dimensional complex bridges, especially those methods based on the finite element models. Example of this application is found with a bridge model which has been discretized into 720 square plate elements using MSc/NASTRAN FE software, and 25 mode shapes with a frequency range from 5 to 305 Hz were used in the moving force identification (Gonzalez et al., 2008). Deng and Cai (2010a) used the solid elements with the ANSYS program for the vehicle–bridge interaction analysis, and the dynamic interaction forces were identified from the measured bridge responses using the superposition principle and the influence surface concept.

11.2.3 *Integrated Bridge Weigh-In-Motion with Structural Health Monitoring*

Recently the B-WIM system has been combined with the structural health monitoring system. The traffic loads and bridge responses can be simultaneously monitored. Data from the B-WIM systems can be used for traffic planning, pavement design, bridge rating and structural health monitoring. Cardini and Dewolf (2009) presented a long-term B-WIM system using an existing structural health monitoring system, and strain measurements from the steel girders were used to determine the truck weights. Liu et al. (2009) presented a method to evaluate the safety of existing bridges based on monitored live load effects. A video-assisted approach for structural health monitoring of highway bridges was also presented by Chen et al. (2006). The videos of vehicle passing by were captured and synchronized with data recordings from accelerometers. The basic information on vehicle types, arrival times and speeds are extracted from the video images to develop a physics-based simulation model of the traffic excitation. Other examples of integrated system identification are with Zhu and Law (2007) who developed an iterative algorithm to identify the moving loads and structural damage simultaneously from the bridge responses. The moving loads and the pre-stressed force were also identified simultaneously from the bridge responses (Law et al., 2008).

References

AASHTO (2002) *Standard Specifications for Highway Bridges*. The American Association of State Highway and Transportation Officials. Washington D.C. 17th Edition.

AASHTO (2007) *LRFD Bridge Design Specifications*. American Association of State Highway and Transportation Officials. 4th Edition.

Abdel Wahab, M.M., Roeck, G.D. and Peeters, B. (1999) "Parameterization of damage in reinforced concrete structures using model updating." *Journal of Sound Vibration*, 228(4), 717–730.

Abu-Hilal, M. and Mohsen, M. (2000) "Vibration of beams with general boundary conditions due to a moving harmonic load." *Journal of Sound and Vibration*, 232(4), 703–717.

Abu-Hilal, M. and Zibdeh, H.S. (2000) "Vibration analysis of beams with general boundary conditions traversed by a moving force." *Journal of Sound and Vibration*, 229(2), 377–388.

Agrawal, O.P., Stanisic, M.M. and Saigal, S. (1988) "Dynamic responses of orthotropic plates under moving masses." *Ingenieur-Archiv*, 58(1), 9–14.

ANON (1992) *Report of the OECD Working Group IR2 on the Dynamic Loading of Pavements*. Paris: OECD.

Ashebo, D.B., Chan, T.H.T. and Yu, L. (2007) "Evaluation of dynamic loads on a skew box girder continuous bridge Part I: field test and modal analysis." *Engineering Structures*, 29, 1052–1063.

Asnachinda, P., Pinkaew, T. and Laman, J.A. (2008) "Multiple vehicle axle load identification from continuous bridge bending moment resposne." *Engineering Structures*, 30, 2800–2817.

Bakht, B. and Jaeger, L.G. (1985) *Bridge Analysis Simplified*. New York: McGraw-Hill.

Bellman, R. (1967). *Introduction to the Mathematical Theory of Control Processes*. Academic Press, New York.

Bendat, J.S. and Piersol, A.G. (1993) *Engineering Application of Correlation and Spectral Analysis*. 2nd Edition, Wiley, New York.

Bilello, C. and Bergman, L.A. (2004) "Vibration of damaged beams under a moving mass: theory and experimental validation." *Journal of Sound Vibration*, 274(3–5), 567–582.

Blair, M.A., Camino, T.S. and Dickens, J.M. (1991) "An iterative approach to a reduced mass matrix." *Proceedings of the 8th International Modal Analysis Conference*, Florence, Italy, April, 510–515.

Briggs, J.C. and Tse, M.K. (1992) "Impact force identification using extracted modal parameters and pattern matching". *International Journal of Impact Engineering* 12, 361–372.

Busby, H.R. and Trujillo, D.M. (1987) "Solution of an inverse dynamic problem using an eigenvalue reduction technique." *Computers and Structures* 25(1), 109–117.

Busby, H.R. and Trujillo, D.M. (1993) "An Inverse Problem for a Plate under Pulse Loading". *Inverse Problems in Engineering: Theory and Practice*, ASME, pp. 155–161.

Busby, H.R. and Trujillo, D.M. (1997) "Optimal Regularization of an Inverse Dynamics Problem." *Computers & Structures*, 63(2), 243–248.

Cantieni, R. (1983) "Dynamic load tests on highway bridges in Switzerland – 60 years of experience." *Report 211, Federal Laboratories for Testing of Materials*, Dubendorf, Switzerland.

Cantieni, R. (1992) "Dynamic behaviour of highway bridges under the passage of heavy vehicles" *Swiss Federal Laboratories for Materials Testing and Research (EMPA)* Report No. 220, 240p.

Cao, X., Sugiyama, Y. and Mitsui, Y. (1998) "Application of artificial neural networks to load identification." *Computers and Structures* 69, 63–78.

Cardini, A.J. and Dewolf, J.T. (2009) "Implementation of a long-term bridge weigh-in-motion system for a steel girder bridge in the interstate highway system." *Journal of Bridge Engineering ASCE*, 14(6), 418–423.

Cebon, D. (1988) "Theoretical road damage due to dynamic tyre forces of heavy vehicles Part 2: Simulated damage caused by a tandem-axle vehicle." *Proceedings of the Institution of Mechanical Engineers Part C: Journal of Mechanical Engineering Science*, 202(2), 109–117.

Chan, T.H.T. and O'Connor, C. (1990) "Wheel loads from highway bridge strains." *Journal of Structural Engineering Division-ASCE*, 116, 1751–1771.

Chan, T.H.T., Law, S.S., Yung, T.H. and Yuan, X.R. (1999) "An interpretive method for moving force identification." *Journal of Sound and Vibration*, 219(3), 503–524.

Chan, T.H.T., Law, S.S. and Yung, T.H. (2000) "Moving force identification using an existing prestressed concrete bridge." *Engineering Structures*, 22(10), 1261–1270.

Chan, T.H.T., Yu, L. and Law, S.S. (2000) "Comparative studies on moving force identification from bridge strains in laboratory." *Journal of Sound and Vibration*, 235(1), 87–104.

Chang, Q. and Fischbach, G.D. (2006) "An acute effect of neuregulin 1 beta to suppress alpha 7-containing nicotinic acetylcholine receptors in hippocampal interneurons" *Journal of Neuroscience*, 25(44), 11295–11303.

Chatterjee, P.K., Datta, T.K. and Surana, C.S. (1994) "Vibration of continuous bridges under moving vehicles." *Journal of Sound and Vibration*, 169(5), 619–632.

Chen, Y., Feng, M.Q. and Tan, C.A. (2006) "Modeling of traffic excitation for system identification of bridge structures." *Computer-aided Civil and Infrastructure Engineering*, 21, 57–66.

Chen, Y.H., Tan, C.A. and Bergman, L.A. (2002) "Effects of boundary flexibility on the vibration of a continuum with a moving oscillator." *Journal of Vibration and Acoustics ASME*, 124, 552–560.

Choi, K.Y. and Chang, F.K. (1996) "Identification of impact force and location using distributed sensors." *AIAA Journal*, 34(1), 136–142.

COST323 (1999) "European specification on weigh-in-motion of road vehicles." *EUCO-COST/323/8/99*, LCPC, Paris, France.

Davis, P. and Sommerville, F. (1987) "Calibration and accuracy testing of weigh-in-motion systems". *Transportation Research Record* 1123, 122–126.

Dempsey, A.T., O'Brien, E.J. and O'Connor, J.M. (1995) "A bridge weigh-in-motion system for the determination of gross vehicle weights." Jacob, B. et al. (Eds.) *Post Proceedings of the First European Conference on Weigh-in-motion of Road Vehicles*, ETH, Zürich, 239–249.

Deng, L. and Cai, C.S. (2010a) "Identification of dynamic vehicular axle loads: theory and simulation." *Journal of Vibration and Control*, available online.

Deng, L. and Cai, C.S. (2010b) "Identification of dynamic vehicular axle loads: demonstration by a field study." *Journal of Vibration and Control*, available online.

Ding L.N., Hao H. and Zhu X.Q. (2009) "Evaluation of dynamic vehicle axle loads on bridges with different surface conditions." *Journal of Sound and Vibration*, 323, 826–848.

Doyle, J.F. (1984) "Further developments in determining the dynamic contact law". *Experimental Mechanics*, 24, 265–270.

Doyle, J.F. (1987a) "An experimental method for determining the dynamic contact law". *Experimental Mechanics*, 27, 229–233.

Doyle, J.F. (1987b) "Experimentally determining the contact force during the transverse impact of an orthotropic plate." *Journal of Sound and Vibration* 118(3), 441–448.

Doyle, J.F. (1994). "A genetic algorithm for determining the location of structural impacts". *Experimental Mechanics*, 34, 37–44.

Druz, J., Crisp, J.D. and Ryall, T. (1991). "Determining a dynamic force on a plate – an inverse problem". *AIAA Journal*, 29, 464–470.

Fafard, M. and Mallikarjuna, S.M. (1993) "Dynamics of bridge–vehicle interaction." *Structural Dynamics, Proc. EURODYN'93*, 951–960.

Fierro, R.D., Golub, G.H., Hansen, P.C. and O'Leary, D.P. (1997) "Regularization by truncated total least squares." *SIAM Journal of Scientific Computing*, 18(4), 1223–1241.

Franklin, G.F., Powell, L.D. and Workman, M.L. (1998) *Digital Control of Dynamic Systems*, Third Edition, Addison-Wesley.

Friswell, M.I., Garvey, S.D. and Penny, J.E.T. (1995) "Model reduction using dynamic and iterated IRS techniques." *Journal of Sound and Vibration*, 186(2), 311–323.

Fryba, L. (1972) *Vibration of Solids and Structures under Moving Loads*. Groningen: Noordhoff International, Groningen, The Netherlands.

Golub, G.H., Heath, M. and Wahaba, G. (1979) "Generalized cross-validation as a method for choosing a good ridge parameter". *Technometrics*, 21(2), 215–223.

Gonzalez, A., Rowley, C., O'Brien, E.J. (2008) "A general solution to the identification of moving vehicle forces on a bridge". *International Journal for Numerical Methods in Engineering*. 75(3), 335–354.

Gordis, J.H. (1992) "An analysis of the improved reduced system (IRS) model reduction procedure." *Proceedings of the 7th International Modal Analysis Conference, San Diego, California, February*, 471–479.

Gorman, D.J. (1975) *Free Vibration Analysis of Beams and Shafts*. New York: John Wiley & Sons.

Gorman, G.H. and Heath, M. (1979) *Generalized cross-validation as a method for choosing a good ridge parameter*. Techno-metrics, 21(2), 215–223.

Grace, N.F. and Kennedy, J.B. (1985) "Dynamic analysis of orthotropic plate structures." *Journal of Engineering Mechanics ASCE*, 111(8), 1027–1037.

Green, M.F. and Cebon, D. (1994) "Dynamic response of highway bridges to heavy vehicle loads: theory and experimental validation." *Journal of Sound and Vibration*, 170(1), 51–78.

Gupta, R.K. and Traill-Nash, R.W. (1980) "Vehicle Braking on highway bridges". *Journal of Engineering Mechanics ASCE*, 106(4), 641–658.

Gutierrez, R.H. and Laura, P.A.A. (1997) "Vibration of a beam of non-uniform cross-section traversed by a time varying concentrated force." *Journal of Sound and Vibration*, 207 (3), 419–425.

Guyan, R.J. (1965) "Reduction of stiffness and mass matrices." *AIAA Journal*, 3(2), 380.

Hamed, E. and Frostig, Y. (2006) "Natural frequencies of bonded and unbonded prestressed beams-pretress force effects." *Journal of Sound and Vibration*, 295, 28–39.

Hanninen, K., Katila, H., Saarela, M., Rontu, R., Mattila, K.M., Fan, M., Hurme, M. and Lehtimaki, T. (2007) "Interleukin-1 beta gene polymorphism and its interactions with neuregulin-1 gene polymorphism are associated with schizophrenia". *European Archives of Psychiatry and Clinical Neuroscience*. 258(1), 10–15.

Hansen, P.C. (1992) "Analysis of discrete ill-posed problems by means of the L-curve." *SIAM Review*, 34(4), 561–580.

Hansen, P.C. and O'Leavy, D.P. (1993) "The use of the L-curve in the regularization of discrete ill-posed problems." *SIAM Sci. Comput.* 14, 1487–1503.

Hayashikawa, T. and Wantanabe, N. (1981) "Dynamic behavior of continuous beams with moving loads." *Journal of the Engineering Mechanics Division ASCE*, 107, 229–246.

Henchi, K., Fafard, M., Dhatt, G. and Talbot, M. (1997) "Dynamic behavior of multi-span beams under moving loads." *Journal of Sound and Vibration*, 199, 33–50.

Henchi, K., Fafard, M., Talbot, M. and Dhatt, G. (1998) "An efficient algorithm for dynamic analysis of bridges under moving vehicles using a coupled modal and physical components approach." *Journal of Sound and Vibration*, 212(4), 663–683.

Heywood, R.J. (1994) "Influence of truck suspensions on the dynamic response of a short span bridge". *International Journal of Vehicle Design*.

Hollandsworth, P.E. and Busby, H.R. (1989) "Impact force identification using the general inverse technique". *International Journal of Impact Engineering*, 8, 315–322.

Hoshiya, M. and Maruyama, O. (1987) "Identification f running load and beam system." *Journal of Engineering Mechanics*, ASCE. 113: 813–824.

Huang, T.C. (1961) "The effect of rotatory inertial and of shear deformation on the frequency and normal mode equations of uniform beams with simple and conditions." *Journal of Applied Mechanics ASME*, 28, 578–584.

Huang, D.Z., Wang, T.L. and Shahawy, M. (1992) "Impact analysis of continuous multi-girder bridges due to moving vehicles." *Journal of Structural Engineering ASCE*, 118, 3427–3443.

Huang, D.Z. and Wang, T.L. (1998) "Vibration of highway steel bridge with longitudinal grades." *Computers and Structures*, 69, 235–245.

Huang, D.Z., Wang, T.L. and Shahawy, M. (1998) "Vibration of horizontally curved box girder bridges due to vehicles." *Computers and Structures*, 68, 513–528.

Huebner, K.H., Dewhirst, D.L., Smith, D.E. and Byrom, T.G. (2001) *The Finite Element Method for Engineers*. 4th Edition, John Wiley & Sons.

Huffington, N.J. and Hoppmann II, W.H. (1958) "On the transverse vibrations of rectangular orthotropic plates." *Journal of Engineering Mechanics Division ASCE*, 25, 389–395.

Humar, J.L. and Kashif, A.H. (1995) "Dynamic response analysis of slab-type bridges." *Journal of Structural Engineering ASCE*, 121, 48–62.

Hwang, E.-S. and Nowak, A.S. (1991) "Simulation of dynamic load for bridges." *Journal of Structural Engineering ASCE*, 117(5), 1413–1434.

Ichikawa, M., Miyakawa, Y. and Matsuda, A. (2000) "Vibration analysis of the continuous beam subject to a moving mass." *Journal of Sound and Vibration*, 230(3), 493–506.

Irons, B. (1965) "Structural eigenvalue problems elimination of unwanted variables." *AIAA Journal*, 3(5), 961–962.

ISO8608:1995(E). *Mechanical vibration – Road surface profiles – Reporting of measured data.*

Jayaraman, G., Chen, P. and Snyder, V.W. (1990) "Free vibrations of rectangular orthotropic plates with a pair of parallel edges simply supported." *Computers and Structures*, 34(2), 203–214.

Jiang, R.J., Au, F.T.K. and Cheung, Y.K. (2003) "Identification of masses moving on multi-span beams based on a genetic algorithm". *Computers and Structures*, 81(22–23), 2137–2148.

Jiang, R.J., Au, F.T.K. and Cheung, Y.K. (2004) "Identification of vehicles moving on continuous bridges with rough surface." *Journal of Sound and Vibration*, 274(3–5), 1045–1063.

Jiang, X.Q. and Hu, H.Y. (2008) "Reconstruction of distributed dynamic loads on an Euler beam via mode-selection and consistent spatial expression". *Journal of Sound and Vibration*, 316(1–5), 122–136.

Jordan, R. and Whiston, G. (1984) "Remote impact analysis of use of propagated acceleration signals, II: Comparison between theory and experiment". *Journal of Sound and Vibration*, 97, 53–63.

Ju, S.H. and Lin, H.T. (2007) "A finite element model of vehicle-bridge interaction considering braking and acceleration." *Journal of Sound and Vibration*, 303(1–2), 46–57.

Juang, J.-N. (1994) *Applied system Identification*. New Jersey.

Kammer, D.C. (1998) "Input force reconstruction using a time domain technique". *Journal of Vibration and Acoustics*, ASME, 120(4), 868–874.

Karoumi, R., Wiberg, J. and Liljencrantz, A. (2005) "Monitoring traffic loads and dynamic effects using a instrumented railway bridge." *Engineering Structures*, 27, 1813–1819.

Khang, N.V., Dien, N.P. and Huong, N.T.V. (2009) "Transverse vibrations of prestressed continuous beams on rigid supports under the action of moving bodies." *Archive of Applied Mechanics*, 79, 939–953.

Kim, S.M. and Roesset, J.M. (1998) "Moving loads on a plate on elastic foundation." *Journal of Engineering Mechanics ASCE*, 124(9), 1010–1017.

Kim, C.W., Kawatani, M. and Kim, K.B. (2005) "Three-dimensional dynamic analysis for bridge–vehicle interaction with roadway roughness." *Computers and Structures*, 83(19–20), 1627–1645.

Kishan, H. and Traill-Nash, R.W. (1977) "A modal method for calculation of highway bridge response with vehicle braking." Institution of Engineers, Australian Civil Engineering Transactions. CE19, 44–50.

Kocaturk, T. and Simsek, M. (2006) "Dynamic analysis of eccentrically prestressed viscoelastic Timoshenko beams under a moving harmonic load." *Computers and Structures*, 84(31–32), 2113–2127.

Koh, C.G., Ong, J.S.Y., Chua, D.K.H. and Feng, J. (2003) "Moving element method for train-track dynamic." *International Journal for Numerical Methods in Engineering*, 56(11), 1549–1567.

Kuhar, E.J. and Stahle, C.V. (1974) "Dynamic transformation method for model synthesis." *AIAA Journal*, 12(5), 672–678.

Kwon, O.B., Longart, M., Vullhorst, D., Hoffman, D.A. and Buonanno, A. (2005) "Neuregulin-1 reverses long-term potentiation at CA1 hippocampal synapses", *Journal of Neuroscience*. 25(41), 9378–9383.

Lane, H., Kettil, P. and Wiberg, N-E. (2008) "Moving finite elements and dynamic vehicle interaction". *European Journal of Mechanics A/Solids* 27(4), 515–531.

Larrondo, H.A., Avalos, D.R. and Laura, P.A.A. (1998) "Transverse vibration of simply supported anisotropic rectangular plates carrying an elastically mounted concentrated mass." *Journal of Sound and Vibration*, 215(5), 1195–1202.

Laura, P.A.A., Avalos, D.R. and Larrondo, H.A. (1999) "Forced vibration of simply supported anisotropic rectangular plates." *Journal of Sound and Vibration*, 220(1), 178–185.

Law, S.S. (1988) *Finite element modeling and parameter study on the effect of boundary conditions on the vibrational response of bridge deck*. Research Report. Civil and Structural Engineering Department, Hong Kong Polytechnic, Hunghom, Kowloon, Hong Kong.

Law, S.S., Chan, T.H.T. and Zeng, Q.H. (1997) "Moving force identification: A time domain method." *Journal of Sound and Vibration*, 201(1), 1–22.

Law, S.S., Chan, T.H.T. and Zeng, Q.H. (1999) "Moving force identification: A frequency and time domains analysis." *Journal of Dynamic Systems, Measurement and Control* ASME, 12(3), 394–401.

Law, S.S. and Fang, Y.L. (2001) "Moving force identification: optimal state estimation approach." *Journal of Sound and Vibration*, 239(2), 233–254.

Law, S.S. and Li, J. (2010) "Updating reliability of a concrete bridge structure based on condition assessment with uncertainties." *Engineering Structures*, 32(1), 286–296.

Law, S.S., Chan, T.H.T., Zhu, X.Q. and Zeng, Q.H. (2001) "Regularization in moving force identification." *Journal of Engineering Mechanics ASCE*, 127(2), 136–148.

Law, S.S., Chan, T.H.T., Zhu, X.Q. and Zeng, H.Q. (2001) "Moving loads identification through regularization." *Journal of Engineering Mechanics*, 127(2), 136–148.

Law, S.S. and Zhu, X.Q. (2004) "Dynamic behavior of damaged concrete bridge structures under moving vehicular loads." *Engineering Structures*, 26(9), 1279–1293.

Law, S.S., Bu, J.Q., Zhu, X.Q. and Chan, S.L. (2004) "Vehicle axle loads identification using finite element method." *Engineering Structures*, 26(8), 1143–1153.

Law, S.S. and Lu, Z.R. (2005) "Time domain responses of a prestressed beam and prestress identification." *Journal of Sound and Vibration*, 288(4–5), 1011–1025.

Law, S.S. and Zhu, X.Q. (2005) "Bridge dynamic responses due to road surface roughness and braking of vehicle." *Journal of Sound and Vibration*, 282(3–5), 805–830.

Law, S.S., Bu, J.Q., Zhu, X.Q. and Chan, S.L. (2007) "Moving load identification on a simply supported orthotropic plate." *International Journal of Mechanical Sciences*, 49, 1262–1275.

Law, S.S., Wu, S.Q. and Shi, Z.Y. (2008) "Moving load and prestress identification using wavelet-based method." *Journal of Applied Mechanics ASME*, 75(2), 141–147.

Lee, H. and Park, Y.S. (1995) "Error analysis of indirect force determination and a regularization method to reduce force determination error." *Mechanical System and Signal processing*, 9(6), 615–633.

Lee, H.P. (1994) "Dynamic response of a beam with intermediate point constraints subject to a moving load." *Journal of Sound and Vibration*, 171, 361–368.

Lee, H.P. and Ng, T.Y. (1994) "Dynamic response of a cracked beam subject to a moving load." *Acta Mechanica*, 106, 221–230.

Lee, H.P. (1996) "The dynamic response of a Timoshenko beam subject to a moving mass." *Journal of Sound and Vibration*, 198(2), 249–256.

Lee, J.W., Kim, J.D., Yun, C.B., Yi, J.H. and Shim, J.M. (2002) "Health-monitoring method for bridges under ordinary traffic loadings." *Journal of Sound and Vibration*, 257(2), 247–264.

Lee, S.Y. and Yhim, S.S. (2004) "Dynamic analysis of composite plates subjected to multi-moving loads based on a third order theory." *International Journal of Solids and Structures*, 41(16–17), 4457–4472.

Leissa, A.W. (1973) "The free vibration of rectangular plates." *Journal of Sound and Vibration*, 31(3), 257–293.

Li, H., Wekezer, J. and Kwasniewski, L. (2008) "Dynamic response of a highway bridge subjected to moving vehicles." *Journal of Bridge Engineering*, 13(5), 439–448.

Li, W.L. (2000) "Free vibration of beams with general boundary conditions." *Journal of Sound and Vibration*, 237(4), 709–725.

Li, X.Y. and Law, S.S. (2010a) "Adaptive Tikhonov Regularization for Damage Detection based on Nonlinear Model Updating." *Mechanical System and Signal Processing*. 24(6), 1646–1664.

Li, X.Y. and Law, S.S. (2010b) "Consistent regularization for damage detection with noise and model errors." *AIAA Journal*. 48(4), 777–787.

Liljencrantz, P., Karoumi, R. and Olofsson, P. (2007). "Implementing bridge weigh-in-motion for railway traffic." Computers and Structures, 85, 80–88.

Lin, Y.H. (1995) Comments on "Dynamic response of a beam with intermediate point constraints subject to a moving load." *Journal of Sound and Vibration*, 180, 809–812.

Lin, Y.H. and Trethewey, M.W. (1990) "Finite element analysis of elastic beams subject to moving dynamic loads." *Journal of Sound and Vibration*, 136(2), 323–342.

Liu, J.J., Ma, C.K., Kung, I.C. and Lin, D.C. (2000) "Input force estimation of a cantilever plate by using a system identification technique." *Computer Methods in Applied Mechanics and Engineering* 190, 1309–1322.

Liu, M., Frangopol, D.M. and Kim, S. (2009) "Bridge safety evaluation based on monitored live load effects." *Journal of Bridge Engineering ASCE*, 14(4), 257–169.

Liu, Y. and Shepard, W.S. (2005) "Dynamic force identification based on enhanced least squares and total least-squares schemes in the frequency domain." *Journal of Sound and Vibration*, 282(1–2), 37–60.

Lu, Z.R. and Law, S.S. (2006) "Force identification based on sensitivity in time domain." *Journal of Engineering Mechanics*, 195(44–47), 1050–1056.

Mabsout, M.E., Tarhini, K.M., Frederick, G.R. and Kesserwan, A. (1999) "Effect of multi-lanes on wheel load distribution in steel girder bridges." *Journal of Bridge Engineering ASCE*, 4, 99–106.

Mackertich, S. (1990) "Moving load on a Timoshenko beam." *Journal of the Acoustic Society America*, 88(2), 1175–1178.

Mahmoud, M.A. (2001) "Effect of cracks on the dynamic response of a simple beam subject to a moving load." *Proceedings of the Institute of Mechanical Engineers Part F: Journal of Rail and Rapid Transit*, 215(3), 207–215.

Mahmoud, M.A. and Abou Zaid, M.A. (2002) "Dynamic response of a beam with a crack subject to a moving mass." *Journal of Sound Vibration*, 256(4), 591–603.

Majumder, L. and Manohar, C.S. (2003) "A time-domain approach for damage detection in beam structures using vibration data with a moving oscillator as an excitation source." *Journal of Sound and Vibration*, 268(4), 699–716.

Marchesiello, S., Fasana, A., Garibaldi, L. and Piombo, B.A.D. (1999) "Dynamics of multi-span continuous straight bridges subject to multi-degrees of freedom moving vehicle excitation." *Journal of Sound and Vibration*, 224, 541–561.

Mason, J.C. and Handscomb, D.C. (2003) *Chebyshev Polynomials*. CRC Press.

Matta, K.W. (1987) "Selection of degrees of freedom for dynamic analysis." *Journal of Pressure Vessel Technology-ASME*, 109(1), 65–69.

Mazurek, D.F. and Dewolf, J.T. (1990) "Experimental study of bridge monitoring techniques." *Journal of Structural Engineering ASCE*, 115(9), 2532–2549.

Michaels, J. and Pao, Y.H. (1986) "Determination of dynamic forces from wave motion measurements." *Journal of Applied Mechanics*, ASME, 53, 61–68.

Millan, M.J. (2005) "N-methyl-D-aspartate receptors as a target for improved antipsychotic agents: novel insights and clinical perspectives." *Psychopharmacology* 179(1), 30–53.

Miller, C.A. (1980) "Dynamic reduction of structural models." *Journal of the Structural Division*, 106(10), 2079–2108.

Moller, P.W. (1999) "Load identification through structural modification." *Journal of Applied Mechanics ASME* 66(1), 236–241.

Morozov, V.A. (1984) *Methods for solving incorrectly posed problems*. Springer, Berlin, 1–64.

Moses F. (1979) "Weigh-in-motion system using instrumented bridges." *Journal of Transportation Engineering ASCE*, 105(3), 233–249

Moses, F. (1984) "Weigh-in-motion system using instrumented bridge." *Journal of Transportation Engineering*, ASCE, 105, (TE3) 233–249.

Mulcahy, N.L. (1983) "Bridge response with tractor-trailer vehicle loading." *Earthquake Engineering and Structural Dynamics*, 11, 649–665.

Naguleswaran, S. (2003) "Transverse vibration of an Euler-Bernoulli uniform beam on up to five resilient supports including ends." *Journal of Sound and Vibration*, 261, 372–384.

Nallasivam, K., Dutta, A. and Talukdar, S. (2007) "Dynamic analysis of horizontally curved thin-walled box-girder bridge due to moving vehicle." *Shock and Vibration*, 14(3), 229–248.

Ng, S.F. and Kulkarni, G.G. (1972) "On the transverse free vibrations of beam-slab type highway bridges." *Journal of Sound and Vibration*, 21(3), 249–261.

Nordberg, T.P. and Gustafsson, I. (2006a) "Dynamic regularization of input estimation problems by explicit block inversion." *Computer Methods in Applied Mechanics and Engineering*, 168(44–47), 5877–5890.

Nordberg, T.P. and Gustafsson, I. (2006b) "Using QR factorization and SVD to solve input estimation problems in structural dynamics." *Computer Methods in Applied Mechanics and Engineering*, 195(44–47), 5891–5908.

Nordstrom, L.J.L. and Nordberg, T.P. (2004) "A time delay method to solve non-collocated input estimation problems." *Mechanical Systems and Signal Processing*, 18(6), 1469–1483.

Nordström, Las, J.L. (2006) "A dynamic programming algorithm for input estimation on linear time-variant systems." *Computational Methods in Applied Mechanics and Engineering*, 195(44–47), 6407–6427.

Norman, O.K. and Hopkins, R.C. (1952) "Weighing vehicles in motion." *Public Road*, 1952: 27(1): 1–17.

Nowak, A.S., Nassif, H. and Defrain, L. (1993) "Effect of truck loads on bridges." *Journal of Engineering Mechanics ASCE* 119, 853–867.

O'Brien, E.J., Znidaric, A. and Dempsey, A.T. (1999) "Comparison of two independently developed bridge weigh-in-motion systems, Heavy Vehicle Systems." *International Journal of Vehicle Design*, 5(1/4) 147–161.

O'Callahan, J.C. (1989) "A procedure for an Improved Reduced System (IRS) Model." *Proceedings of the Seventh International Modal Analysis Conference*, Las Vegas, Nevada.

O'Callahan, J.C. (1992) "A procedure for an improved reduced system (IRS) model." *Proceedings of the 7th International Modal Analysis Conference, Las Vegas*, 17–21.

O'Connor, C. and Chan, T.H.T. (1988) "Dynamic wheel loads from bridge strains." *Journal of Structural Engineering*, ASCE. 114(ST8): 1703–1723.

Pan, T.C. and Li, J. (2002) "Dynamic vehicle element method for transient response of coupled vehicle-structure systems." *Journal of Structural Engineering*, 128(2), 214–223.

Parhi, D.R. and Behera, A.K. (1997) "Dynamic deflection of a cracked beam with moving mass." *Proceedings of the Institution of Mechanical Engineers: C—Journal of Mechanical Engineering*, 211, 77–87.

Pesterev, A.V. and Bergman, L.A. (1997) "Response of elastic continuum carrying moving linear oscillator." *Journal of Engineering Mechanics ASCE*, 123(8), 878–884.

Pesterev, A.V. and Bergman, L.A. (1998) "Response of a non-conservative continuous system to a moving concentrated load." *Journal of Applied Mechanics ASME*, 65, 436–444.

Pesterev, A.V., Yang, B., Bergman, L.A. and Tan, C.A. (2001) "Response of elastic continuum carrying multiple moving oscillators." *Journal of Engineering Mechanics ASCE*, 127(3), 260–265.

Peters, R.T. (1984) "AXWAY – a system to obtain vehicle axle weight." *Proceeding of 12th ARRB Conference*, Hobart, Australia, Vol. 2, 10–18.

Peters, R.T. (1986) "CULWAY – an unmanned and undetectable highway speed vehicle weighting system." *Proceedings of 13th ARRB and 5th REAAA Combined Conference*, Australia, Vol. 6, 70–83.

Petrovski, J. and Naumovski, N. (1979) "Processing of strong motion accelerograms. Part 1: Analytical methods." *Report 66. I211S*, Stronjze.

Pinkaew, T. and Akarawittayapoom, T. (2003) "Determination of truck weight from response of bridges." *16th ASCE Engineering Mechanics Conference, July 16–18, 2003, University of Washington*.

Pinkaew, T. (2006) "Identification of vehicle axle loads from bridge responses using updated static component technique." *Engineering Structures*, 28(11), 1599–1608.

Pinkaew, T. and Asnachinda, P. (2007) "Experimental study on the identification of dynamic axle loads of moving vehicles from the bending moments of bridges." *Engineering Structures*, 29(9), 2282–2293.

Qian, C.Z. and Tang, J.S. (2008) "A time delay control for a nonlinear dynamic beam under moving load." *Journal of Sound and Vibration*, 309(1–2), 1–8.

Rao, K.C. and Mirza, S. (1989) "A note on vibrations of generally restrained beams." *Journal of Sound and Vibration*, 130(3), 53–465.

Rao, Z.S., Shi, Q.Z. and Hagiwara, I. (1999) "Optimal estimation of dynamic loads for multiple input system." *Journal of Vibration and Acoustics ASME* 121(3), 397–401.

Rowley, C.W., O'Brien, E.J., Gonzalez, A. and Znidaric, A. (2009) "Experimental testing of a moving force identification bridge weigh-in-motion algorithm." *Experimental Mechanics*, 49, 743–746.

Saiidi N., Douglas B. and Feng S. (1994) "Prestress force effect on vibration frequency of concrete bridges." Journal of Structural Engineering ASCE, 120(7), 2233–2241.

Santantamarina, J.C. and Fratta, D. (1998) *Introduction to Discrete Signals and Inverse Problems in Civil Engineering*. ASCE Press, 200–238.

Schueller, J.K. and Wall, T.M.P. (1991) "Impact of fruit on flexible beams to sense modulus of elasticity". *Experimental Mechanics*, 31, 118–121.

Shan, V.N. and Raymund, M. (1982) "Analytical selection of masters for the reduced eigenvalue problem." *International Journal for Numerical Methods in Engineering*, 18(1), 89–98.

Simonian, S.S. (1981a) "Inverse problems in structural dynamics – I. Theory." *International Journal of Numerical Method in Engineering*, 17, 357–365.

Simonian, S.S. (1981b) "Inverse problems in structural dynamics – II. Applications." *International Journal of Numerical Method in Engineering*, 17, 367–386.

Simsek, M. and Kocaturk, T. (2006) "Dynamic analysis of eccentrically prestressed damped beam moving harmonic force using higher order shear deformation theory." *Journal of Structural Engineering*, 133(12), 1733–1741.

Sniady, P. (2008) "Dynamic response of a Timoshenko beam to a moving force." *Journal of Applied Mechanics*, 75, Paper No. 024503.

Stevens, E.K. (1987) "Force identification problems – an overview." *Proceedings of SEM Spring Conference on Experimental Mechanics*, FL, USA. 838–844.

Taheri, M.R. and Ting, E.C. (1989) "Dynamic response of plate to moving loads: Structural impedance method." *Computers and Structures*, 33(6), 1379–1393.

Taheri, M.R. and Ting, E.C. (1990) "Dynamic response of plates to moving loads: Finite element method." *Computers and Structures*, 34(3), 509–521.

Thite, A.N. and Thompson, D.J. (2003a) "The quantification of structure-borne transmission paths by inverse methods. Part 1: Improved singular value rejection methods." *Journal of Sound and Vibration*, 264(2), 411–431.

Thite, A.N. and Thompson, D.J. (2003b) "The quantification of structure-borne transmission paths by inverse methods. Part 2: Use of regularization techniques." *Journal of Sound and Vibration*, 264(2), 433–451.

Tikhonov, A.N. (1963) "On the solution of ill-posed problems and the method of regularization." *Soviet Mathematics*, 4, 1035–1038.

Tikhonov, A.N. and Arsenin, V.Y. (1977). *Solution of ill-posed problems*. John Wiley & Sons, Inc., New York, N.Y.

Trujillo, D.M. (1978) "Application of Dynamic Programming to the General Inverse Problem." *International Journal of Numerical Methods in Engineering*, 12, 613–624.

Trujillo, D.M. and Busby, H.R. (1983) "Investigation of a technique for the differentiation of empirical data." *Journal of Dynamic Systems, Measurement and Control ASME*, 105(3), 200–202.

Trujillo, D.M. and Busby, H.R. (1997) *Practical Inverse Analysis in Engineering*. CRC Press LLC, Boca Raton, FL.

Wang, R.T. and Chou, T.H. (1998) "Non-linear vibration of Timoshenko beam due to a moving force and the weight of beam." *Journal of Sound and Vibration*, 218(1), 117–131.

Wang, T.L., Huang, D.Z. and Shahawy, M. (1992) "Dynamic response of multigirder bridges." *Journal of Structural Engineering*, 118(8), 2222–2238.

Wang, T.L., Huang, D.Z., Shahawy, M. and Huang, K.Z. (1996) "Dynamic response of highway girder bridge." *Computers and Structures*, 60(6), 1020–1027.

Wang, R.T. and Lin, T.Y. (1996) "Vibration of multispan Mindlin plates to a moving load." *Journal of Chinese Institute of Engineering*, 19(4), 467–477.

Wang, R.T. (1997) "Vibration of multi-span Timoshenko beams to a moving force." *Journal of Sound and Vibration*, 207, 731–742.

Whiston, G. (1984) "Remote impact analysis by use of propagated acceleration signals, I: Theoretical methods." *Journal of Sound and Vibration* 97, 35–51.

Williams, D. and Jones, R.P.N. (1948) "Dynamic loads in aeroplanes under given impulsive loads with particular reference to landing and gust loads on a large flying boat." *Aeronautic Research Council, TR#2221*.

Wu, J.J. (2005a) "Dynamic analysis of an inclined beam due to moving loads." *Journal of Sound and Vibration*, 288, 107–131.

Wu, J.J. (2005b) "Vibration analysis of a portal under the action of a moving distributed mass using moving mass element." *International Journal for Numerical Methods in Engineering*, 62(14), 2028–2052.

Wu, J.J. (2007) "Use of moving distributed mass element for the dynamic analysis of a flat plate undergoing a moving distributed load." *International Journal for Numerical Methods in Engineering*, 71(3), 347–362.

Wu, J.S. and Dai, C.W. (1987) "Dynamic response of multi-span non-uniform beams due to moving loads." *Journal of Structural Engineering ASCE*, 113, 458–474.

Wu, S.Q. and Law, S.S. (2010a) "Dynamic analysis of bridge-vehicle system with uncertainties based on finite element model." *Probabilistic Engineering Mechanics*, 25(4), 425–432.

Wu, S.Q. and Law, S.S. (2010b) "Moving force identification based on stochastic finite element model." *Engineering Structures*, 32, 1016–1027.

Wu, S.Q. and Law, S.S. (2010c) "Dynamic Analysis of Bridge with Non-Gaussian Uncertainties under a Moving Vehicle." *Probabilistic Engineering Mechanics* (in press).

Wu, J.S., Lee, M.L. and Lai, T.S. (1987) "The dynamic analysis of a flat plate under a moving load by the finite element method." *International Journal for Numerical Methods in Engineering*, 24(4), 743–762.

Wu, S.Q. and Shi, Z.Y. (2006) "Identification of vehicle axle loads based on FEM-Wavelet-Galerkin method." *Journal of Vibration Engineering*, 19(4), 494–498 (in Chinese).

Wu, J.S. (2008) "Transverse and longitudinal vibrations of a frame structure due to a moving trolley and the hoisted object using moving finite element." *International Journal of Mechanical Sciences*, Volume 50, Issue 4, 613–625.

Xiao, Z.-G., Yamada, K., Inoue, J. and Yamaguchi, K. (2006) "Measurement of truck axle weights by instrumenting longitudinal ribs of orthotropic bridge." *Journal of Bridge Engineering ASCE*, 11(5), 526–532.

Yamaguchi, E., Kawamura, S., Matuso, K., Matsuki, Y. and Naito, Y. (2009) "Bridge-weigh-in-motion by two-span continuous bridge with skew and heavy-truck flow in Fukuoka area, Japan." *Advances in Structural Engineering*, 12(1), 115–125.

Yang, Y.B., Liao, S.S. and Lin, B.H. (1995) "Impact formulas for vehicles moving over simple and continuous beams." Journal of Structural Engineering ASCE, 121(11), 1644–1650.

Yang, Y.B. and Yau, J.D. (1997) "Vehicle-bridge interaction element for dynamic analysis." *J of Structural Engineering, ASCE*, **123**(11), 1512–1518.

Yang, Y.B. and Wu, Y.S. (2001) "A versatile element for analyzing vehicle-bridge interaction response." *Engineering Structure*, 23(5), 452–469.

Yang, B., Tan, C.A. and Bergman, L.A. (2000) "Direct numerical procedure for solution of moving oscillator problems." *Journal of Engineering Mechanics ASCE*, 126(5), 462–469.

Yang, J., Chen, Y., Xiang, Y. and Jia, X.L. (2008) "Free and forced vibration of cracked inhomogeneous beams under an axial force and a moving load." *Journal of Sound Vibration*, 312(1–2), 166–181.

Yau, J.D., Wu, Y.S. and Yang, Y.B. (2001) "Impact response of bridges with elastic bearings to moving loads." *Journal of Sound and Vibration*, 248(1), 9–30.

Yen, C.S. and Wu, E. (1995a) "On the inverse problem of rectangular plates subject to elastic impact, Part I: Method development and numerical verification." *Journal of Applied Mechanics*, ASME 62(4), 692–698.

Yen, C.-S., and Wu, E. (1995b) "On the Inverse Problem of Rectangular Plates Subjected to Elastic Impact, Part II: Experimental Verification and Further Application," *Journal of Applied Mechanics*, ASME 62(3), 699–705.

Yener, M. and Chompooming, K. (1994) "Numerical method of lines for analysis of vehicle–bridge dynamic interaction." *Computers and Structures*, 53, 709–726.

Yu, L. and Chan, T.H.T. (2003) "Moving force identification based on the frequency-time domain method." *Journal of Sound and Vibration*, 261(2), 329–349.

Yu, L., Chan, T.H.T. and Zhu, J.H. (2008a) "A MOM-based algorithm for moving force identification: Part I – Theory and numerical simulation." *Structural Engineering and Mechanics*, 29(2), 135–154.

Yu, L., Chan, T.H.T. and Zhu, J.H. (2008b) "A MOM-based algorithm for moving force identification: Part II – Experiment and comparative studies." *Structural Engineering and Mechanics*, 29(2), 155–169.

Zheng, D.Y., Cheung, Y.K., Au, F.T.K. and Cheng, Y.S. (1998) "Vibration of multi-span non-uniform beams under moving loads by using modified beam vibration functions." *Journal of Sound and Vibration*, 212(3), 455–467.

Zhong, W.X. and Williams, F.W. (1994) "A precise time step integration method." *Proceedings of Institution of Mechanical Engineers, Part C: Journal of Mechanical Engineering Science*, 208, 427–430.

Zhou, D. (1994) "Eigenfrequencies of line supported rectangular plates." *International Journal of Solids and Structures*, 31, 347–358.

Zhu, X.Q. and Law, S.S. (1999) "Moving forces identification on a multi-span continuous bridge." *Journal of Sound and Vibration*, 228, 377–396.

Zhu, X.Q. and Law, S.S. (2000) "Identification of vehicle axle loads from bridge dynamic responses." *Journal of Sound and Vibration* 236(4), 705–724.

Zhu, X.Q. and Law, S.S. (2001a) "Orthogonal function in moving loads identification on a multi-span bridge." *Journal of Sound and Vibration*, 245(2), 329–345.

Zhu, X.Q. and Law, S.S. (2001b) "Identification of moving loads on an orthotropic plate." *Journal of Vibration and Acoustics*, ASME, 123(2), 238–244.

Zhu, X.Q. and Law, S.S. (2001c) "Precise time-step integration for the dynamic response of a continuous beam under moving loads." *Journal of Sound and Vibration*, 240(5), 962–970.

Zhu, X.Q. and Law, S.S. (2002a) "Dynamic load on continuous multi-lane bridge deck from moving vehicles." *Journal of Sound and Vibration*, 251(4), 697–716.

Zhu, X.Q. and Law, S.S. (2002b) "Moving loads identification through regularization." *Journal of Engineering Mechanics ASCE*, 128(9), 989–1000.

Zhu, X.Q. and Law, S.S. (2002c) "Practical aspects in moving load identification." *Journal of Sound and Vibration*, 258(1), 123–146.

Zhu, X.Q. and Law, S.S. (2003a) "Dynamic axle and wheel loads identification: laboratory studies." *Journal of Sound and Vibration*, 268, 855–879.

Zhu, X.Q. and Law, S.S. (2003b) "Dynamic behavior of orthotropic rectangular plates under moving loads." *Journal of Engineering Mechanics ASCE*, 129(1), 79–87.

Zhu, X.Q. and Law, S.S. (2003c) "Time domain identification of moving loads on bridge deck." *Journal of Vibration and Acoustics*, ASME, 125(2), 187–198.

Zhu, X.Q. and Law, S.S. (2006) "Moving load identification on multi-span continuous bridges with elastic bearings." *Mechanical Systems and Signal Processing*, 20, 1759–1782.

Zhu, X.Q., Law, S.S. and Bu, J.Q. (2006) "A state space formulation for moving loads identification." *Journal of Vibration and Acoustics ASME*, 128(4), 509–520.

Zhu, X.Q. and Law, S.S. (2007) "Damage detection in simply supported concrete bridge structures under moving vehicular loads." *Journal of Vibration and Acoustics ASME*, 129(1), 58–65.

Znidaric, A. (2010) "Bridge-WIM as an efficient tool for optimized bridge assessment." *Proceedings of the Workshop "Civil Structural Health Monitoring 2": WIM (Weigh In Motion), Load Capacity and Bridge Performance*, Taormina, Italy.

Subject Index

Structures and Infrastructures Series

Book Series Editor: Dan M. Frangopol

ISSN: 1747–7735

Publisher: CRC/Balkema, Taylor & Francis Group

1. Structural Design Optimization Considering Uncertainties
   Editors: Yiannis Tsompanakis, Nikos D. Lagaros & Manolis Papadrakakis
   2008
   ISBN: 978-0-415-45260-1 (Hb)

2. Computational Structural Dynamics and Earthquake Engineering
   Editors: Manolis Papadrakakis, Dimos C. Charmpis,
   Nikos D. Lagaros & Yiannis Tsompanakis
   2008
   ISBN: 978-0-415-45261-8 (Hb)

3. Computational Analysis of Randomness in Structural Mechanics
   Christian Bucher
   2009
   ISBN: 978-0-415-40354-2 (Hb)

4. Frontier Technologies for Infrastructures Engineering
   Editors: Shi-Shuenn Chen & Alfredo H-S. Ang
   2009
   ISBN: 978-0-415-49875-3 (Hb)

5. Damage Models and Algorithms for Assessment of Structures
   under Operating Conditions
   Siu-Seong Law & Xin-Qun Zhu
   ISBN: 978-0-415-42195-9 (Hb)

6. Structural Identification and Damage Detection using Genetic Algorithms
   Chan Ghee Koh & Michael John Perry
   ISBN: 978-0-415-46102-3 (Hb)

7. Design Decisions under Uncertainty with Limited Information
   Efstratios Nikolaidis, Zissimos P. Mourelatos & Vijitashwa Pandey
   ISBN: 978-0-415-49247-8 (Hb)

8. Moving Loads – Dynamic Analysis and Identification Techniques
   Siu-Seong Law & Xin-Qun Zhu
   ISBN: 978-0-415-87877-7 (Hb)